Semiconductor Photochemistry and Photophysics

MOLECULAR AND SUPRAMOLECULAR PHOTOCHEMISTRY

Series Editors

V. RAMAMURTHY

Professor
Department of Chemistry
Tulane University
New Orleans, Louisiana

KIRK S. SCHANZE

Professor
Department of Chemistry
University of Florida
Gainesville, Florida

1. Organic Photochemistry, *edited by V. Ramamurthy and Kirk S. Schanze*
2. Organic and Inorganic Photochemistry, *edited by V. Ramamurthy and Kirk S. Schanze*
3. Organic Molecular Photochemistry, *edited by V. Ramamurthy and Kirk S. Schanze*
4. Multimetallic and Macromolecular Inorganic Photochemistry, *edited by V. Ramamurthy and Kirk S. Schanze*
5. Solid State and Surface Photochemistry, *edited by V. Ramamurthy and Kirk S. Schanze*
6. Organic, Physical, and Materials Photochemistry, *edited by V. Ramamurthy and Kirk S. Schanze*
7. Optical Sensors and Switches, *edited by V. Ramamurthy and Kirk S. Schanze*
8. Understanding and Manipulating Excited-State Processes, *edited by V. Ramamurthy and Kirk S. Schanze*
9. Photochemistry of Organic Molecules in Isotropic and Anisotropic Media, *edited by V. Ramamurthy and Kirk S. Schanze*
10. Semiconductor Photochemistry and Photophysics, *edited by V. Ramamurthy and Kirk S. Schanze*

ADDITIONAL VOLUMES IN PREPARATION

Semiconductor Photochemistry and Photophysics

edited by

V. Ramamurthy
Tulane University
New Orleans, Louisiana
U.S.A.

Kirk S. Schanze
University of Florida
Gainesville, Florida
U.S.A.

CRC Press
Taylor & Francis Group
Boca Raton London New York

CRC Press is an imprint of the
Taylor & Francis Group, an **informa** business

First published 2003 by Marcel Dekker, Inc.

Published 2019 by CRC Press
Taylor & Francis Group
6000 Broken Sound Parkway NW, Suite 300
Boca Raton, FL 33487-2742

© 2003 by Taylor & Francis Group, LLC
CRC Press is an imprint of Taylor & Francis Group, an Informa business

First issued in paperback 2019

No claim to original U.S. Government works

ISBN 13: 978-0-367-44679-6 (pbk)
ISBN 13: 978-0-8247-0958-7 (hbk)

Visit the Taylor & Francis Web site at
http://www.taylorandfrancis.com

and the CRC Press Web site at
http://www.crcpress.com

Library of Congress Cataloging-in-Publication Data
A catalog record for this book is available from the Library of Congress.

Preface

Semiconductor photochemistry and photophysics play an important role in the broad field of supramolecular photochemistry. The unique properties of nanocrystalline semiconductor particles—which include quantum size effects on the bandgap, high surface area which is optimal for interfacial reactions, good photo- and thermal stability, and compatibility with the environment (i.e., "green" chemistry)—have led to an explosion of interest in the field. This volume of the Molecular and Supramolecular Photochemistry series provides chapters, authored by experts in the field, that discuss the area of semiconductor photochemistry and photophysics and highlight recent important advances in the area.

The breakthrough development of dye-sensitized solar cells (DSSCs) based on nanocrystalline semiconductor particles (the "Grätzel cell") has stimulated a surge of interest in research aimed at understanding the operation of the cells and optimization of light-to-electrical conversion efficiency. At the present time, DSSCs exhibit incident photon-to-current efficiencies approaching unity and overall solar-to-electrical power conversion efficiency in excess of 10%. Five chapters in this volume describe fundamental work aimed at understanding the underlying mechanisms by which DSSCs operate, as well as molecular and device-engineering efforts being undertaken to improve their light-to-electrical power conversion efficiency.

Another major theme is the application of nanocrystalline TiO_2 as a photocatalyst for environmental remediation. During the past several decades a broad research community composed of chemists, engineers, and materials scientists has developed technologies that use TiO_2 as the light absorber and primary oxidant to mineralize organic pollutants in water- and air-streams. Three chapters in this volume highlight recent advances in this field. Several other chapters describe additional aspects of semiconductor photochemistry and photophysics, including applications of luminescent nanocrystalline semiconductor particles to chemical sensor development and fundamental studies aimed at understanding the structure–property relationships for nanocrystalline semiconductors composed of layered materials.

We are optimistic that the field of semiconductor photochemistry will lead to new materials, devices, and techniques to solve many important problems facing our society, including development of renewable energy sources, methods for environmental remediation, and advanced sensor and monitoring technologies. The chapters herein are up-to-date, comprehensive, and authoritative. The book is a valuable resource for both students and scientists working in the broad area of optically responsive materials based on nanostructured semiconductors.

Kirk S. Schanze
V. Ramamurthy

Contents

v

Contributors

Masakazu Anpo   Department of Applied Chemistry, Graduate School of Engineering, Osaka Prefecture University, Osaka, Japan

Hironori Arakawa   Photoreaction Control Research Center, National Institute of Advanced Industrial Science and Technology, Tsukuba, Japan

Matteo Biancardo   Department of Chemistry, University of Ferrara, Ferrara, Italy

Carlo A. Bignozzi   Department of Chemistry, University of Ferrara, Ferrara, Italy

Michel Che   Laboratoire de Reactive de Surface, Université Pierre et Marie Curie, Paris, France

Ann Davidson   Laboratoire de Reactive de Surface, Université Pierre et Marie Curie, Paris, France

Arthur B. Ellis   Department of Chemistry, University of Wisconsin–Madison, Madison, Wisconsin, U.S.A.

Dennis A. Gaal   Department of Chemistry, Northwestern University, Evanston, Illinois, U.S.A.

Michael Grätzel   Laboratory for Phototonics and Interfaces, Institute of Molecular and Biological Chemistry, School of Basic Science, Swiss Federal Institute Technology, Lausanne, Switzerland

Brian A. Gregg   National Renewable Energy Laboratory, Golden, Colorado, U.S.A.

Kohjiro Hara   Photoreaction Control Research Center, National Institute of Advanced Industrial Science and Technology, Tsukuba, Japan

Joseph T. Hupp   Department of Chemistry, Northwestern University, Evanston, Illinois, U.S.A.

David F. Kelley   Department of Chemistry, Kansas State University, Manhattan, Kansas, U.S.A.

Satoru Kishiguchi   Department of Applied Chemistry, Graduate School of Engineering, Osaka Prefecture University, Osaka, Japan

Thomas F. Kuech   Department of Chemical Engineering, University of Wisconsin–Madison, Madison, Wisconsin, U.S.A.

Michael Lewandowski   Department of Chemical Engineering, North Carolina State University, Raleigh, North Carolina, U.S.A.

Valter Maurino   Department of Analytical Chemistry, University of Torino, Torino, Italy

Luke J. Mawst   Department of Electrical and Computer Engineering, University of Wisconsin–Madison, Madison, Wisconsin, U.S.A.

Claudio Minero   Department of Analytical Chemistry, University of Torino, Torino, Italy

Mohammad K. Nazeeruddin   Laboratory for Phototonics and Interfaces, Institute of Molecular and Biological Chemistry, School of Basic Science, Swiss Federal Institute of Technology, Lausanne, Switzerland

Anne-Marie L. Nickel   Department of Physics and Chemistry, Milwaukee School of Engineering, Milwaukee, Wisconsin, U.S.A.

David F. Ollis   Department of Chemical Engineering, North Carolina State University, Raleigh, North Carolina, U.S.A.

Kevin E. O'Shea   Department of Chemistry, Florida International University, Miami, Florida, U.S.A.

Ezio Pelizzetti   Department of Analytical Chemistry, University of Torino, Torino, Italy

Peter F. H. Schwab   Department of Chemistry, University of Ferrara, Ferrara, Italy

Gordon A. Shaw   Department of Chemistry, University of Wisconsin–Madison, Madison, Wisconsin, U.S.A.

Masato Takeuchi   Department of Applied Chemistry, Graduate School of Engineering, Osaka Prefecture University, Osaka, Japan

Hiromi Yamashita   Department of Applied Chemistry, Graduate School of Engineering, Osaka Prefecture University, Osaka, Japan

Jeng-Ya Yeh   Department of Electrical and Computer Engineering, University of Wisconsin–Madison, Madison, Wisconsin, U.S.A.

Contents of Previous Volumes

1

Heterosupramolecular Devices Based on Nanocrystalline Semiconductors

Carlo A. Bignozzi, Matteo Biancardo, and Peter F. H. Schwab

University of Ferrara, Ferrara, Italy

I. INTRODUCTION

It has been recognized for some time that for the successful design of molecular devices, the addressability and organization of the molecular components have to be considered. Many studies on photochemically driven supramolecular assemblies have been carried out in solution, but although the results obtained from these studies have been very exciting, addressability and organization issues cannot be solved in the solution phase. For this reason, there is an increasing interest in the investigation of photochemically active molecular assemblies bound to active solid semiconductor substrates [1–4].

Over the last 10 years, our group has been working in the field of supramolecular photochemistry at semiconductor interfaces. In particular, we have been interested in the design, synthesis, and characterization of supramolecular coordination compounds (i.e., inorganic coordination compounds with suitable built-in light-induced and electron-transfer functions) that can be integrated into semiconductor devices. These studies provide a general molecular approach for the controlled conversion of light into an electrical response and thus have potential applications in displays, sensors, electrochromics, and other molecular photonic devices.

Part of the research described here has been motivated by the development of photovoltaic cells based on sensitized nanocrystalline TiO_2 [5]. These studies have shown that although the basic photophysical properties of the molecular components are maintained upon immobilization of a molecular component on a semiconducting surface, the interaction with the surface can greatly change the rate of the individual photophysical processes. For example, when bound to TiO_2, ruthenium polypyridyl complexes that are intrinsically photolabile in solution become photostable. This change in behavior is explained by the very fast, subpicosecond charge injection from the excited state of the surface-bound compound into the valence band of the semiconductor [6–9]. In addition to this fast injection, the back-electron-transfer process is several orders of magnitude slower than the forward-electron-transfer reaction, and as a result, effective charge separation is observed [10]. This ability of the TiO_2 surface to assist in a long-lived charge separation has been one of the driving forces behind the work carried out in this area [11,12].

Other interesting aspects, arising from the possibility of binding a molecular system to high-surface wide-band-gap nanocrystalline semiconductors, are related to the production of new types of electrochromic device which appear to be promising for display applications. In the past years, four major classes of materials have evolved for incorporation into electrochromic devices, namely liquid crystals, molecular dyes, metal oxide films, and conductive polymers. All of these systems have virtues and limitations. Molecular dyes can be deposited as thin films onto an optically transparent electrode or dispersed through an electrolyte between two transparent electrodes. They normally exhibit response times in the millisecond to second range, depending on the rate of diffusion of counterions in the film or of direct diffusion of the electroactive species toward the electrodes. Metal oxide films, generally based on tungsten, nickel, or molybdenum, show a pH-dependent electrochromism and slower switching rates with respect to molecular dyes. Conductive polymers such as functionalized polypyrroles, polythiophenes, or combinations of low-band-gap and high-band-gap polymers deposited on transparent electrodes have been found to exhibit extremely high coloration efficiencies and subsecond switching times. High-surface-area nanocrystalline TiO_2 electrodes have been recently proposed for supporting electroactive species which can be bound to the semiconductor surface by means of suitable anchoring units. Depending on the molecular design, high-contrast electrochromic devices with sharp colors can be produced. In addition, short response times are obtained because all molecules bound to the semiconductor surface are in contact with the solution containing the counterions, ensuring for a fast charge compensation during the redox process. Our research in this field has focused on the design and preparation of charge transfer and mixed-valence compounds which can be adsorbed on the surface of the semiconductor, giving rise to fast color changes as a function of the redox properties of the component molecular units.

The aim of this contribution is to use representative experimental results to exemplify a supramolecular approach to the functionalization of wide-band-gap semiconductors for the realization of photoelectrochemical and electrochromic devices. In this context, we consider it useful to present a brief overview of alternative electrochromic devices in order to give some perspective to our concept.

II. PHOTOINDUCED CHARGE SEPARATION ACROSS DYE-SENSITIZED TiO₂ INTERFACES

The inherent electronic nature of semiconductor metal oxides can directly interact with molecular excited states in a manner not energetically possible with insulators. More specifically, an excited sensitizer, S*, may transfer an electron to the semiconductor forming a charge separated pair [Eq. (1)]:

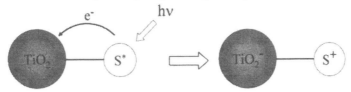

For artificial photosynthetic applications, the importance of this charge-separation process is that it provides a molecular basis for the conversion of photons into potential energy. If the interfacial charge-separation pair has a sufficiently long lifetime, it may undergo subsequent bimolecular redox processes forming useful chemical products. For these types of applications, it is desirable to prevent the energy-wasting charge recombination process that yield ground-state products [Eq. (2)]:

A fundamental goal, therefore, is to form interfacial charge-separated states that have long lifetimes.

A particular advantage of interfacial charge-separated states at semiconductor materials is that the injected electrons can be collected as an electrical response. This forms the basis for new applications that exploit both electronic and optical properties of the sensitized materials such as charge storage, displays, and optical switching.

To date, one of the most promising applications has been the direct conversion of light into electricity with sensitized nanocrystalline semiconductor films.

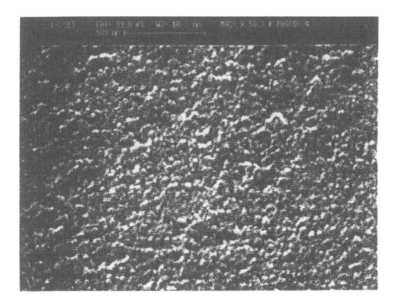

Figure 1 Transmission electronic microscope picture of a porous nanocrystalline TiO$_2$ film.

In order to take advantage of nanometer-sized semiconductor clusters, one must provide an electron pathway for conduction between the particles. This has been achieved by sintering colloidal solutions deposited on conductive glasses. The resulting material is a porous nanostructured film, like that shown in Fig. 1, which retains many of the characteristics of colloidal solutions, but is in a more manageable form and may be produced in a transparent state. Furthermore, the Fermi level within each semiconductor particle can be controlled potentiostatically, a feature which is fundamental for the functioning of the electrochromic devices described in Section III.

The attention devoted to supramolecular sensitizers containing multifold chromophoric and electroactive centers arises from the construction of molecular devices based on nanometric and well-defined molecular architectures [4]. The use of these species for sensitization of titanium dioxide has provided fundamental insights into interfacial electron-transfer processes.

In general, a supramolecular species possesses the following attributes: (1) The intrinsic properties of the molecular components are not significantly perturbed and (2) the properties of the supramolecular system are not a simple superposition of the properties of the molecular components; there is a supramolecular function [13]. Upon substitution of one of the molecular components by a semiconductor, a heterosupramolecular system is formed. In addition to interfacial electron transfer, heterosupramolecular systems have been designed to support intramolecular electron transfer or energy-transfer functions. In order for these heterosupramolecular systems to perform these and other desired functions, several nontrivial issues must be addressed. The first area of concern is the molecular architecture at the semiconductor surface where the appropriate molecular components must be assembled with suitable connectors in the right sequence such that each step is thermodynamically allowed. In addition, delicate issues of a kinetic nature must be taken into account. As a matter of fact, the kinetics of each of the electron-transfer steps should be optimized, such that absorbed photons are quantitatively converted into an interfacial charge-separated state and, hence, into an efficient electrical response.

The discussion which follows is divided into two main sections. The first termed "antenna sensitizers" presents studies of polynuclear compounds with a surface bound unit that can accept energy from covalently linked chromophoric groups and inject electrons into the semiconductor from its excited state. The second describes supramolecular assemblies designed to promote intramolecular and interfacial electron transfer upon light absorption.

A. Antenna Sensitizer Molecular Devices

The requisites for the supramolecular antenna sensitizers are (1) an efficient antenna effect, vectorially translating absorbed energy toward a molecular component and (2) the capability of the molecular component bound to the semiconductor surface to inject electrons into the semiconductor from its excited state. Antenna sensitizers can increase the fraction of light harvested by a sensitized semiconductor surface. Two simple prototypes, following the "branched" or "one-dimensional" design, are shown schematically in Fig. 2. An a priori evaluation

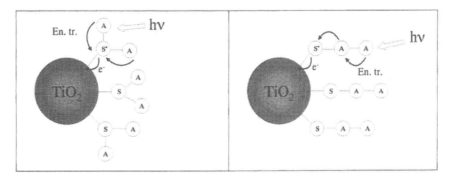

Figure 2 Example of antenna-sensitizer molecular devices on TiO_2.

of the two types of design is difficult, as both have virtues and limitations. For example, when a finite driving force for each energy-transfer step is necessary, a branched design, where extensive use of parallel processes is made, is energy saving relative to the one-dimensional design, where all the processes are in series. If, however, fast isoenergetic energy hopping between components can take place, followed by trapping at the lowest-energy component sitting on the surface, then the difference between the two types of design becomes much less critical. In the branched design, the dimensions of the antenna system cannot be increased without introducing energy losses. In the one-dimensional design, quite large antenna systems could be envisioned.

There are important antenna–semiconductor issues that are specific to the system of interest. For a supramolecular system considered as an independent photochemical molecular device, an obvious expectation is that "the larger the antenna system, the larger the light–harvesting efficiency." The same is not necessarily true for light-to-electrical energy conversion on a semiconductor. A highly branched supramolecular system projects a much larger area than a simple molecular sensitizer onto the semiconductor surface. At saturation coverage, this would strongly reduce the potential gain represented by the antenna effect. From this point of view, the one-dimensional design would be superior to the branched one, as one could think of increasing the nuclearity of the supramolecular system indefinitely without substantially increasing the occupied area. These arguments, however, should be taken with caution, as the semiconductor surface available for absorption is far from being an idealized flat surface. In the state-of-the-art nanocrystalline photoanodes, an extremely rough surface is present, with nanometer-sized pores and cavities (Fig. 1). As such, the dimensions may be comparable to those of the antenna systems and geometric considerations may, therefore, be critical in determining the optimal design and the degree of nuclearity for a heterosupramolecular antenna sensitizer.

The idea of using sensitizer-antenna molecular devices in the sensitization of semiconductors stems from the need of increasing the light-harvesting efficiency of nanoporous thin electrodes and from the low efficiency observed with multilayers of sensitizers deposited on electrode surfaces. In a first attempt to investigate the feasibility of such an approach, the $(cis\text{-}[(NC)Ru(bpy)_2(CN)]_2$ $Ru(dcbH_2)_2)$ complex [bpy $=$ 2,2′-bipyridine and dcbH$_2$ $=$ 4,4′-$(CO_2H)_2$-2,2′-bipyridine] was synthesized [14]. Efficient energy funneling from the peripheral chromophores to the central —Ru(dcbH$_2$)$_2$ unit was demonstrated by conventional photophysical experiments and by time-resolved resonance Raman spectroscopy [15]. Experiments carried out with this branched antenna sensitizer on TiO$_2$ in aqueous solution at pH 3.5 resulted in significant photocurrents. Plots of the photocurrent efficiency versus the excitation wavelength were similar to the absorption spectrum of the trinuclear compound, indicating that the efficiency with which absorbed light was converted to electrons in the external circuit was constant and did not depend on whether the incident light was absorbed by the central unit or by the terminal ones. Subsequent experiments on this complex anchored to nanocrystalline TiO_2 gave a global conversion efficiency of ~7% under simulated sunlight conditions with turnover numbers of at least five million without decomposition [16]. High photocurrent efficiencies were also observed with related compounds based on the same $Ru(dcb)_2(CN)_2$ core and a lateral $Ru(1,10$-phenanthroline) antenna [17].

The charge-injection process from the photoexcited (cis-$[(NC)Ru(bpy)_2$ $(CN)]_2Ru(dcbH_2)_2$) to the TiO_2 semiconductor was investigated by monitoring the initial decay of the time-resolved luminescence [18]. From the lifetime of the fast-emission component (170 ps), an injection rate of ~6 $\times$ 10^9 s^{-1} was estimated. As will be discussed in Section II.C, the possibility of remote interfacial electron transfer from the metal-to-ligand change-transfer (MLCT) states localized on the lateral —$Ru(bpy)_2$ units cannot be ruled out, and it might represent an important pathway for charge injection into the conduction band. In the case of the trinuclear complex, however, remote injection would have to compete with the fast intracomponent energy-transfer processes which have been observed in the picosecond time domain for this class of polynuclear compounds [19,20].

Interesting photoelectrochemical properties were also observed for the trinuclear complex ($[trans$-$(NC)Ru(py)_4(CN)]_2Ru(dcbH_2))^{2+}$ (where py = pyridine), with incident photocurrent conversion efficiencies (IPCE) in the order of 60% following excitation in the lowest lying $Ru^{II} \rightarrow (dcbH_2)$ MLCT band at 540 nm (Fig. 3). The absorption difference spectra of the complex bound to TiO_2, recorded at different delay times after pulsed 355-nm excitation (Fig. 4) are characteristic of the one-electron oxidized form of the complex, ($[(NC)Ru^{II}$ $(py)_4(CN)]_2Ru^{III}(dcbH_2))^{3+}$. Considering the spectroscopic features of the complex with Ru $\rightarrow$ py MLCT transitions at higher energy (360 nm) compared to the $Ru(II) \rightarrow (dcbH_2)$ MLCT (540 nm) (Fig. 4a) and considering its cyclic voltammetry with three reversible waves in acetonitrile ($E_{1/2}$ at 0.85 V E_1, 1.10 V E_2, and 1.40 V E_3) that are assigned to $Ru(II)/(III)$ processes involving the central (E_1) and to the two external units (E_2, E_3), it can be concluded that the final interfacial charge-separated state can either be formed through direct population of the $Ru(II) \rightarrow (dcbH_2)$ MLCT state and energy transfer to this state from the $trans(NC)Ru(py)_4(CN)$ units (process 1 in Fig. 5) or via remote injection from the exited $trans$-$(NC)Ru(py)_4(CN)$ unit followed by intramolecular electron transfer (processes 2 and 3 in Fig. 5). This mechanism is currently under investigation by using ultrafast laser spectroscopy.

B. Charge-Separating Sensitizers

Two simple supramolecular dyad systems, containing a chromophoric component called sensitizer (S) and a covalently linked acceptor (A) or donor (D) component,

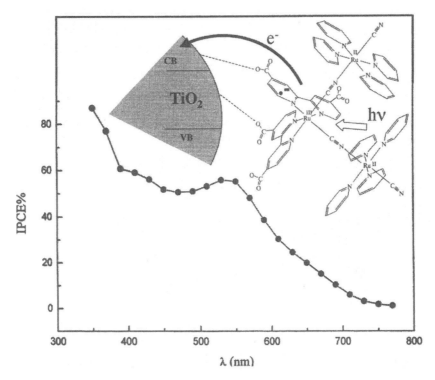

Figure 3 Incident photocurrent conversion efficiencies (IPCE) observed for the tri-nuclear complex ($[trans\text{-}(NC)Ru(py)_4(CN)_2]Ru(dcbH_2))^{2+}$ on a 6-μm-thick TiO_2 pho-toanode in the presence of 0.3 M I_2 and 0.03 M LiI.

are shown bound to a semiconductor in Fig. 6. When anchored to a semiconductor surface that can act as an electron acceptor, these dyads may be referred to as "heterotriads." In principle, an extension from heterotriads to larger systems can be envisioned keeping in mind that each additional electron-transfer step may reduce the free energy that can be stored in the final interfacial charge-separated state. The heterotriads shown are designed to efficiently perform the processes indicated by steps 1 and 2 in Fig. 6. The goal is to quantitatively photocreate an interfacial charge-separated pair with an electron in the semiconductor and a "hole" localized on a molecular unit away from the semiconductor surface. The rate constant for electron–hole recombination may be inhibited relative to that observed with a simple molecular sensitizer. The kinetics must be optimized in order to achieve efficient long-lived charge separation. In the heterotriad shown in Fig. 6a, interfacial excited-state electron transfer is expected to be quantitative

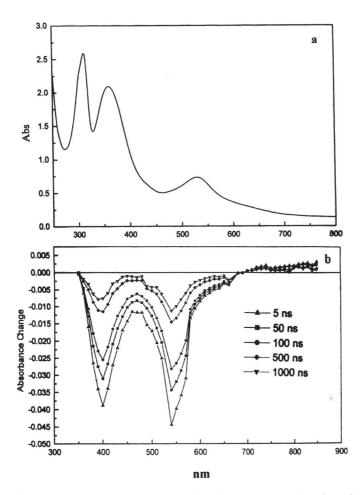

Figure 4 Absorption spectrum in CH$_3$CN solution (a) and transient absorbance difference spectra on TiO$_2$ film (b) of ([*trans*-(NC)Ru(py)$_4$(CN)$_2$]$_2$Ru(dcbH$_2$))$^{2+}$

and ultrafast, so the key to the overall efficiency is likely to be determined by the kinetic competition between the secondary electron transfer process (step 2) and the primary charge-recombination process (step 3). On the other hand, as shown in Fig. 6b, the relative kinetics for intramolecular and interfacial electron transfer are expected to control the efficiency (steps 2 and 4, respectively). In both cases, it is evident that considerable control of the factors that govern electron-transfer rate constants, such as driving force, reorganizational barriers, and

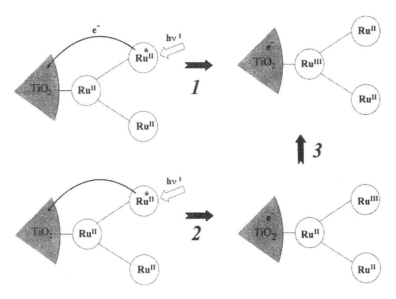

Figure 5   Photoinduced direct and remote charge-injection process leading to a common interfacial charge-separated state.

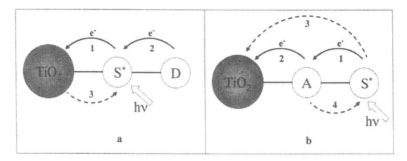

Figure 6   Photoinduced stepwise electron-transfer process in TiO_2–sensitizer–donor and TiO_2 acceptor–sensitizer heterotriads.

electronic coupling, must be attained before successful supramolecular devices of this type can be developed. The studies described below represent the first steps toward the required molecular precision.

The first dyad synthesized to perform the function is Ru(dcbH$_2$)$_2$(4-CH$_3$,4'-CH$_2$–PTZ-2,2'-bipyridine)$^{2+}$, where dcb is 4,4'-(CO$_2$H)$_2$-2,2'-bipyridine and PTZ is the electron donor phenothiazine [11,12]. The resulting heterotriad with TiO$_2$ is shown schematically in Fig. 7. Irradiation of the dyad with visible light results in the creation of a MLCT excited state, which is quenched by electron transfer from the PTZ group in fluid solution. The reductive excited-state quenching is moderately exergonic ($<$ 0.25 eV) and has a rate constant of 2.5 $\times$ 10^8 s^{-1} in methanol, as estimated from the lifetime of the residual *Ru(II) emission. The corresponding charge-recombination step is faster than the forward one so that there is no appreciable transient accumulation of the electron-transfer product.

When the dyad system is attached to TiO$_2$, MLCT excitation can result in a new charge-separated state with an electron in TiO$_2$ and an oxidized PTZ group, abbreviated TiO$_2$(e^-)–Ru–PTZ$^+$. In principle there are two possible pathways available to reach this charge separation. In the first pathway, charge injection is followed by oxidation of the phenothiazine donor by the oxidized sensitizer unit, TiO$_2$–Ru(II)*–PTZ $\rightarrow$ TiO$_2$(e^-)–Ru(III)–PTZ $\rightarrow$ TiO$_2$(e^-)–Ru(II)–PTZ$^+$. In an alternative pathway, reductive quenching by the PTZ group is followed by charge injection into the semiconductor, TiO$_2$–Ru(II)*–PTZ $\rightarrow$ TiO$_2$–Ru(I)–PTZ$^+$ $\rightarrow$ TiO$_2$(e^-)–Ru(II)–PTZ$^+$. Note that the "Ru(I)" intermediate does not refer to ruthenium in a +1 oxidation state but, rather, to Ru(II) coordinated to a reduced dcbH$_2$ ligand [i.e., Ru(II)dcb$^-$].

A flash photolysis study of the heterotriad shown in Fig. 7 was undertaken [11,12]. With nanosecond time resolution, it was not possible to determine

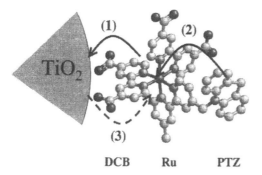

Figure 7   Interfacial and intramolecular electron-transfer process for Ru(dcbH$_2$)$_2$(4-CH$_3$, 4'-CH$_2$–PTZ-2-2'bpy)$^{2+}$ on TiO$_2$.

whether the $TiO_2(e^-)$–Ru(II)–PTZ$^+$ state was formed by interfacial electron transfer from the excited or reduced state (i.e. pathway 1 or pathway 2, respectively). However, electron injection into TiO_2 from MLCT excited states occurs on a femtosecond time scale, so pathway 1 is the most probable under the experimental conditions employed. After electron injection, electron transfer from PTZ to the Ru(III) center ($-\Delta G \sim 0.36$ eV) produces the charge-separated state $TiO_2(e^-)$–Ru(II)–PTZ$^+$. Recombination of the electron in TiO_2 with the oxidized PTZ to yield the ground state occurs with a rate constant of 3.6×10^3 s^{-1}. Under otherwise identical conditions, excitation of a model compound that does not contain the PTZ donor, Ru(dmb)(dcb)$_2^{2+}$ [where dmb is 4,4'-(CH$_3$)$_2$-2,2'-bipyridine), gave rise to the immediate formation of a charge-separated state, $TiO_2(e^-)$–Ru(III), whose recombination kinetics were complex and analyzed by a distribution model with an average rate constant of 3.9×10^6 s^{-1}. Therefore, translating the "hole" from the Ru center to the pendant PTZ moiety inhibits recombination rates by about three orders of magnitude.

Grätzel and co-workers have recently reported an interesting study of heterotriads of this type and have emphasized their potential application in photochromic devices [21].

The possibility of using polynuclear metal complexes as charge-separating sensitizers was tested for binuclear Ru(II)–Os(II) species. The binuclear compound [Ru(dcbH$_2$)$_2$(Cl)–BPA–Os(bpy)$_2$Cl]$^{2+}$ [where BPA is 1,2-bis(4-pyridyl) ethane and bpy is 2,2'-bipyridine] shown in Fig. 8 was prepared to study processes like that shown in Fig. 6a [22]. A key difference between this binuclear compound and the Ru(II)–PTZ dyads is that the Os(II) donor is a chromophore that absorbs visible light. The binuclear compound binds to TiO_2 thin films with about half of the surface coverage of a model compound, [Ru(dcbH$_2$)$_2$(Cl)(py)]$^+$ (where py is pyridine). This suggests that the binuclear complex lies on the nanocrystalline TiO_2 surface in a more or less extended conformation.

Transient absorbance difference spectra measured following 532-nm laser excitation, where both Ru(II) and Os(II) chromophores absorb, reveal the typical

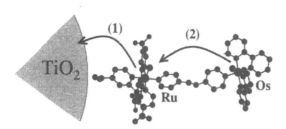

Figure 8 Interfacial and intramolecular electron-transfer process for [Ru(dcbH$_2$)$_2$ (Cl)–BPA–Os(bpy)$_2$Cl]$^{2+}$ on TiO_2.

bleaching of the spin-forbidden MLCT transition localized on the Os(II) unit. Spectral and kinetic analysis of the transient signals are consistent with the formation of the charge-separated state $TiO_2(e^-)$–Ru(II)–Os(III). This state can be formed through charge injection from the excited Ru chromophore followed by intramolecular Os(II) → Ru(III) electron transfer or via remote electron transfer from the MLCT excited state localized on the Os(II)(bpy)$_2$ unit. Comparative actinometry revealed that the desired intramolecular electron-transfer process occurs efficiently with 532- or 417-nm light excitation. The occurrence of remote transfer from the excited Os unit was confirmed by time-resolved experiments after selective laser excitation of the Os(II) chromophore at 683 nm. The lifetime of the $TiO_2(e^-)$–Ru(II)–Os(III) was not significantly different than that of model compounds, presumably because of the semiconductor–dyad orientation.

Recently, two dinuclear ruthenium and osmium containing polypyridyl complexes immobilized on nanocrystalline TiO$_2$ surfaces were studied. The compounds are based on the bridging ligand 3,5-bis-(pyridin-2-yl)-1,2,4-triazole (Hbpt) (Chart 1) [23]. Both homonuclear and heteronuclear ruthenium and osmium complexes of this ligand have been studied in solution. Electrochemical studies indicate that in the presence of the deprotonated bpt$^-$ bridge, the interaction between the two metal centers is strong. Photophysical measurements have shown that for the mixed-metal ruthenium–osmium compound efficient energy transfer takes place from the Ru center to the Os moiety. An important difference between the bpt-based compounds and the Ru–BPA–Os dimer is the rigidity of the bpt bridge which prevents rotation around the linker. The results reported in the study show that upon immobilization of these dinuclear compounds onto a solid substrate, a substantial difference is observed between the photophysical processes of the heterotriad and of the dyad in solution. The data also strongly indicate that direct injection from moieties not directly bound to the oxide surface can be efficient. Importantly, for the TiO$_2$–RuOs triad studied, no osmium-based emission was detected and injection from both the ruthenium and the osmium centers was faster than the laser pulse. Also, time-resolved absorbance measurements revealed the formation of only one final product, $(e^-)TiO_2$–Ru(II)Os(III), upon irradiation of TiO$_2$–RuOs.

C. Remote and Stepwise Interfacial Electron-Transfer Processes

Precedence for remote electron transfer like that observed in the TiO$_2$–Ru(II)–L–Os(II) triads exists. In previous work, a supramolecular approach for designing a molecular sensitizer with controlled orientation of the component units on the semiconductor surfaces was reported (Fig. 9). The binuclear compound is based on a fac-Re(I)(dcbH$_2$)(CO)$_3$ surface anchoring unit and a —Ru(II)(bpy)$_2$ chromophore linked through an ambidentate cyanide ligand [24]. Due to the facial

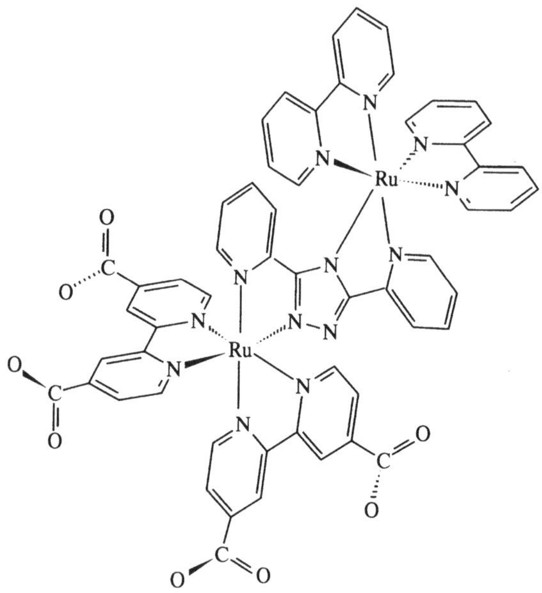

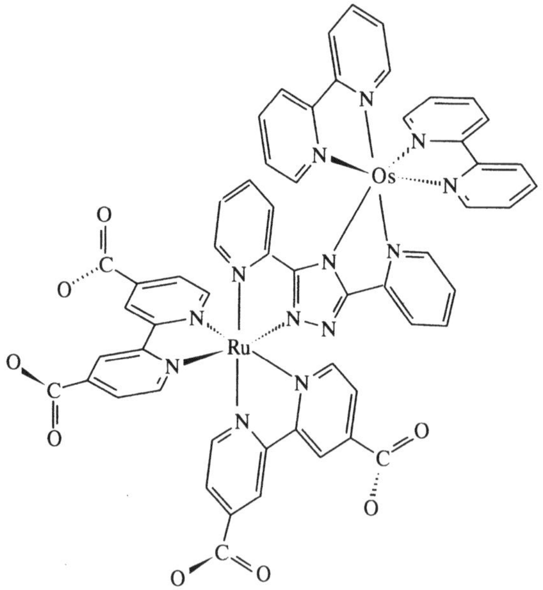

Chart 1

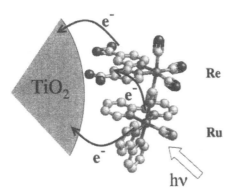

Figure 9 Interfacial and remote electron-transfer process for the binuclear [fac-Re$^{(I)}$ (dcb)(CO)$_3$(CN)Ru$^{(II)}$(bpy)$_2$(CN)] complex on TiO$_2$.

geometry of the surface-bound Re- group, the -Ru(bpy)$_2$ unit is forced to be proximate to the semiconductor surface. Visible-light excitation of TiO$_2$ photoanodes, loaded with Re(dcbH$_2$)(CO)$_3$(CN)Ru(bpy)$_2$(CN)$^+$ or its linkage isomer, in a regenerative solar cell resulted in efficient light-to-electrical energy conversion. Plots of the photocurrent efficiency versus excitation wavelength demonstrated that the Ru–polypyridine group absorbs the visible light and converts it efficiently into an electrical current. The transient absorption difference spectra for these sensitizers bound to TiO$_2$ showed a broad bleach in the region from 400 to 600 nm, expected for the TiO$_2(e^-)$–Re(I)–Ru(III) state. The formation of this state is promptly observed after the laser pulse. This indicates that either remote electron injection into TiO$_2$ or intraligand (bpy$^-$ → dcb) electron hopping from Ru(III) (bpy$^-$) to Re(I)(dcbH$_2$) occurs within the laser pulse ($k > 5 \times 10^8$ s^{-1}). The data demonstrate a rapid and efficient injection process from a chromophoric group which is not directly coupled to the semiconductor surface. In fact, the solar-energy conversion efficiencies realized with TiO$_2$–Re(I)–Ru(II) materials in regenerative photoelectrochemical cells were the same, within experimental error, of those measured for materials sensitized with cis-Ru(dcbH$_2$)$_2$(CN)$_2$.

In order to test the performances of the second type of heterotriad shown in Fig. 6b, two Ru(II)–Rh(III) polypyridine dyads, containing carboxylic acids groups at the Rh(III) unit, Rh(dcb)$_2$–(BL)–Ru(dmp)$_2$ and Rh(dcb)$_2$–(BL)– Ru(bpy)$_2$ (where BL is 1,2-bis [4 (4′-methyl-2,2′-bipyridyl)]ethane) and dmp is 4,7-dimethyl-1,10-phenanthroline) were prepared [25]. In both species, the Ru unit was designed to play the role of the chromophoric donor and the Rh unit that of the acceptor. Their photophysical behavior in solution indicates the occurrence of fast (10^8–10^9 s^{-1}) and efficient (>95%) photoinduced electron transfer from the excited Ru chromophore to the covalently bound Rh(III) center.

   The Rh–Ru dyads anchored to TiO_2 could be used as efficient sensitizers, giving rise to maximum incident photon-to-current efficiency values of ~40–50% with iodide electron donors in acetonitrile. The photoinduced electron–transfer properties in the heterosupramolecular systems TiO_2–Rh–Ru(dmb)$_2$ and TiO_2–Rh–Ru(dmp)$_2$ can be summarized as follows. Upon light excitation of the Ru(II) unit, one-third of the surface-bound dyads undergo direct electron injection from the excited state of the Ru chromophore, TiO_2–Rh(III)–Ru(II)* → $TiO_2(e^-)$–Rh(III)–Ru(III), with a rate constant $>10^8 \text{ s}^{-1}$. The dyads that undergo this remote injection process probably have different surface orientations or accidental contacts in small cavities within the nanocrystalline TiO_2 film. The remaining dyads display stepwise charge-injection processes [i.e., TiO_2–Rh(III)–Ru(II)* → TiO_2–Rh(II)–Ru(III) → $TiO_2(e^-)$–Rh(III)–Ru(III)]. The first process has comparable rates and efficiencies as the process for the free dyads in solution. The second step is 40% efficient, because of primary charge recombination between the reduced Rh unit and the oxidized Ru, TiO_2–Rh(II)–Ru(III) → TiO_2–Rh(III)–Ru(II). When the recombination of the injected electron and oxidized Ru(III) sites was studied, a remarkable slowdown was observed relative to a simple mononuclear sensitizer, TiO_2–Ru(dcbH$_2$)$_2$dmb^{2+}. Perhaps more importantly, these studies demonstrate a "stepwise" interfacial electron-transfer process, like that shown in Fig. 6b, for the first time.

III. ELECTROCHROMIC DEVICES (ECDs)

Classically speaking, electrochromic materials are chemical species whose color can be interchanged electrochemically due to the generation of a different electronic absorption band in the visible region [26]. In a more general sense, the terminology does not only include visible color changes but also electrochemically induced changes in the near-infrared (NiR), thermal infrared, and microwave regions. This broader definition is reasonable considering the more recent progress toward a multispectral energy modulation of radiation by absorbance and reflection; however, we will focus predominantly on the "classical" field and our use of the term should be understood in this context.

   Electrochromic materials of interest can be classified according to three types of color change: (1) from a bleached state (transparent) to a colored state, (2) between two colored states, and (3) between several colored states, if more than two redox states are accessible. This behavior is called polyelectrochromic [27].

   The appeal of electrochromic materials is at least twofold. From a fundamental point of view, the field encompasses a wide range of systems and presents a challenge to the chemists' understanding of their redox operation and the accompanying physicochemical structure changes [26,28]. From a commercial standpoint, even though only very few devices have been successfully brought to the market, electrochromic devices hold much promise and several areas of interest have been proposed and are currently under investigation and development [29]. However, it is prudent to caution that several practical hurdles still have to be overcome and intense competition is present in some important markets such as in display technology, where liquid crystal displays (LCDs) have been dominating for more than two decades and are well entrenched. Nonetheless, the distinct disadvantages of LCDs [30], namely poor contrast in the absence of back lighting, rapidly vanishing legibility under decreasing viewing angle, narrow temperature operation range, and severe size limitations, leave enough room for improvement for a new technology to emerge.

Among the electrochromic products currently under development are switchable windows and mirrors, electromagnetic shutters, slow display technologies, earthtone chameleon materials, protective eyewear, controllable light-reflective or light-transmissive devices for optical information storage, controllable aircraft canopies, and glare-reduction systems for offices [29]. Car rearview mirrors have already achieved commercial success. They act as safety devices by preventing mirror-reflected glare which causes an "after image" to stay on the eye's retina. Prototypes of window glass that is almost infinitely variable in its degree of tinting have been demonstrated and are very close to commercialization. Those smart windows offer control of thermal conditions within a building for energy conservation at low projected cost [31].

From these desired applications, several key elements and properties of the electrochromic materials can be derived, their respective importance depending on their use. Among them are a high contrast ratio (sharp colors), high coloration efficiency (adsorbance change/charge injected per unit area), long cycle life, low response time, and good write–erase efficiency (percentage of originally formed coloration that may be subsequently electrobleached) [32].

For their characterization, electrochromic compounds are initially tested at a single working electrode under potentiostatic control using a three-electrode arrangement. Traditional characterization techniques such as cyclic voltammetry, coulometry, chronoamperometry, all with in situ spectroscopic measurements, are applied to monitor important properties [27]. From these results, promising candidates are selected and then incorporated into the respective device.

Not surprisingly, the research field of electrochromic compounds is very broad and includes inorganic, organic, and polymeric as well as several hybrid materials. We shall therefore attempt to give a brief overview of past and current achievements before we present our own contributions. Due to its scope, this chapter cannot be considered comprehensive, and the reader is referred to recent review articles for a more detailed description. [26,27,33].

A. Metal Oxide Thin Films

By applying an outside potential, many metal oxide films are switchable to a nonstoichiometric redox state, creating an intervalence charge transfer with an intense electronic absorption band. The change is usually from transparent to colored. Depending on the metal oxide, the intensely colored absorption state can be produced upon reduction (cathodic ion insertion) or upon oxidation (anodic ion insertion). Detailed coloration mechanisms are not fully elucidated, but it is widely accepted that injection and extraction of electrons and metal cations (Li^+, H^+, etc.) in the cathodic case and proton extraction or anion insertion in the anodic case are involved.

1. Examples

a. WO_3. The electrochromism of tungsten trioxide was reported over 30 years ago and has therefore been investigated extensively [34,35]. To explain this behavior, the proposition is widely accepted that WO_3 is transparent as a thin film with all metal sites at oxidation state VI, but upon reduction, sites at oxidation state V are generated and a blue color is observed. The intercalation of small metal cations, such as Li^+, is likely involved in the electrochromic effect, as can be rationalized in the following equation [34,35]:

$$WO_3 + x(M^+ + e^-) \rightarrow M_x W^{VI}_{(1-x)}) W^V_x O_3$$

In the nonstochiometric redox state, intervalence charge transfer between adjacent W^{VI} and W^V sites leads to the observed intense blue color. Apparently, if the ratio of W^{VI} to W^V is increased further, the ion insertion becomes irreversible, resulting in a permanent color change to red or golden [34,35]. It should be noted that in spite of the apparent progress in explaining the electrochromism on a fundamental level, questions remain and alternate mechanisms have been proposed which suggest the involvement of other oxidation states [36].

b. TiO_2. Nanocrystalline TiO_2 films show a similar behavior, switching from transparent to blue upon electrochemical reduction. Application of a negative potential triggers the rapid accumulation of electrons in the TiO_2 electrode which is compensated by small ions in the vicinity of the electrode–electrolyte interface. The migration of protons and lithium ions within the pores of the metal oxide is reasonably fast and the accompanying coloration is moderately intense (absorbance change from 0.2 to 2.2 at 780 nm within 14 s upon potential sweep from -0.64 to -1.64 V versus Ag/AgCl) and reversible at a comparable rate, but not sufficient for use in devices on its own [37]. The suggested mechanism is based on the formation of Ti^{3+} ions and the presence of electrons in the conduction band:

$$TiO_2 + nxe^- + xM^{n+} \rightarrow xMTiO_2$$

c. Other Metal Oxides. Several other metal oxides show coloration on electrochemical reduction, such as MoO_3, V_2O_5, and Nb_2O_5, and similar mechanisms have been proposed [27].

Examples for electrochromic behavior upon electrochemical oxidation can be found among group VIII metal oxides. Thin films of transparent hydrated iridium oxide turn blue-black, whereas nickel oxide switches from pale green to brown-black, possibly due to the absorbance of Ni^{3+} centers [26]. The systems are much less thoroughly investigated and a detailed mechanistic explanation is not known. However, proton extraction and anion insertion have been suggested.

Metal-oxide-based electrochromic systems are especially interesting for the development of electrochromic windows because they mostly switch from a transparent state to a dark colored state [38,39]. In addition, their relatively slow response times are acceptable for this kind of application, possibly even preferable from an aesthetic point of view. Again, WO_3 has seen the most use in the development of actual devices. Several different deposition techniques have been applied. For example, a prototype electrochromic window based on WO_3 with reasonable dimensions (0.7 × 1 m) has been assembled that reduces light transmission by a factor of 4 in its colored state [28].

B. Conducting Polymers

Conjugated conducting polymers consist of a backbone of resonance-stabilized aromatic molecules. Most frequently, the charged and typically planar oxidized form possesses a delocalized π-electron band structure and is doped with counteranions (p-doping). The band gap (defined as the onset of the $\pi-\pi^*$ transition) between the valence band and the conduction band is considered responsible for the intrinsic optical properties. Investigations of the mechanism have revealed that the charge transport is based on the formation of radical cations delocalized over several monomer units, called polarons [27].

Reduction of conducting polymers leads to a loss of electronic conjugation and an exit of anions, resulting in an electrically insulating "undoped" and neutral form. In the neutral form, the extended delocalization of the π-electrons along the polymer backbone generates an optical absorption band due to $\pi-\pi^*$ transitions in the visible region, but at higher energies than the electronic transitions of the oxidized polymer [40]. In principle, the opposite process leading to radical anions doped with countercations for charge balancing (n-doping) is possible as well. However, the lower stability of this cathodic doping has limited its applicability and reports are much less common [27].

The very large number of conducting polymers provides an immense pool for potential electrochromic devices. Not only are there many classes of conducting polymers, but their variations due to monomer derivatization are virtually endless. As thin films, potentially all conducting polymers show electrochromic behavior as the redox processes are accompanied by changes in the optical absorption bands due to the replacement of the $\pi-\pi^*$ transitions by electronic transitions at lower energies in the charged form. The changes in the optical absorption

bands can be tuned by modification of the monomeric unit or use of copolymerization. Key to the color tailoring is the control of the band gap which is directly
related to the relative energies of the HOMO and LUMO [41]. By determining
the effective conjugation length and the electron density along the polymer backbone, the substitution pattern on the monomeric units allows the adjustment of
the gap in a controlled direction through the introduction of electronic and steric
effects. Other parameters of interest are the bond-length alteration, resonance
energy, interchain effects, and interring torsion angles. Small changes can have
a noticeable impact [42]. Another, less deliberate approach for the control of
color is the copolymerization of distinct monomers or homopolymerization of
hybrid monomers containing several distinct units. Several examples have demonstrated that color changes are observed upon variation of the ratios of monomers
used in copolymerizations of carbazoles, pyrrols, and thiophenes [43–46].

The diffusion of the charge-compensating counterions through the thin films
determines the response time of the systems during redox switching. A more
open polymer morphology therefore enhances the ionic mobility and yields a
faster response [40].

In addition to their immense variability, conducting polymers are characterized by several other key advantages, such as fast (subsecond) switching times,
high coloration efficiency, durability, mechanical flexibility, and facile processability.

1. Examples

a. Polypyrroles. Polypyrroles can be efficiently prepared as thin films
by electrochemical polymerization of pyrrol (the oxidation potential of pyrrol is
$+0.9$ V versus Fc/Fc$^+$) from organic solvents, such as acetonitrile or propylene
carbonate, using multiple-scan cyclic voltammetry [47]. By controlling the charge
passed, the thickness of the film can be conveniently modulated. The oxidative
electropolymerization begins with the formation of a radical cation species from
the monomer. Both a radical cation–radical cation coupling or an attack of the
radical cation on a neutral monomer have been proposed as likely mechanistic
pathways for the formation of polypyrrole (Scheme 1) [48].

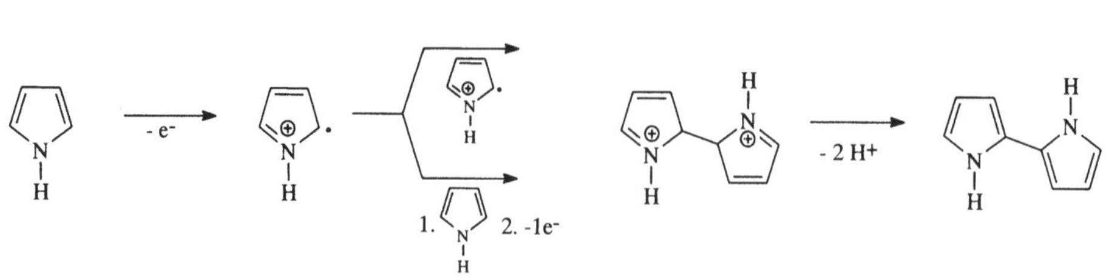

Scheme 1

The blue-violet (λ_{max} = 670 nm) color of the doped (oxidized) polypyrrole changes to yellow-green (λ_{max} = 420 nm) upon electrochemical reduction [49]. Polypyrrole itself has two distinct disadvantages that prevent its use in devices [33]. First, the reversibility of the doping process is facile only in very thin films. As a result, in films thick enough to ensure mechanical stability, the contrast ratio becomes very low. Second, the polypyrroles have been shown to be highly sensitive to overoxidation where the conjugated polymer backbone is degraded through reactions involving positions 3 and 4 of the pyrrole ring [50]. However, when these positions are blocked by suitable substituents, this type of degradation can be avoided and polymers of high electrochemical stability can be obtained [51,52]. Furthermore, the substituents can be chosen to reduce the oxidation potentials and improve specific solubility requirements [53]. Cyclic alkylenedioxy moieties, for example, have been reported to produce excellent results (Scheme 2) [54].

The substituents have an electron-donating effect, but by varying the length of the alkyl bridge, small variations on the magnitude of the effect can modulate the color due to the induced strain and the impact on the planarity of the pyrrole rings. The oxidation potentials of the corresponding polymers are lowered to between -0.6 V and -0.4 V versus Fc/Fc$^+$ and the HOMO energies are raised, resulting in a lower band gap [52]. As a consequence of the modifications, the obtained conducting polymers display greatly improved stability against overoxidation, long-lasting reversibility between their red/orange neutral states and their blue/gray doped states, high contrast ratios, and very fast switching times (less than 200 ms) [52].

Furthermore, additional positions for further derivatization are available. Solubilizing groups can be attached to the alkyl bridge, further increasing solvent compatibility, whereas N-alkylation allows the facile introduction of substituents, ranging from completely nonpolar to ionic [53].

b. Polythiophenes. Similar to polypyrroles, polythiophenes can be efficiently prepared by multiple-scan electropolymerization usually performed in

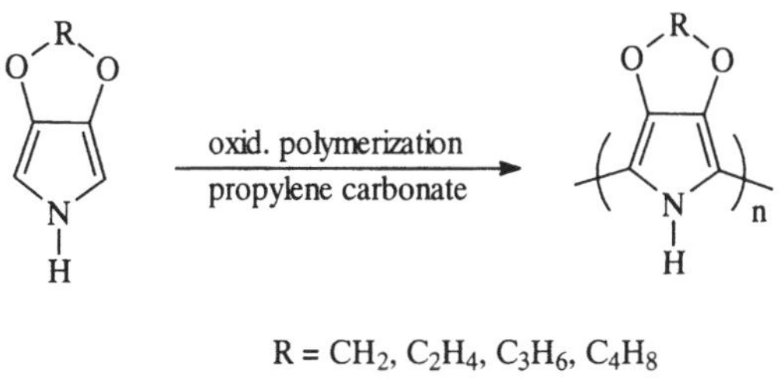

$$R = CH_2,\ C_2H_4,\ C_3H_6,\ C_4H_8$$

Scheme 2

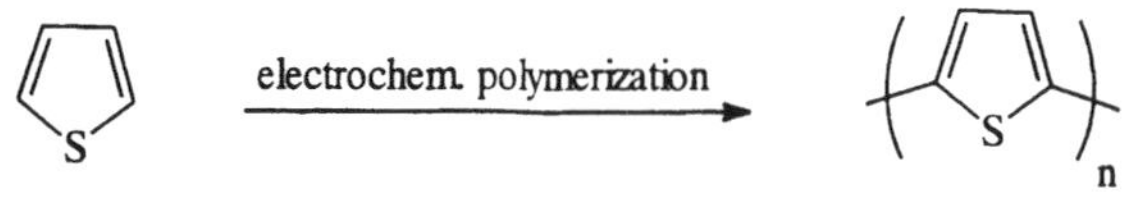

electrochem. polymerization

Scheme 3

acetonitrile at potentials around $+ 1$ V versus Ag/Ag$^+$ (Scheme 3) [55]. Again, the polymerization is initiated by the formation of the monomer radical cation. The parent polymer is chemically significantly more stable than its pyrrole-based counterpart and easily processable [49]. Upon reduction of the doped polymer, a color change from blue ($\lambda_{max} = 730$ nm) to red ($\lambda_{max} = 470$ nm) is observed. The derivatization of the monomer is facile and allows for further improvements of the materials. Interestingly, the substituent effects depend significantly on their relative position on the polymer backbone. For example, 3-methylthiophene-based polymers show a wide variety of colors in the reduced form when either the monomer or various oligomers derived from it are used in the polymerization process [56]. Apparently, subtle changes in the effective conjugation length of the polymer have very noticeable effects. The oxidized forms of the example shown are all blue (Scheme 4).

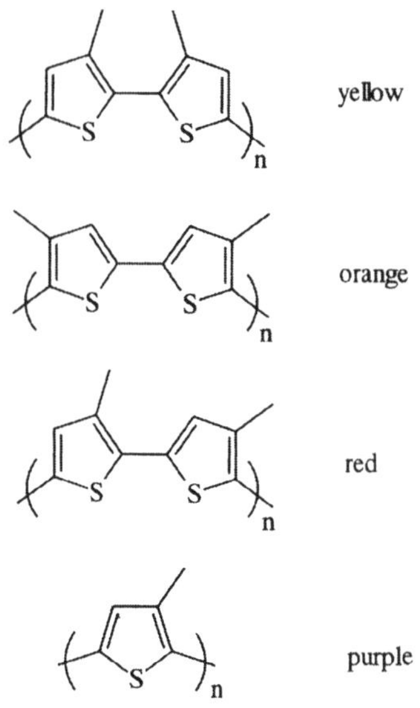

yellow

orange

red

purple

Scheme 4

Further research on the substitution of the thiophene 3-position with phenyl groups containing electron-withdrawing or electron-donating groups (such as methyl, methoxy, fluoro, chloro, bromo, trifluoromethyl, sulfoxy) in the para position have lead to polymers with unique features [57]. The electron-withdrawing groups allow the formation of a radical anion and thus stabilize the n-doped state. As a result, such conducting polymers can be reversibly oxidized and reduced and electrochromic devices can be built with identical anode and cathode materials [58].

Recently, a new and very useful class of substituted polythiophenes has been synthesized and investigated (Scheme 5) [55]. The introduction of cyclic alkylenedioxy moieties provides additional electron density to the polymers and effectively lowers their band gap, shifting the absorption maximum to the NIR region in the doped state. The variation of the bridging alkyl chain of the substituents showed very little additional effect on the conjugation of the polymer backbone, and the band gaps for all studied polymer derivatives were essentially the same at 1.7 eV [55]. Thus, thin films of the polymers exhibit virtual transparency in the visible region, but turn blue-black upon electrochemical reduction to the neutral form. Practically, a cathodically coloring electrochromic material is obtained which complements the anodically coloring behavior of most other common conjugated polymers and allows the preparation of dual-polymer electrochromic devices [40]. In addition, poly(3,4-ethylenedioxythiophene)s are characterized by a low oxidation potential, high chemical stability even under elevated temperatures, and high conductivity. As a result, 300-nm thin films of the material can be fully switched with relatively fast response times of 0.8 to 2.2 s with a relatively high electrochromic contrast ($\Delta\%T$ of 44–63%) [55].

c. Polyaniline. Another class of conducting polymers is based on polyanilines [27]. Again, the polymer can be formed by oxidative electropolymerization which likely initiates from a resonance-stabilized radical cation [59]. The elucidation of the electrochemistry of polyaniline is complicated by the fact that the redox processes are accompanied by protonation/deprotonation steps and sev-

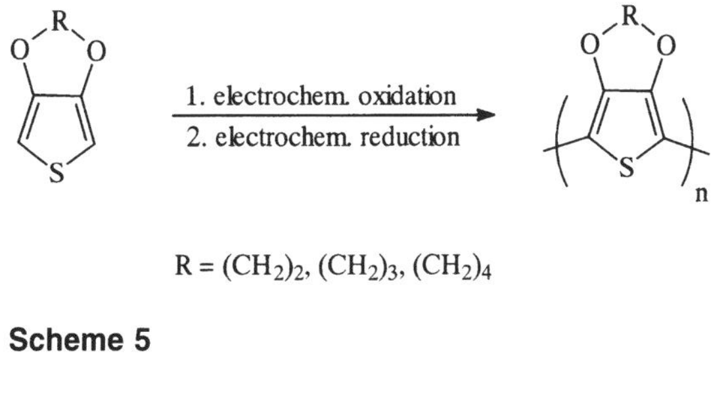

R = $(CH_2)_2$, $(CH_2)_3$, $(CH_2)_4$

Scheme 5

eral mechanisms have been proposed [60]. Depending both on the oxidation state and the protonation state, polyaniline-based thin films display polyelectrochromic behavior ranging from transparent to black [61]. However, only the yellow-green transition is repetitively reversible (Scheme 6) [62]. Changes in the absorption bands of polyaniline below 0.3 V are attributed to changes in the relative intensities between a $\pi-\pi^*$ transition related to the extend of the backbone conjugation and the contribution of radical cations, which appear with increasing potential [63,64]. Above 0.3 V, the polymer becomes conducting accompanied by a broad free-carrier electron band. Alkyl substitution, such as in poly(o-toluidine) and poly(m-toluidine), has improved the stability of the oxidation states compared to

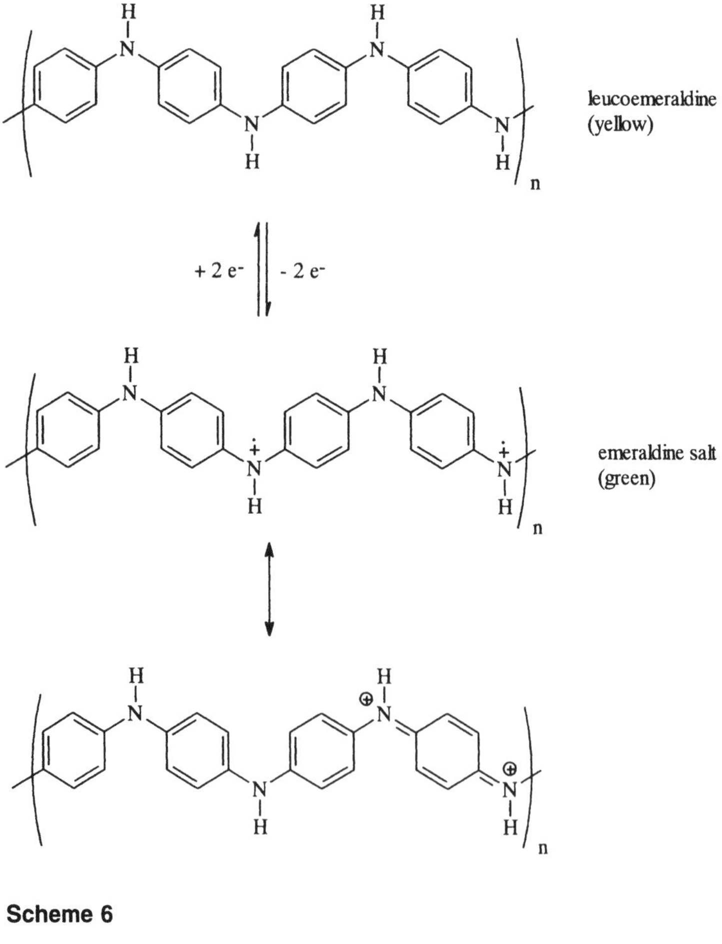

leucoemeraldine
(yellow)

$+ 2 e^-$ | $- 2 e^-$

emeraldine salt
(green)

Scheme 6

the parent, but also increased the response times [65]. In all cases, the reduction process is faster than the oxidation. Electrochemical quartz-crystal microbalance measurements have also revealed that the redox switching in poly(o-toluidine) is very complex and depends on various external parameters [66].

Several processing techniques for coatings with polyanilines have been developed allowing the spray or brush coating of large areas [67]. In electrochromic devices, however, polyaniline has mainly been used in combination with other materials, such as in complementary ECDs with Prussian blue [27] or in composite conducting polymers prepared by electropolymerization of monomers in the presence of additives. Thus, styrenes, cellulose acetate, nitrilic rubber, and indigo dyes have been combined with polyanilines to form new materials [27].

d. Copolymers. The synthesis of copolymers provides a potentially infinite number of conducting polymers and, therefore, electrochromic materials, especially considering that the ratio of the monomers and the type and position of the substituents influence the properties to a large extent. In order to maintain better control over the exact relative positions of substituents within the polymer, but also guided by synthetic considerations, it is often advantageous to build a mixed monomeric unit prior to polymerization. In addition, several important physical properties can be measured on these mixed oligomers in advance and tailored according to desired values.

The tuning of the band gap can be achieved by using alternating donor–acceptor units along the backbone [68,69]. Thiophene–silole copolymers fall into that category. We have already discussed the advantages of the thiophene monomer earlier. The incorporation of silol derivatives is of interest because of their unique electronic structure, including the low-lying LUMO [46]. Its cisoid butadiene structure can also interact with the aromatic heterocycles affecting the band gap. In fact, it has been shown that the band gap is tunable as a function of the component ratios [70]. A very recently reported poly [2,5-bis(2-(3,4-ethylenedioxy)thienyl)silole] shows a very low oxidation potential and a narrow band gap of 1.3–1.4 eV (Scheme 7) [46]. Upon oxidation, the conducting polymer changes from blue to transmissive yellow-green as the π–π* transition is depleted, creating a cathodically coloring electrochromatic behavior [46].

Copolymers of thiophenes and didodecyloxybenzenes have been synthesized as well, exhibiting multicolor electrochromism coupled with high contrast ratios and fast response times (Scheme 8) [45]. Since all three basic colors in the RGB (red–green–blue) system can be obtained, the material is a potential candidate for display applications.

In poly[2,5-bis(3,4-ethylene-dioxy-2-thienyl)pyridine], a copolymer of thiophene and pyridine in a 2:1 ratio, the effective HOMO and LUMO energy levels of the π-system are controlled in such a way that both p-doped and n-doped states are accessible (Scheme 9) [43]. The thin films display multicolor

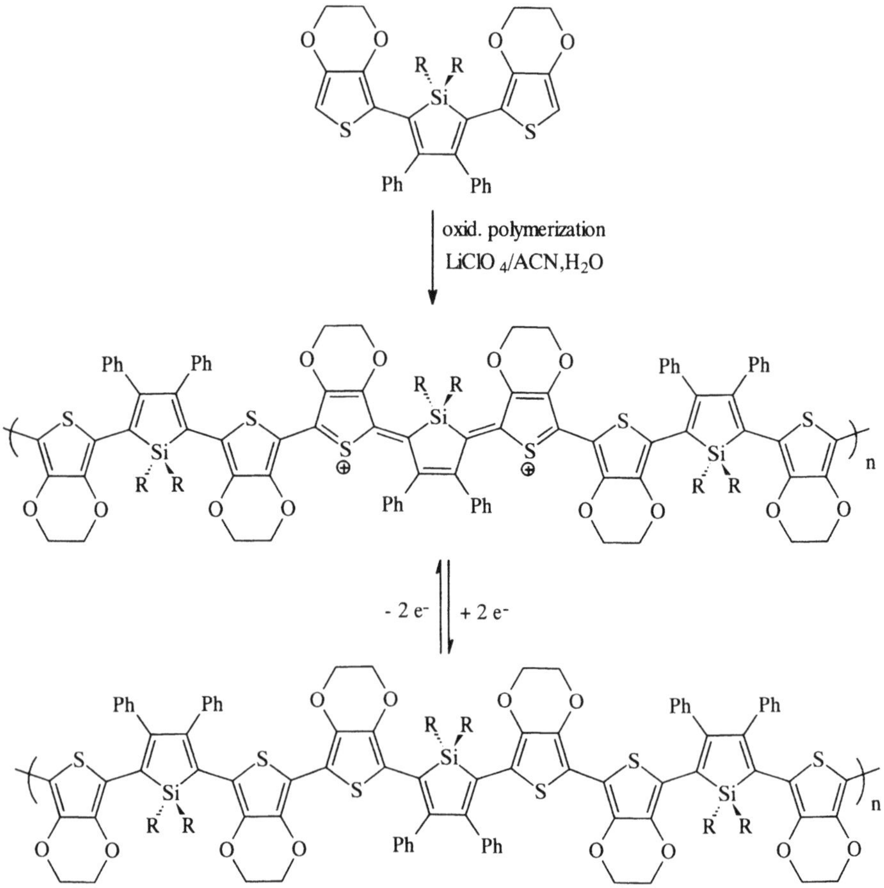

Scheme 7

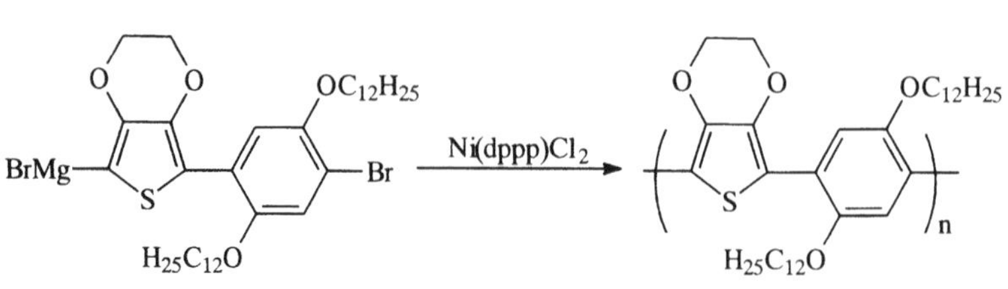

Scheme 8

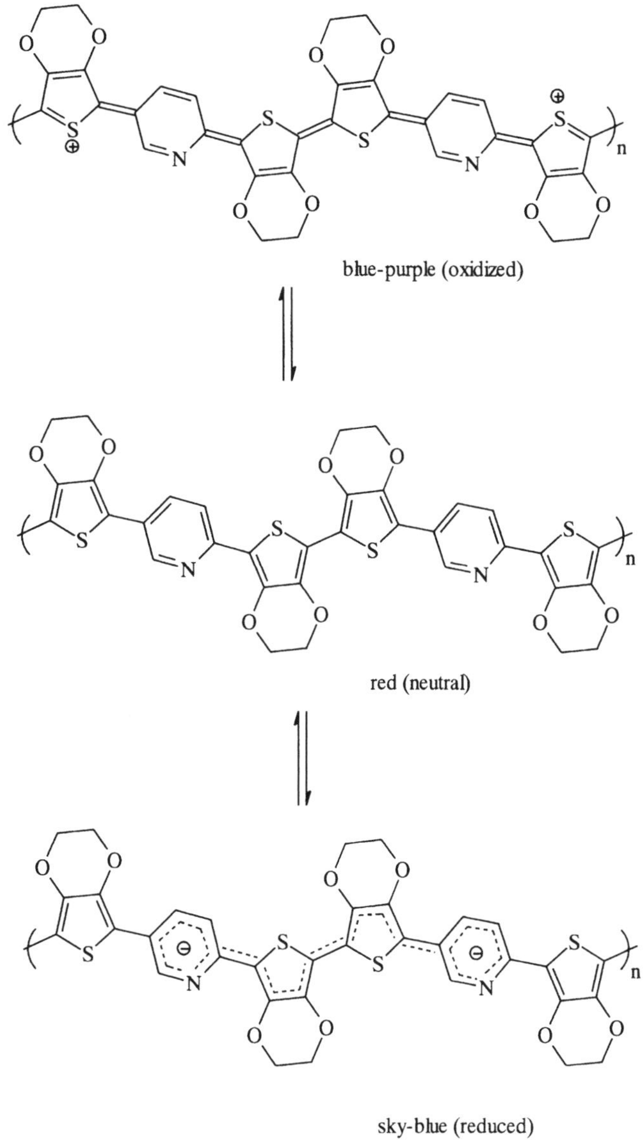

blue-purple (oxidized)

red (neutral)

sky-blue (reduced)

Scheme 9

Scheme 10

electrochromism and can be switched between a pale blue reduced state, a red neutral state, and a blue-purple oxidized state. In addition, the polymer contains reversibly quarternizable sites that result in pH sensitivity. At pH below 2, the neutral polymer is protonated, switches from red to indigo blue, and its band gap is significantly lowered [43].

      e.   Nonconjugated Polymers Containing Pendant π-Groups.   All of the polymers used in electrochromic materials discussed so far belong to the class of conjugated polymers. Recently, the group of nonconjugated polymers containing pendant π-groups has gained some increased attention due to the distinct color changes that result from definitive structures of the radical cations formed upon doping [71,72]. Another reported advantage is their facile processability [44].

      Several such polymers have shown electrochromic behavior, among them poly(*n*-vinylcarbazole) [73] which switches from colorless in the neutral state to green in the doped state (Scheme 10) and poly(*N*-phenyl-2-(2′-thienyl)-5-(5″-vinyl-2″-thienyl)pyrrole) [74], which changes from yellow to reddish brown upon oxidation (Scheme 11). A study of the electrochromic properties of blends consist-

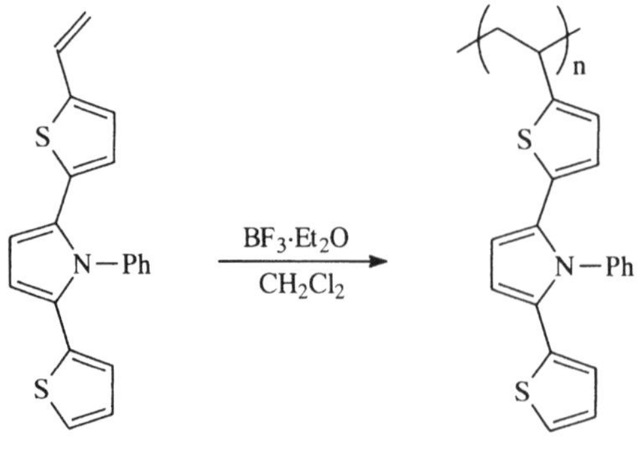

Scheme 11

ing of these nonconjugated polymers has revealed a dependence of the observed color on the ratio of the individual components of the blends. This technique provides a very simple way of generating a polyelectrochromic system and tailoring its colors in the range between different shades of brown, tan and green [44].

f. Nonconjugated Polymers Containing Complexed Transition Metals. Nonconjugated polymers with pendant ligands capable of complexing transition metals, so-called metallopolymers [75], have been prepared as electrochromic materials using both reductive and oxidative electropolymerisation techniques [76]. The electrochromic effect is due to the changes in the absorption bands of the complexes upon oxidation and reduction of the metal.

Attachment of the chromophores to a polymer backbone facilitates the preparation of all-solid-state systems. Iron, ruthenium, and osmium have shown interesting results, usually in combination with polypyridyl ligands, such as bipyridyl, terpyridyl, or phenanthroline. For the preparation via reductive electropolymerization, the ligands are fitted with a suitable substituent, predominantly vinyl, but also chloro, that polymerizes upon electrochemically induced radical formation [33]. Beyond a certain chain length, the oligomer becomes insoluble in the electrolyte and starts to form a thin film on the electrode surface. The film thickness can be controlled through the number of repetitive cyclic voltammograms [76]. For oxidative electropolymerization, the polypyridyl ligands are substituted with amino [77] or anilino [78] groups which polymerize following a similar pathway as was reported for the formation of polyaniline. Variation of both the metal and the ligands can be used to obtain a desired color of the film when the metal is its reduced form. Upon oxidation, the film usually switches to a transparent state. Other modifications, such as the addition of secondary pendant groups like crown ethers for the controlled binding of alkaline and earthalkaline metals have also demonstrated significant impact on the electrochromic behavior [76].

A very interesting effect can be accomplished by incorporation of photolabile substituents on the metal complex in the polymer [79]. One example are thin films of poly[$Ru^{II}(vbpy)_2(py)_2$]Cl_2 on indium-tin-oxide (ITO) glass which substitute chloride for pyridine in two steps upon irradiation (Scheme 12). Each loss of a photolabile ligand is accompanied by a distinct color change of the metal complex in the reduced state and a change in the oxidation potential. With the use of contact lithography, films can be created with any desired pattern of the different colored species and with image resolution below 10 μm. Application of various potentials can then trigger the electrochromic response of each domain [79].

C. Metallophthalocyanines

Phthalocyanines, which are porphyrin derivatives with highly delocalized π-systems, can form two types of metal complexes: one with the metal in the center

orange $E_{1/2} = +1.27$ V vs SCE

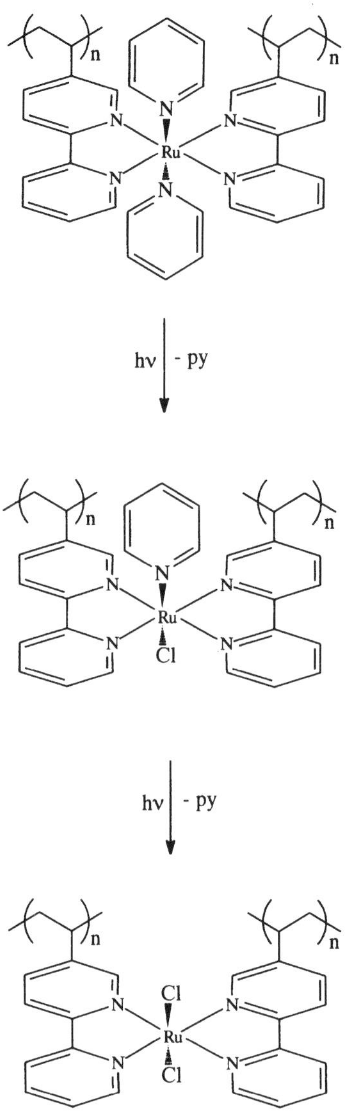

hv | - py

red $E_{1/2} = +0.77$ V vs SCE

hv | - py

purple $E_{1/2} = +0.35$ V vs SC

Scheme 12

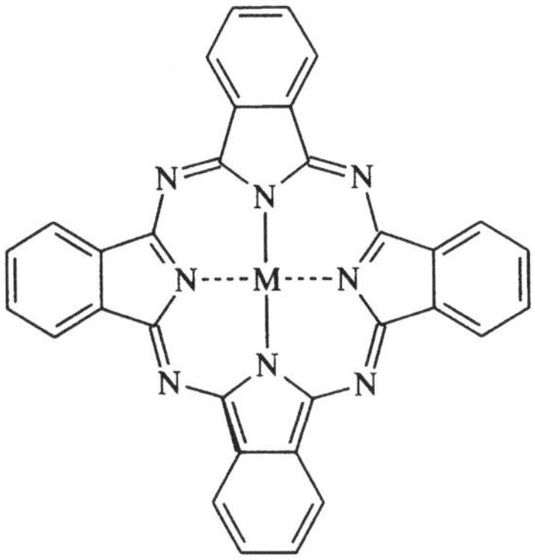

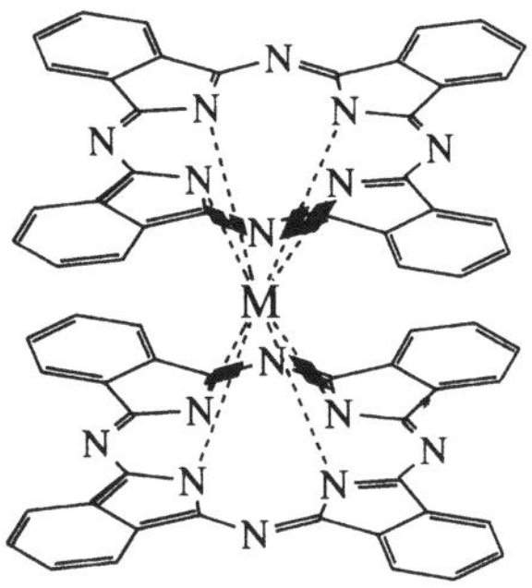

Chart 2

of the macrocycle and the other one with two rings sandwiched around the metal (Chart 2) [80]. Due to their very unique and interesting optical features, as well as their chemical versatility and stability, these metallophthalocyanines are widely used industrial pigments and have also been thoroughly investigated in many other areas, including electrochromism [81]. In fact, the observation of electrochromic behavior on a thin film consisting of bis(phthalocyaninato)lutetium(III) sandwich complexes dates back to 1970 [26].

Sublimation of the compound under vacuum produces a green film which shows polyelectrochromic behavior to which both ligand-and metal-based redox processes contribute [82]. In principle, a total of five colors are accessible upon oxidation and reduction, but only the color switch to blue upon reduction is utilizable. Despite fast response times, good stability, and reversibility, difficult processability of the films have limited any potential application [83]. Derivatives with pendant hydroxy and amino groups have been subjected to oxidative electro-polymerization, yielding several electrochromic thin films with fast and reversible electrochromic response at negative potentials. Again, the variation of the metal (such as Ni, Co) has been used to tune the color [84]. Multilayer Langmuir–Blodgett (LB) films of alkyloxy-substituted metalphthalocyanins have been prepared which display polyelectrochromic behavior [85,86] and very facile redox processes attributable to the formation of channels in the layer structure, allowing ion transport and better diffusion of counterions into the film. Studies have shown that the phthalocyanine units stack with their large faces perpendicular to the surface. Deposition of multilayers onto ITO-coated glass using LB techniques produced an electrode capable of polyelectrochromic behavior between blue–green–yellow and red. The display was characterized by fast response times and very high stability [87,88].

D. Viologen-Based Systems

1,1′-Disubstituted 4,4′-bipyridyl derivatives are very widely known under their trivial name "viologens" [89]. A recent book gives a detailed overview over their long history and still growing importance in many areas of chemistry and biology [90]. Its remarkable importance in the area of electrochromic devices arises from the accessibility of three common redox states (Scheme 13). The dication, being the most oxidized and most stable of the three, is colorless in the absence of any charge transfer with a counteranion. Upon reduction of the dication, a radical cation is formed which is stabilized by delocalization through the π-framework. Intramolecular charge transfer between the formally charged and uncharged nitrogen atoms produces an intense absorption band in the visible region with an exceptionally high absorption coefficient. Depending on the substituents on the nitrogen atoms and their effect on the molecular orbital levels, the observed colors range from blue/violet for various alkyl groups to green for aryl/groups with electron-withdrawing substituents. Further reduction leads to a neutral, colorless state of the molecule. This clean redox picture of viologens is complicated by the fact that, in solution, the radical cations have the tendency to spin-pair, forming a diamagnetic dimeric species, especially in aqueous solution and in organic solvents at low temperature [91]. The optical properties of the dimer are significantly altered, leading to a red color for the methyl-substituted viologen, affecting

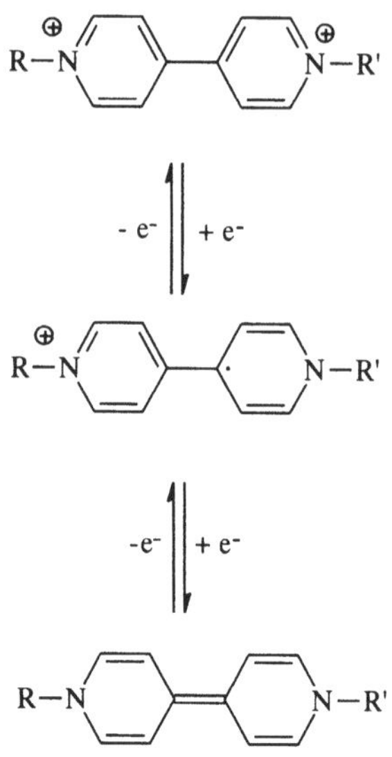

Scheme 13

the purity of the color. In addition, the electro-oxidation of the dimer is slower and the bleaching of the colored state is delayed with a negative impact on its switching time.

The development of display technology has been a major driving force behind the interest in the electrochromism of viologens. In the 1970s, a heptyl viologen was successfully used in an alphanumeric character display with a very fast response time of 10–50 ms and a long lifetime of over 10^5 switching cycles [92]. However, the technology ultimately did not prove competitive against LCDs. Commercially, has been much more successful the development of a rearview mirror for automobiles that automatically dims any dazzling incoming light [93,94]. The system is based on solution electrochromism and consists of a highly reflective metal surface (mirror) and a ITO-glass surface as the electrodes of the cell separated by a fraction of a millimeter. The electrolyte contains both a cathodically (a substituted viologen) and an anodically electrochromic compound. When the cell is switched on by a photosensitive detector, the ions migrate to the electrodes and switch to their colored states, effectively absorbing the incident light. Over time in the absense of an outside current, the colored species diffuse back into the bulk of the solution where they undergo a mutual redox reaction resulting in the quenching of the color.

More recently, substituted viologens containing —PO_3H_2 or —COOH functions have found application in electrochromic windows that try to combine the advantages of metal oxide films with those based on redox chromophores [95–98]. A monolayer of diphosphonoethyl-substituted viologen is hereby chemisorbed onto a nanostructured TiO_2 film on an ITO electrode. Because the redox potential of the viologen lies above the conduction band edge of TiO_2 at the liquid–solid interface, electrons can be transferred reversibly from the conduction band to the molecule and the semiconductor becomes conducting for the adsorbed species (see Section III.E). The diphosphono groups are chosen because, as Lewis bases, they react readily with subvalent, Lewis acidic titanium atoms present on the surface of the electrode, forming strong covalent bonds. The window is assembled with a second conducting glass electrode, and the space in between is filled with an electrolyte and a redox promotor, such as ferrocene. Upon application of a negative potential of about − 1 V, the viologen molecules are reduced and the electrode turns deeply blue at a rapid rate [99]. The high coloration can be explained by the surface roughness of the nanostructured film which effectively stacks the redox chromophores, leading to 100–1000-fold amplification compared to a flat monolayer [98]. Due to the direct linking of the redox chromophore to the electrode surface, the rate of the reduction is not limited to the diffusion rate of solution-phase chromophores. Furthermore, the accumulated negative charge on the electrode is easily compensated by ions adsorbed at the surface of the nanostructured film, making the color switch independent of the slow ion intercalation characteristic of pure metal oxide systems. The window is self-erasing due to the reaction of the redox promoter with the reduced viologen. Modification of the substituents on the viologen have led to the accessibility of other colors, such as green and pink in similar systems [100].

In order to obtain a chromophoric window with an extended memory, the reaction of the redox promoter with the viologen chromophor must be avoided [99]. This can be achieved by substituting the redox promoter with a second electrochromophoric compound which has an appropriate oxidation potential and which is deposited on the surface of the anode. For example, a modified phenothiazine has been chemisorbed on a SnO_2:Sb electrode in combination with the described viologen TiO_2 electrode. Upon applying an external potential biasing the TiO_2 electrode 1.2 V negative of the SnO_2:Sb electrode, the viologen is reduced, whereas the phenothiazine is oxidized to their respective radical cations [99]. The net result is a fast color change from transparent to blue-red. The steady-state current is low assuring the preservation of the color in the absence of the external potential. A high stability both under electrochromic cycling and under prolonged coloration is achieved [98,99].

E. Polynuclear Mixed-Valence Metal Complexes

When two colorless solutions of $Fe^{2+}(aq)$ and $Fe(CN)_6^{3-}$ are combined, the mixture turns rapidly to a dark blue color. The same effect is observed upon blending the solutions of Fe^{3+} and $Fe(CN)_6^{4-}$. However, any other combination of the solutions does not show a blue coloration at all. Not surprisingly, this beautiful phenomenon had attracted much interest in chemistry and its elucidation has created a research area of major importance in transition metal chemistry and beyond. The appearance of the blue color is due to the formation of a new complex anion, namely Prussian blue $Fe^{III}[Fe^{II}(CN)_6]$, in which visible-light absorption induces an electron transfer from Fe^{II} to Fe^{III}, a process apparently only possible in a material where two oxidation states are simultaneously present and sufficiently linked [101]. In order to generalize the conditions for the occurrence of such optical transitions, so-called mixed-valence compounds have been the subject of intense experimental investigation [101,102]. Relying on a systematic review of experimental results, Hush [103,104] has developed a theoretical model linking the physical properties of the mixed-valent compounds with the subject of electron-transfer reactions in solution. His model predicts that mixed-valence compounds should display a so-called intervalence transfer (IT) (or metal-to-metal charge transfer) band only if moderately coupled and that the properties of this transition are closely linked to the kinetic and thermodynamic factors governing the corresponding thermal electron-transfer process. In particular, the absorption energy is related to the energetic barrier of the thermal electron transfer. The theory also allows to calculate the degree of electronic coupling (H_{AB}) between the metal centers from the position and band intensity of the observed

IT band. The correlation of these properties can be illustrated in a diagram plotting potential energy versus nuclear configuration for a symmetrical mixed-valence complex (Fig. 10).

Based on the strength of the electronic coupling between the metals, Robin and Day [105] have developed a system in which mixed-valence compounds are broadly distinguished in three classes. In a very weakly coupled or Class I material, only the properties of the individual mononuclear species are observed due to the lack of communication ($H_{AB} = 0$). In the other extreme, a Class III com-

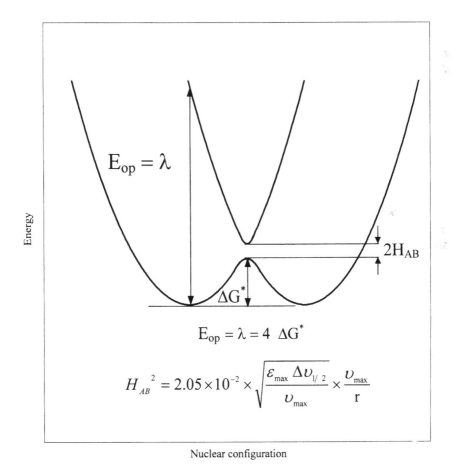

$$E_{op} = \lambda$$

$$E_{op} = \lambda = 4\ \Delta G^{*}$$

$$H_{AB}^{2} = 2.05 \times 10^{-2} \times \sqrt{\frac{\varepsilon_{max}\ \Delta\upsilon_{1/2}}{\upsilon_{max}}} \times \frac{\upsilon_{max}}{r}$$

Nuclear configuration

Figure 10 Optical intervalence transfer in a symmetric mixed-valence compound. Correlation between spectroscopic and kinetic parameters.

pound is so strongly coupled that the observed properties are entirely new and unique for the whole new system, whereas contributions from the individual units are absent. Complexes belonging to Class II display a mixture of new properties due to communication between the metals, whereas other properties of the individual units are retained. The effect of the changes in the electronic coupling on the energy surface of the ground state can be illustrated in the same potential energy versus nuclear configuration diagram. For the existence and position of the inter-valence transfer band, some important conclusions can be drawn from the classification. Because the band intensity depends on the electronic coupling, members of Class I are not expected to display IT bands. Class III complexes possess a single minimum on the ground-state potential energy surface; therefore the energy of the IT band is directly proportional to H_{AB} and the transition should most properly be identified as an HOMO–LUMO transition. Finally, compounds classified in II are anticipated to show the metal-to-metal charge transfer (MMCT) band, with its position related to the activation energy of adiabatic thermal electron transfer ($E_{op} = \lambda = 4\Delta G^*$) [101]. The situation in unsymmetrical complexes can be similarly explained if the energy difference between the two states is taken into account ($E_{op} = \lambda + \Delta E$) (Fig. 11).

Even though the Hush model was originally developed to describe MMCT transitions in mixed-valence binuclear or polynuclear complexes, its principles are applicable to any kind of transition between localized redox centers in a complex molecule, including ligands whose redox potentials are at reasonable energies. In this context, the very common (MLCT) [102] and the less frequent ligand-to-ligand chart transfer (LLCT) [106–108] bands have been discussed in terms of the Hush formalism for IT transitions. As a result of this extension of the theory, it is possible to use the relative intensities for the corresponding IT spectral transitions to predict the relative "degree of adiabaticity" of various unimolecular electron-transfer paths in a complex molecule [109].

The demonstration that MLCT and LLCT bands can, in principle, be treated with the same theoretical model has a tremendous impact on practical aspects and, therefore, the search and the development of electrochromic compounds in the area as well. Especially, the inclusion of organic ligands with their own redox properties along with their interaction with the metal cores and their mediation of the interactions between metals opens up a virtually limitless supply of possible combinations with potential electrochromic behavior.

Following the discovery of the polyelectrochromic behavior of Prussian blue in 1978, its properties have been thoroughly investigated and several other polynuclear transition metal hexacyanometallates have been developed, forming a new class of mixed-valence compounds with the general formula $M_1[M_2(CN_6)]$ [110]. Due to its insolubility, thin films of Prussian blue are obtained by electrode-position during electrochemical reduction of a solution containing iron(III) and hexacyanoferrate(III) ions. Hereby, the soluble complex Prussian brown [iron(III)

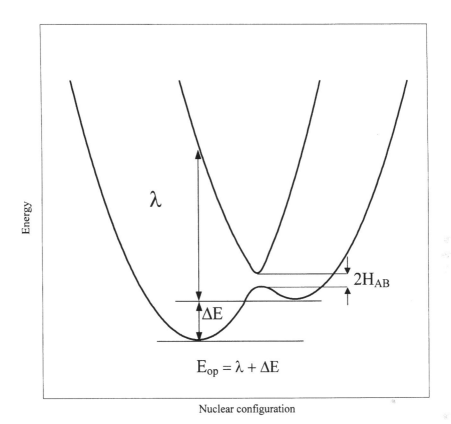

Figure 11 Optical intervalence transfer in a unsymmetric mixed-valence compound. Correlation between spectroscopic and thermodynamic parameters.

hexacyanoferrate(III)] is converted in a single-electron-transfer process to the insoluble Prussian blue, accompanied by the deposition of charge-compensating cations such as K^+ into the film. The deeply blue color of Prussian blue (λ_{max} = 690 nm) has been attributed to an intervalence charge-transfer band, involving Fe(III) and Fe(II) centers [27].

Prussian blue films undergo several color changes both upon oxidation and reduction, accompanied by the respective counterion migration. Partial electrochemical oxidation results in a gradual change to a green color ascribable to the formation of a continuous series of mixed-valence species, such as Berlin green $[Fe^{III}3(Fe^{III}(CN)_6)_2(Fe^{II}(CN)_6)]^-$. Further oxidation leads to the yellow color of the neutral $[Fe^{III}Fe^{III}(CN)_6]$ complex (λ_{max} = 425 nm) [111]. On the other hand, reduction of Prussian blue abruptly produces a transparent state, known as Prussian white $[Fe^{II}Fe^{II}(CN)_6]^{2-}$. The cyclic voltammogram (CV) of a Prussian blue (PB)-modified electrode shows a sharp reversible wave in reduction and a broad peak in oxidation, illustrating the fact that one transition is between two distinct species and the other between a continuous series of compounds. From an application point of view, the reduction of Prussian blue is more promising due to its sharpness and reversibility.

Several devices based solely on the compound have been prepared [27], usually involving a solid phase sandwiched between two optically transparent electrodes. An example is an arrangement of two ITO electrodes with a membrane of a solid polymer electrolyte (Nafion) [112] in between on which Prussian blue has been chemically generated.

A more promising apparatus was devised in which the complementary electrochromic behavior of a cathodically coloring material was combined with the anodically coloring capability of Prussian blue. The system was constructed from two optically transparent electrodes (OTEs) separated by a thin transparent ionic conductor $[KCF_3SO_3$ in poly(ethyleneoxide)] [113]. On one electrode, a thin film of Prussian blue was deposited; on the other, a film of tungsten trioxide was deposited. Upon application of a sufficient potential with the appropriate polarity, both compounds can be simultaneously colored (blue) and bleached (transparent), increasing the overall effect. Prussian blue was also used in a similar fashion in combination with a substituted viologen [30]. The device was built from an OTE covered with a nanocrystalline TiO_2 film on which the viologen was chemisorbed, a second electrode on which the Prussian blue was electrodeposited, and the resulting cell was filled with a liquid electrolyte consisting of $LiN(SO_2CF_3)_2$ in glutaronitrile. Under a sufficient potential biasing the TiO_2 electrode negative of the Prussian blue electrode, the viologen is in its reduced form and the Prussian blue is in its oxidized from, both being deeply blue colored. Reversal of the polarity switches the device to its transparent state. The switching time of the system is low because the rate of Prussian blue oxidation is limited by hole diffusion into the bulk film.

Not surprisingly, the very interesting and practically useful properties of Prussian blue have resulted in several studies of similar hexacyano derivatives [114] but only a few of them (e.g., ruthenium- and osmium-based systems) have found use in electrochromic devices. More systematic studies should, indeed, be called for in order to harvest the full potential of this group of compounds.

Other mixed-valence metal complexes have been published that display electrochromic behavior, among them bipyridyl ruthenium complexes linked through dioxolene containing bridging ligands, such as 3,3′,4,4′-tetrahydroxybiphenyl, 3,3″,4,4″-tetrahydroxy-*p*-terphenyl, and 9-phenyl-2,3,7-trihydoxy-6-fluorone [115,116] (Chart 3). Many of the redox processes that lead to frequently

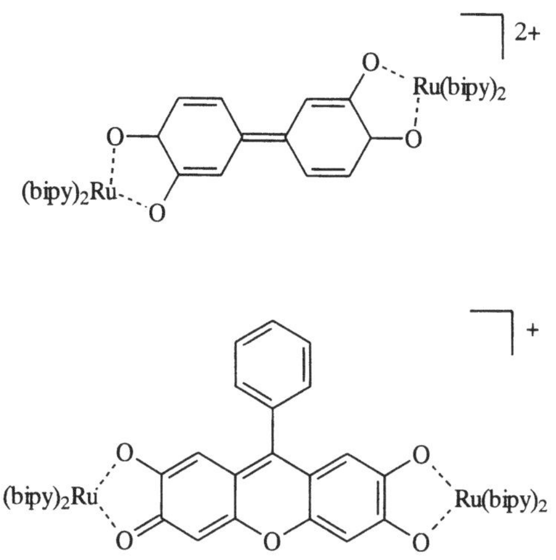

Chart 3

quite dramatic changes in the absorption spectra streching even into the NIR region are, in fact, predominantly ligand based.

In addition, several ruthenium trisbipyridine complexes with electron-withdrawing substitutents exhibit multicolor electrochromism in solution [117,118]. When their spectroelectrochemical behavior is studied in an optically transparent thin-layer electrode (OTTLE), up to six stable reductions are observable. Clearly, with only one metal present in the complex, the accompanying transitions all involve ligand-to-metal interactions. In contrast, the same compounds with electron-donating substitutents are entirely devoid of the electrochromism, illustrating again how relatively minor changes in the molecular structure can have significant effects on the electrochromic behavior [118].

As shown, optical electron-transfer transitions in polynuclear complexes are certainly not a new phenomenon and have been studied very broadly and thoroughly [119–122]. However, most of the examples are strictly solution based and the attachment to an electrode has not been addressed. As a result, their applicability to devices has not been investigated. Overall, it is quite obvious that the tremendous potential of the field for the discovery of electrochromic materials has hardly been touched.

One of the key issues for the further exploitation and development of the area is the choice of the electrode material and the anchoring of the mixed-valence species. It is apparent from the above examples that the choice of the metal periphery and even the substituents on those ligands have material effects on the

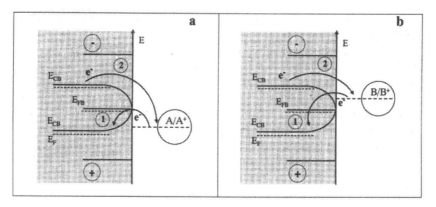

Figure 12 Schematic representation of thermodynamic and kinetic parameters influencing interfacial electron-transfer processes between the semiconductor and an adsorbed redox specie.

behavior of the overall system. For the electrode material, similar considerations apply as shown above for the viologen-based systems. In fact, nanostructured wide-gap semiconductor particles, such as TiO_2 or SnO_2, offer an appealing choice. Application of an external potential to the semiconductor particle, in contact with a conductive electrode, allows one to modify the space charge in the solid. As shown in Fig. 12a, for an n-type semiconductor such as titanium dioxide, the depletion layer, which is present at positive potentials with respect to the potential of zero charge (flat-band situation), can be transformed in an accumulation layer by negative polarization of the electrode. Thus, the consequent tuning of the conduction-band energy allows to modify the thermodynamics and kinetics of the heterogeneous electron-transfer process with the molecular species absorbed on the semiconductor. For the molecular species (A) having the redox level indicated in Fig. 12, the electron-transfer process (1) is expected to be slow, due to an activation barrier given by $E_{FB}-E_{1/2}(A/A^+)$, whereas the activationless process (2) is expected to be fast. Upon changing the redox level of the absorbed species, both oxidative and reductive interfacial electron-transfer processes can be fast. This is illustrated in Fig. 12b for a molecular species (B) with a redox potential slightly negative with respect to the flat-band potential of the semiconductor. The arguments are relevant for the design of electrochromic molecular species which may undergo fast electron transfer on the semiconductor films with accompanying chromatic changes.

Several strategies for the attachment of redox components onto a host surface are, in principle, feasible, among them chemisorption, electrostatic association, hydrogen-bonding, physisorption, and physical entrapment [32]. Because of the mentioned durability requirements, covalent linking is the method of choice for electrochromic devices. Metal oxides are characterized by the availability of hydroxyl groups on the surface, which provide convenient reactive sites for the anchoring of molecular species. Various functional groups are capable of spontaneously forming stable covalent bonds with the surface layer, among them carboxylic acids (carboxy ester linkage), phosphonic acids (phosphonate ester linkage), boronic acids (boronic ester linkage), and silanes (siloxy linkage). Many common polypyridyl-based ligands, such as bipyridine and terpyridyl, have been prepared containing these functional groups (Chart 4) [32,123]. The general procedure is to prepare the metal complex with these activated substituents and expose the nanocrystalline semiconductor surface to a solution in a nonaqueous solvent. Water is to be avoided because some of the listed linkages are labile in its presence. Another less often encountered procedure is to treat the surface first with a binding ligand, creating selective ion-binding sites and to complex the metal cation in the second step [124].

We have shown that our concept works in principle and have realized a few promising examples. They are bimetallic and trimetallic systems, each metal with a distinct periphery, linked through a bridging ligand and containing one or more of the anchoring groups described earlier (Fig. 13). The electrochromic behavior is tested using conventional three-electrode arrangements after their chemisorption onto an optically transparent working electrode layered with TiO_2 or SnO_2/Sb.

The general principle that we have followed in the molecular design of the polynuclear species is based on the introduction of a metal-containing moiety which can be directly bound to the surface of the semiconductor and can be interconverted between two oxidation states at a potential close to 0 V versus SCE. This should allow to maximize the electronic coupling with the semiconductor and promote color changes by applying a small potential difference between the

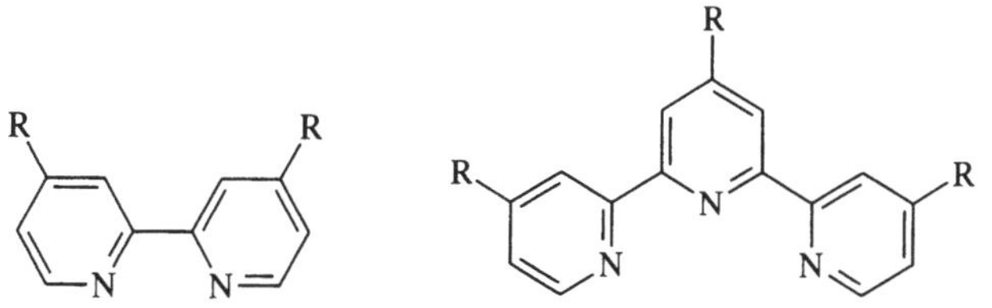

$R = CO_2H, PO_3H_2, BO_2H_2$

Chart 4

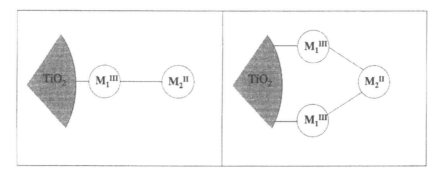

Figure 13   Binuclear and trinuclear mixed-valence species on TiO_2 films.

electrodes, a very important feature in regard to low energy consumption. Following this idea, a series of binuclear complexes of the type shown in Chart 5 have been prepared and tested [125]. All of these species display a distinct metal-to-metal $Ru^{II} \rightarrow Ru^{III}$ charge-transfer band in the one electron oxidized form rendering them green (**1**) or blue (**2,3**) in color. Upon reduction, the intervalence band is quenched and the color of the complex is now governed by the $d\pi-\pi^*$ ($Ru \rightarrow py$) MLCT band resulting in a color change to red. The absorption spectra measured following the reduction and oxidation of the $(HOOCpy)CH_4C_5Ru^{III}$ $(NH_3)4(NC)Ru^{II}(bpy)]^{3+}$ complex in solution (Fig. 14a) and on transparent TiO_2 films (Fig. 14b) clearly show a distinct color change in a narrow potential range (-0.5 to $+0.5$ V versus SCE), with switching times in the order of milliseconds.

Stability tests performed in sandwich-type cells containing the dyes adsorbed on SnO_2/Sb electrodes demonstrated a high stability, with optical density changes lower than 2% after cycling the electrochromic device 20000 times between -0.5 and $+0.5$ V.

Owing to the presence of the amine and cyanide ligands, known to give rise to specific donor–acceptor interaction with solvents [126–130], an interesting solvatochromic behavior is observed for these species. For complex **1** the spectral changes are dominated by amine interactions with the solvents as shown by the linear correlation of the solvent donor number* [131] with the IT band maxima and with the half-wave potential of the ruthenium amine moiety.

It can be appreciated that this type of polynuclear system in general possesses a high degree of flexibility thanks to the possibility of tuning its spectroscopic properties through changes of metal centers, of coordinated or bridging

---

* The donor number (DN) of a solvent is defined as the enthalpic change at 25°C for formation of the adduct between $SbCl_5$ and the solvent in 1,2-dichloroethane.

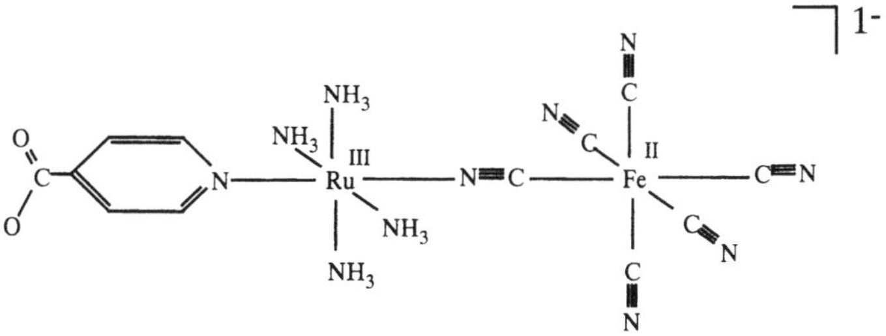

Chart 5

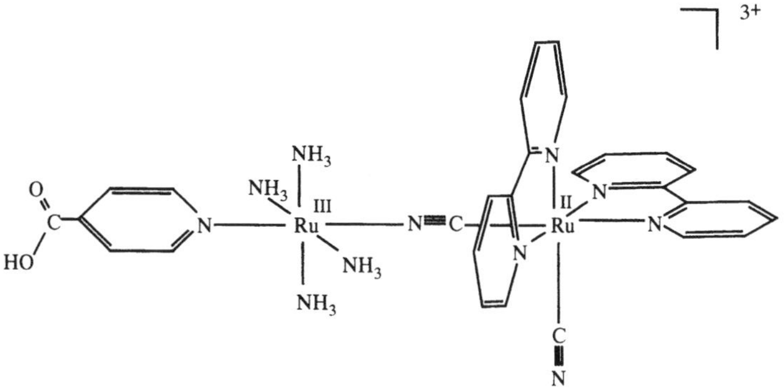

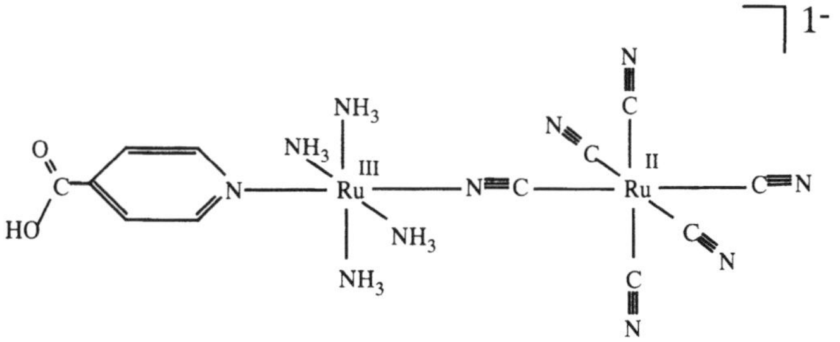

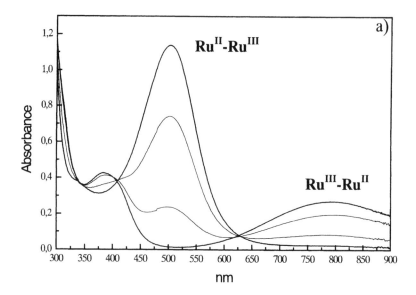

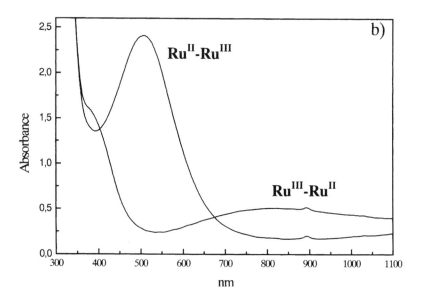

Figure 14 (a) Absorption spectral changes of [HOOCpyRu(NH$_3$)$_4$(NC)Ru-(bpy)$_2$(CN)](PF$_6$)$_3$ in water totally reduced of the Ru^{III} center. (b) Absorption spectral changes of [HOOCpyRu(NH$_3$)$_4$(NC)Ru(bpy)$_2$(CN)](PF$_6$)$_3$ on TiO$_2$ electrodes in the range of applied potential from $+0.5$ V to -0.5 V versus SCE.

ligands and solvents. In addition, the availability of a fast-screening protocol to test suitable ligand–metal combinations for modifying their properties should be of immense utility.

IV. CONCLUSIONS

In the last two decades, much has been learned about fundamental aspects of electron transfer in organic and inorganic systems in homogeneous solution. More recently, the attention of many laboratories has been attracted by the extraordinary potential applications of these fundamental concepts for building real devices which operate on a molecular level.

   This chapter provides some examples of the work developed in the field of heterosupramolecular chemistry toward the realization of artificial photosynthetic materials for the conversion of solar energy into electricity and of electrochromic devices based on nanostructured semiconductors functionalized with charge-transfer polynuclear complexes. In both cases, the molecular design is closely correlated to specific features expected for the device and should meet requirements such as light absorption, vectorial energy and electron transfer, and assembling of suitable redox components. The study of these heterogeneous systems is essential to a better understanding of the surface photochemical or electrochemical processes and can provide the basis for the design of novel microheterogeneous assemblies with great potential for a variety of applications.

ACKNOWLEDGMENTS

This work is dedicated to Dario Bignozzi and Sofia Bignozzi. Financial support from a EU contract is gratefully acknowledged.

REFERENCES

1. Hagfeldt, A.; Grätzel, M. *Chem. Rev.* **1995**, *95*, 49.
2. Bignozzi, C. A.; Schoonover, J. R.; Scandola, F. *Molecular Level Artificial Photosynthetic Materials*, Progress in Inorganic Chemistry **1997**, Vol. 44, p. 1.
3. Hagfeldt, A.; Grätzel, M. *Acc. Chem. Res.* **2000**, *33*, 296.
4. Bignozzi, C. A.; Argazzi, R.; Kleverlaan, C. J. *Chem. Soc. Rev.* **2000**, *29*, 87.
5. O'Regan, B.; Grätzel, M. *Nature* **1991**, *353*, 737.
6. Tachibana, Y.; Moser, J. E.; Grätzel, M.; Klug, D. R.; Durrant, J. R. *J. Phys. Chem. B.* **1996**, *100*, 20,056.
7. Hannappel, T.; Burfeindt, B.; Storck, W.; Willig, F. *J. Phys. Chem. B* **1997**, *101*, 6799.
8. Ellingson, R. J.; Asbury, J. B.; Ferrere, S.; Ghosh, H. N.; Sprague, J. R.; Lian, T.; Nozik, A. J. *J. Phys. Chem. B* **1998**, *102*, 6455.

9. Heimer, T. A.; Heilweil, E. J.; Bignozzi, C. A.; Meyer, G. J. *J. Phys. Chem. A* **2000**, *104*, 4256.

10. Kuciauskas, D.; Freund, M. S.; Gray, H. B.; Winkler, J. R.; Lewis, N. S. *J. Phys. Chem. B* **2001**, *105*, 392.

11. Argazzi, R.; Bignozzi, C. A.; Heimer, T. A.; Castellano, F. N.; Meyer, G. J. *J. Am. Chem. Soc.* **1995**, *117*, 11,815.

12. Argazzi, R.; Bignozzi, C. A.; Heimer, T. A.; Castellano, F. N.; Meyer, G. J. *J. Phys. Chem. B* **1997**, *101*, 2591.

13. Balzani, V.; Scandola, F. *Supramolecular Photochemistry*; Horwood: Chichester, 1991.

14. Amadelli, R.; Argazzi, R.; Bignozzi, C. A.; Scandola, F. *J. Am. Chem. Soc.* **1990**, *112*, 7099.

15. Bignozzi, C. A.; Argazzi, R.; Indelli, M. T.; Scandola, F.; Schoonover, J. R.; Meyer, G. J. *Proc. Indian Acad. Sci. (Chem. Sci.)* **1997**, *109*, 397.

16. Smestad G.; Bignozzi C. A.; Argazzi R. *Solar Energy Mater. Solar Cells* **1994**, *32*, 259.

17. Nazeeruddin, M. K.; Liska, P.; Moser, J.; Vlachopoulos, N.; Grätzel M. *Helv. Chim. Acta* **1990**, *73*, 1788.

18. Willig, F.; Kietzmann, R.; Schwarzburg, K. *Proceedings of the SPIE Conference on Energy Efficiency and Solar Energy Conversion XI*; SPIE: Bellingham, WA, 1992.

19. Bignozzi, C. A.; Argazzi, R.; Garcia, C. G.; Scandola, F.; Schoonover, J. R.; Meyer, T. J. *J. Am. Chem. Soc.* **1992**, *114*, 8727.

20. Schoonover, J. R.; Gordon, K. C.; Argazzi, R.; Woodruff, W. H.; Peterson, K. A.; Bignozzi, C. A.; Dyer, R. B.; Meyer, T. J. *J. Am. Chem. Soc.* **1993**, *115*, 10,996.

21. Bonhôte, P.; Moser, J.-E.; Humphry-Baker, R.; Vlachopoulos, N.; Zakeeruddin, S. M.; Walder, L.; Grätzel, M. *J. Am. Chem. Soc.* **1999**, *121(6)*, 1324–1336.

22. Kleverlaan, C.; Alebbi, M.; Argazzi, R.; Bignozzi, C. A.; Hasselmann, G. M.; Meyer, G. J. *Inorg. Chem.* **2000**, *39*, 1342.

23. Lees, A. C.; Kleverlaan, C. J.; Bignozzi, C. A.; Vos, J. G. *Inorg. Chem.* **2001**, *40*, 5343.

24. Argazzi, R.; Bignozzi, C. A.; Heimer, T. A.; Meyer, G. J. *Inorg. Chem.* **1997**, *36*, 2.

25. Kleverlaan, C. J.; Indelli, M. T.; Bignozzi, C. A.; Pavanin, L.; Scandola, F.; Hasselman, G. M.; Meyer, G. J. *J. Am. Chem. Soc.* **2000**, *122*, 2840.

26. Monk, P. M. S.; Mortimer, R. J.; Rosseinsky, D. R. *Electrochromism: Fundamentals and Applications*; VCH: Weinheim, 1995.

27. Mortimer, R. J. *Chem. Soc. Rev.* **1997**, *26*, 147.

28. Green, M. *Electrochim. Acta* **1999**, *44*, 2969.

29. Green, M. *Chem. Ind.* **1996**, *17*, 641.

30. Bonhôte, P.; Gogniat, E.; Campus, F.; Walder, L.; Grätzel, M. *Display* **1999**, *20*, 137.

31. Bechinger, C.; Ferrer, S.; Zaban, A.; Sprague, J.; Gregg, B. A. *Nature* **1996**, *383*, 608.

32. Kalyanasundaram, K.; Grätzel, M. *Coord. Chem. Rev.* **1998**, *77*, 347.

33. Mortimer, R. J. *Electrochim. Acta* **1999**, *44*, 2971.
34. Granqvist, C. G. *Handbook of Inorganic Electrochromic Materials*; Elsevier: Amsterdam, 1995.
35. Granqvist, C. G. *Electrochim. Acta* **1999**, *44*, 3005.
36. Zhang, J.-G.; Benson, D. K.; Tracy, C. E.; Deb, S. K.; Czanderna, A. W.; Bechinger, C. In *Electrochromic Materials* III; Ho, K. C.; Greenberg, C. B.; MacArthur, D. M., Eds.; Electrochemical Society Proceedings Series; Electrochemical Society: Pennington, LNJ, 1997, pp. 251–259.
37. Hagfeldt, A.; Vlachopoulos, N.; Grätzel, M. *J. Electrochem. Soc.* **1994**, *14*, L82.
38. Rauh, R. D. *Electrochim. Acta* **1999**, *44*, 3165.
39. Rauh, R. D. *Solar Energy Mater. Solar Cells* **1995**, *39*, 145.
40. Sapp, S. A.; Sotzing, G. A.; Reynolds, J. R. *Chem. Mater.* **1998**, *10*, 2101.
41. Thompson, B. C.; Schottland, P.; Zong, K.; Reynolds, J. R. *Chem. Mater.* **2000**, *12*, 1563.
42. Roncali, J. *Chem. Rev.* **1997**, *97*, 173.
43. Irvin, D. J.; DuBois, C. J.; Reynolds, J. R. *Chem. Commun.* **1999**, 2121.
44. Meeker, D. L.; Mudigonda, D. S. K.; Osborn, J. M.; Loveday, D. C.; Ferraris J. P. *Macromolecules* **1998**, *31*, 2943.
45. Wang, F.; Wilson, M. S.; Rauh, R. D.; Schottland, P.; Thompson, B. C.; Reynolds, J. R. *Macromolecules* **2000**, *33*, 2083.
46. Lee, Y.; Sadki, S.; Tsuie, B.; Reynolds, J. R. *Chem. Mater.* **2001**, *13*, 2234.
47. Ouyang, J.; Li, Y. *Polymer* **1997**, *38*, 1971.
48. Evans, G. P. In *Advances in Electrochemical Science and Engineering*; Gerischer, H.; Tobias, C. W., Eds.; VCH: Weinheim, 1990.
49. Garnier, F.; Tourillon, G.; Gazard, M. Dubois, J. C. *J. Electroanal. Chem.* **1983**, *148*, 299.
50. Pfluger, P.; Street, G. B. *J. Chem. Phys.* **1984**, *80*, 544.
51. Zotti, G.; Zecchin, S.; Schiavon, G.; Groenendaal, L. B. *Chem. Mater.* **2000**, *12*, 2996.
52. Schottland, P.; Zong, K.; Gaupp, C. L.; Thompson, B. C.; Thomas, C. A.; Giurgiu, I.; Hickman, R.; Abboud, K. A.; Reynolds, J. R. *Macromolecules* **2000**, *33*, 7051.
53. Zong, K.; Reynolds, J. R. *J. Org. Chem.* **2001**, *66*, 6873.
54. Gaupp, C. L.; Zong, K.; Schottland, P.; Thompson, B. C.; Thomas, C. A.; Reynolds, J. R. *Macromolecules* **2000**, *33*, 1132.
55. Kumar, A.; Welsh, D. M.; Morvant, M. C.; Piroux, F.; Abboud, K. A.; Reynolds, J. R. *Chem. Mater.* **1998**, *10*, 896.
56. Mastragostino, M. In *Applications of Electroactive Polymers*, Scrosati, B., Ed.; Chapman & Hall: London, 1993.
57. Guerrero, D. J.; Ren, X. M.; Ferraris, J. P. *Chem. Mater.* **1994**, *6*, 1437.
58. Ferraris, J. P.; Henderson, C.; Torres, D.; Meeker, D. *Synth. Met.* **1995**, *72*, 147.
59. Ray, A.; Richter, A. F.; MacDiarmid, A. G.; Epstein, A. J. *Synth. Met.* **1989**, *29*, 151.
60. Rourke, F.; Crayston; J. A. *J. Chem. Soc., Faraday Trans.* **1993**, *89*, 295.
61. Diaz, A. F.; Logan, J. A. *J. Electroanal. Chem.* **1980**, *111*, 111.
62. Kobayashi, T.; Yoneyama, H.; Tamura, H. *J. Electroanal. Chem.* **1984**, *161*, 419.

63. Stilwell, D. E.; Park, S.-M. *J. Electrochem. Soc.* **1989**, *136*, 427.
64. Wei, Y.; Focke, W. W.; Wnek, G. E.; Ray, A.; MacDiarmid, A. G. *J. Phys. Chem.* **1989**, *93*, 495.
65. Mortimer, R. J. *J. Mater. Chem.* **1995**, *5*, 969.
66. Ramirez, S.; Hillman, A. R. *J. Electrochem. Soc.* **1998**, *145*, 2640.
67. Jang, G.-W.; Chen, C. C.; Gumbs, R. W.; Wei, Y.; Yeh, J.-M. *J. Electrochem. Soc.* **1996**, *143*, 2591.
68. Akoudad, S.; Roncali, J. *J. Chem. Soc., Chem. Commun.* **1998**, 2081.
69. Lee, B. L.; Yamamoto, T. *Macromolecules* **1999**, *32*, 1375.
70. Yamaguchi, S.; Goto, T.; Tamao, K. *Angew. Chem. Int. Ed.* **2000**, *39*, 1695.
71. Kitamura, C.; Tanaka, S.; Yamashita, Y. J. *J. Chem. Soc., Chem. Commun.* **1994**, 1585.
72. Nawa, K.; Imae, I.; Noma, N.; Shirota, Y. *Macromolecules* **1995**, *28*, 723.
73. Ferraris, J. P.; McMackin, C.; Torres, D.; Meeker, D.; Rudge, A.; Gottesfeld, S. In *Electrochromic Materials II/1994*; Ho, K. C., MacArthur, D. A., Eds.; Electrochemical Society Proceedings Series; Electrochemical Society: Pennington, NJ, 1994.
74. Ferraris, J. P.; Mudigonda, D. S. K.; Loveday, D. C.; Barashkov, N. N.; Hmyene, M.; Henderson, C. R. In *Electrochromic Materials III/1996*; Ho, K. C., Greenberg, C. B., MacArthur, D. A., Eds.; Electrochemical Society Proceedings Series; Electrochemical Society: Pennington, NJ, 1997.
75. Mortimer, R. J. In *Research in Chemical Kinetics*; Compton, R. G., Hancock, G., Eds.; Elsevier: Amsterdam, 1994.
76. Beer, P. D.; Kocian, O.; Mortimer, R. J.; Ridgway, C. *J. Chem. Soc., Faraday Trans.* **1993**, *89*, 333.
77. Ellis, C. D.; Margerum, L. D.; Murray, R. W.; Meyer, T. J. *Inorg. Chem.* **1983**, *22*, 1283.
78. Horwitz, C. P.; Zuo, Q. *Inorg. Chem.* **1992**, *31*, 1607.
79. Leasure, R. M.; Ou, W.; Linton, R. W.; Meyer, T. J. *Chem. Mater.* **1996**, *8*, 264.
80. Ercolani, C. *J. Porphyrins Phthalocyanines* **2000**, *4*, 340.
81. Leznoff, C. C.; Lever, A. B. P *Phthalocyanins: Properties and Applications*; Wiley: New York, 1996.
82. Collins, G. C. S.; Schiffrin, D. J. *J. Electrochem. Soc.* **1985**, *132*, 1835.
83. Moore, D. J.; Guarr, T. F. *J. Electroanal. Chem.* **1991**, *314*, 313.
84. Kimura, M.; Horai, T.; Hanabusa, K.; Shirai, H. *Chem. Lett.* **1997**, *7*, 653.
85. Goldenberg, L. M. *J. Electroanal. Chem.* **1994**, *379*, 3.
86. Besbes, S.; Plichon, V.; Simon, J.; Vaxiviere, J. *J. Electroanal. Chem.* **1987**, *237*, 61.
87. Granito, C.; Goldenberg, L. M.; Bryce, M. R.; Monkman, A. P.; Troisi, L.; Pasimeni, L.; Petty, M. C. *Langmuir* **1996**, *12*, 472.
88. Rodríguez-Méndez, M. L.; Souto, J.; de Saja, J. A.; Aroca, R. *J. Mater. Chem.* **1995**, *5*, 639.
89. Bird, C. L.; Kuhn, A. T. *Chem. Soc. Rev.* **1981**, *10*, 49.
90. Monk, P. M. S. *The Viologens: Physicochemical Properties, Synthesis and Applications of the Salts of 4,4'-Bipyridine*; Wiley: Chichester, 1998.

91. Monk, P. M: S.; Fairweather, R. D.; Duffy, J. A.; Ingram, M. D. *J. Chem. Soc., Perkin Trans. II* **1992**, 2039.

92. Schoot, C. J.; Ponjee, J. J.; van Dam, H. T.; van Doorn, R. A.; Bolwijn, P. J. *Appl. Phys. Lett.* **1973**, *23*, 64.

93. Byker, H. J. U.S. Patent 4,902,108, 1990 (to Gentex Corp.).

94. Byker, H. J. U.S. Patent 5,128,799, 1992 (to Gentex Corp.).

95. Marguerettaz, X.; O'Neill, R.; Fitzmaurice, D. *J. Am. Chem. Soc.* **1994**, *116*, 2629.

96. Cinnsealach, R.; Boschloo, G.; Rao, S. N.; Fitzmaurice, D. *Solar Energy Mater. Solar Cells* **1998**, *55*, 215.

97. Cinnsealach, R.; Boschloo, G.; Rao, S. N.; Fitzmaurice, D. *Solar Energy Mater. Solar Cells* **1999**, *57*, 107.

98. Grätzel, M. *Nature* **2001**, *409*, 575.

99. Cummins, D.; Boschloo, G.; Ryan, M.; Corr, D.; Rao, S. N.; Fitzmaurice, D. *J. Phys. Chem. B* **2000**, *104*, 11449.

100. Boehlehn, R.; Felderhoff, M.; Michalek, R.; Walder, L. *Chem. Lett.* **1998**, *8*, 815.

101. Creutz, C. *Prog. Inorg. Chem.* **1983**, *30*, 1.

102. Meyer, T. J. *Prog. Inorg. Chem.* **1983**, *30*, 389.

103. Hush, N. S. *Electrochim. Acta* **1968**, *13*, 1005.

104. Hush, N. S. *Prog. Inorg. Chem.* **1967**, *8*, 391.

105. Robin, M. B.; Day, P. *Adv. Inorg. Chem. Radiochem.* **1967**, *10*, 247.

106. Vogler, A.; Kunkely, H. *J. Am. Chem. Soc.* **1981**, *103*, 1559.

107. Vogler, A.; Kunkely, H. *Angew. Chem. Int. Ed. Eng.* **1982**, *21*, 77.

108. Crosby, G. A.; Highland, R. G.; Truesdell; K. A. *Coord. Chem. Rev.* **1985**, *64*, 41.

109. Scandola, F.; Bignozzi, C. A.; Balzani, V. In *Homogeneous and Heterogeneous Photocatalysis*; Pelizzetti, E., Serpone, N., Eds.; D. Reidel: Boston, 1986.

110. Sharpe, A. G. *The Chemistry of Cyano Complexes of the Transition Metals*; Academic Press: New York, 1976.

111. Mortimer, R. J.; Rosseinsky, D. R. *J. Chem. Soc., Dalton Trans.* **1984**, 2059.

112. Honda, K.; Ochiai, J.; Hayashi, H. *J. Chem. Soc., Chem. Commun.* **1986**, 168.

113. Ho K.-C.; Rukavina T. G.; Greenberg C. B., *Electrochromic Materials II*; Ho, K. C.; MacArthur, D. A., Eds.; Electrochemical Society Proceedings Series, Electrochemical Society: Pennington, NJ, 1994.

114. Itaya, K.; Uchida, I.; Neff, V. D. *Acc. Chem. Res.* **1986**, *19*, 162.

115. Joulié, L. F.; Schatz, E.; Ward, M. D.; Weber, F.; Yellowlees, L. J. *J. Chem. Soc., Dalton Trans.* **1994**, 799.

116. Bartham, A. M.; Ward, M. D. *New J. Chem.* **2000**, *24*, 501.

117. Elliott, C. M.; Hershenhart, E. J. *J. Am. Chem. Soc.* **1982**, *104*, 7519.

118. Pichot, F.; Beck, J. H.; Elliott, C. M. *J. Phys. Chem. A* **1999**, *103*, 6263.

119. Taube, H. *Ann. NY Acad. Sci.* **1978**, *313*, 418.

120. Meyer, T. J. *Acc. Chem. Res.* **1978**, *11*, 94.

121. Bignozzi, C. A.; Roffia, S.; Scandola, F. *J. Am. Chem. Soc.* **1985**, *107*, 1644.

122. Bignozzi, C. A.; Paradisi, C.; Roffia, S.; Scandola, F. *Inorg. Chem.* **1988**, *27*, 408.

123. Gillaizeau-Gauthier, I.; Odobel, F.; Alebbi, M.; Argazzi, R.; Costa, E.; Bignozzi, C. A.; Qu, P.; Meyer, G. J. *Inorg. Chem.* **2001**, *40*, 6073.

124. Li, Z. Y.; Mallouk, T. E.; Lai, C. W. *Inorg. Chem.* **1989**, *28*, 178.

125. Biancardo, M.; Argazzi, R.; Costa E.; Schwab, P. F. H.; Bignozzi, C. A. Unpublished.

126. Curtis, J. C.; Sullivan, B. P.; Meyer, T. J. *Inorg. Chem.* **1983**, *22*, 224.

127. Bignozzi, C. A.; Chiorboli, C.; Indelli, M. T.; Rampi Scandola, M. A.; Varani, G.; Scandola, F. *J. Am. Chem. Soc.* **1986**, *108*, 7872.

128. Indelli, M. T.; Bignozzi, C. A.; Marconi, A.; Scandola, F. *J. Am. Chem. Soc.* **1988**, *110*, 7381.

129. Davila, J.; Bignozzi, C. A.; Scandola, F. *J. Phys. Chem.* **1989**, *93*, 1373.

130. Timpson, C. J.; Bignozzi, C. A.; Sullivan, B. P.; Kober, E. M.; Meyer, T. J. *J. Phys. Chem.* **1996**, *100*, 2915.

131. Gutmann, V.; Resch, G.; Linert, W. *Coord. Chem. Rev.* **1982**, *43*, 133.

2

The Essential Interface: Studies in Dye-Sensitized Solar Cells

Brian A. Gregg

National Renewable Energy Laboratory, Golden, Colorado, U.S.A.

I. INTRODUCTION: EXCITONIC SOLAR CELLS

Dye-sensitized solar cells [1–4] are the most promising alternative to conventional solar cells conceived in recent years. They convert light to electricity by a mechanism that is different from conventional cells. These differences provide an opportunity to further our understanding of the essential requirements for all solar cells and for this reason, among many others, the study of dye-sensitized solar cells (DSSCs or dye cells) is a fruitful area of research. It also brings us closer to the ultimate goal of designing and producing highly efficient, inexpensive solar cells. A key difference between dye cells and conventional solar cells, epitomized by silicon p–n junction cells, is the relative importance of interfacial processes. Conventional solar cells are minority carrier devices: Their efficiency is determined by the ability of photogenerated minority carriers (say, electrons in a p-type material) to escape from that side of the device before recombining with the majority carriers. Thus, properties such as the minority carrier lifetime and diffusion length are essential to device function. Although interfaces are also important in these devices, the crucial charge-carrier processes of photogeneration, separation, and recombination, all occur primarily in the bulk material. Therefore, bulk semiconductor properties such as crystallinity and chemical purity often control the efficiency of conventional solar cells, and optimizing these properties can be expensive.

Dye cells (see also Chapters 4 and 10), on the other hand, belong to a special class of majority carrier devices in which electrons are found almost exclusively in one phase and holes in another. Most organic photovoltaic (OPV) cells also belong to this class. Charge carriers in these "excitonic" solar cells are generated at the interface between the electron-conducting and hole-conducting phases via exciton dissociation. This interfacial mode of carrier generation is fundamentally different from the bulk generation occurring in conventional cells and is responsible for many of the unusual features of excitonic cells. In most OPV cells, photogenerated excitons (mobile excited states) must diffuse to the interface before dissociating, but in DSSCs, the excited states are created right at the interface. All important charge carrier processes—photogeneration, separation, and recombination—occur primarily, or exclusively, at the interface in excitonic solar cells, and thus the properties of these interfaces are of paramount importance, bulk properties are less critical. This allows the use of less pure and, therefore, less expensive materials. It also requires a somewhat different conceptual framework because, although superficially similar to conventional cells, excitonic solar cells are mechanistically distinct. The mechanism of the photoconversion process in DSSCs is described briefly in Section II and more thoroughly elaborated in the following sections.

One of the most obvious differences between dye cells and conventional photovoltaic (PV) cells is the nanoporous nature of the dye cell and the presence of an electrolyte solution permeating its entire thickness. Conventional cells, of course, are solid state, contain no mobile ions and usually have planar interfaces. Not surprisingly, this difference has a profound influence on the potential distribution throughout DSSCs and, therefore, on the interpretation of any behavior involving the applied or photogenerated potential. Some of the complexity of DSSCs also derives from the multiplicity of mobile species. Whereas a conventional PV cell has only electrons and holes, a dye cell has these *plus* inert (e.g., Li^+) and electroactive (I^-, I_2, and I_3^-) species. Section III presents a simple model of distributed resistance that describes one aspect of the potential distribution in DSSCs caused simply by the geometry of interpenetrating conducting phases. Experimental work designed to elucidate the effect of mobile electrolyte on the potential distribution in dye cells is then described. How these two independent phenomena affect functioning dye cells and other types of excitonic solar cells is treated in the final parts of Section III.

Although bulk processes like transport are important in all solar cells, the profound influence of the high-surface-area interface is one distinguishing feature of DSSCs and, to a lesser extent, of some OPV cells. The relative energy levels of the sensitizing dye, the TiO_2, and the redox couple determine the driving forces for the interfacial electron-transfer processes of photoinjection and recombination. Section IV describes these energetic relationships at the TiO_2/dye/electrolyte interface and shows how these forces can be affected by seemingly innocuous changes like altering the size of the electrolyte ions or the pH of the surface. The driving forces in a working cell are difficult to determine exactly because they are affected by so many factors. However, the qualitative changes expected to occur in the energy levels upon changing the electrolyte, for example, or upon illuminating the cell can be derived from basic principles and are illustrated by simple experiments.

The photoinduced generation of an electron and a hole in DSSCs is simultaneous with, and identical to, the initial separation of the electron from the hole across the interface. This is an important mechanistic difference relative to conventional solar cells, in which generation and separation are two spatially and temporally distinct processes. It results in different limiting factors to the photovoltages of the two cell types. Conventional solar cells *require* a gradient of electrical potential to separate the photogenerated charge carriers. Thus, the "built-in" electrical potential difference, Φ_{bi}, sets an upper limit to the achievable photovoltage. In dye cells, on the other hand, electrons are generated in (injected into) one phase while the holes remain in the other; thus, the two carrier types are already separated across a phase boundary upon generation. Therefore, Φ_{bi} does not usually determine the photovoltage in a DSSC, which is mainly controlled by the photoinduced chemical potential difference across the interface, $\Delta\mu$. Section V provides the basic equations describing these processes and some relevant experimental results. This section also includes a critical analysis of the view that DSSCs can be described in direct analogy to conventional p–n junction solar cells.

Energy-wasting charge-carrier recombination processes are the bane of solar cells and prevent them from reaching their thermodynamic efficiency limits. In contrast to conventional PV cells in which recombination occurs primarily in the bulk material, recombination is almost entirely an interfacial process in DSSCs. Because the interfacial area is huge (500–1000 times the geometrical surface area) [2], it is essential that the rate of the interfacial recombination process be extremely low. So far, efficient cells have been based on only a single redox couple, I^-/I_2, because of its uniquely slow kinetics for recombination. Use of other redox couples diminishes or eliminates the photovoltaic effect. However, the interfacial nature of the recombination process allows us novel methods for decreasing its rate that are not applicable to conventional cells. Two examples are described in Section VI. When such passivation techniques are perfected, it may be feasible to employ any desired redox couple in dye cells, making it far simpler to produce an efficient solid-state version of the DSSC.

It is not yet clear if the DSSC will become the dominant type of solar cell in the future, but its ability to achieve a high photoconversion efficiency by a mechanism that is fundamentally different from conventional solar cells makes it a perfect system for exploring those features that are indispensable to all solar cells. It is now clear, for example, that a p–n junction is not a necessary requirement for an efficient solar cell and neither is any form of built-in electrical potential gradient or a crystalline semiconductor with high carrier mobilities. All of these once were believed to be indispensable for efficient photoconversion. Our nascent understanding of dye cells and OPV cells, combined with knowledge of conventional solar cells, has revealed the enlightening fact that there is more than one viable photoconversion mechanism. Our challenge is to understand the quintessence of photoconversion processes and their materials requirements well enough to design and build the optimum solar cell.

II. PHOTOCONVERSION MECHANISM IN DYE CELLS

Unless otherwise noted, our experiments and discussion pertain to the conventional DSSC [2,5]. It consists of a nanoporous TiO_2 film sensitized with an adsorbed ruthenium complex like "N3" [Ru(4,4'-dicarboxy-2,2'-bipyridine)$_2$(NCS)$_2$]. The TiO_2 film is sintered onto a F-doped SnO_2 substrate electrode and the counterelectrode consists of F–SnO_2 with a thin film of Pt sputtered on it to catalyze the reduction of I_2/I_3^-. A drop of a liquid electrolyte containing 0.5 M LiI, 0.05 M I_2, and 4-*tert*-butylpyridine in 3-methoxypropionitrile is introduced between the two electrodes immediately before the cells are tested. For the most part, we will ignore the complexities of the I^-/ I_2/I_3^- redox "couple" and refer to the couple as I^-/I_2, and to the oxidized half of it as I_2/I_3^-. For a good description of this couple and its use in dye cells, see Refs. 6–8. Other dyes and electrolytes were sometimes used as described below. We did not employ a blocking layer of TiO_2 on the SnO_2 substrate [9,10] nor did we treat the nanoporous films with $TiCl_4$ solution [2], although these procedures probably would have improved the efficiencies. Our discussion mainly concerns the general features of these cells rather than particular details. As such, it is also relevant to other types of nanoporous devices [9,11–13] and, to some degree, to organic semiconductor devices [14–16].

The convoluted, high-surface-area interface between the TiO_2 and the electrolyte solution is an essential characteristic of DSSCs [1–3,5,17,18]. The photoconversion process begins at this interface when the adsorbed dye, D, absorbs a photon and the resulting excited state, D*, injects an electron, e_{TiO2}, into the nanocrystalline TiO_2 semiconductor:

$$D^* \rightarrow D^+ + e_{TiO_2} \tag{1}$$

In the conventional DSSC, this reaction is extremely rapid, occurring in the subpicosecond regime [19–23] and often exhibits near-unit efficiency. Note that the electron and the "hole" (D^+ at first, and then R^+) are *created* on opposite sides of the interface by the photoinjection process: They never coexist in the same phase and are already separated from each other upon creation. This fundamental difference is overlooked by those who would try to model dye cells as if they were p–n junctions.

The photoinjection reaction is followed by regeneration of the dye by the redox species, R:

$$R + D^+ \rightarrow R^+ + D \tag{2}$$

This reaction occurs in about 10 ns when R is an iodide ion in the 0.5 M concentration range [5]. Diffusion of e_{TiO2} through the nanocrystalline TiO_2 film to the substrate SnO_2 electrode and diffusion of the oxidized redox species, R^+, through the solution to the counterelectrode allow both charge carriers to be transferred to the external circuit where useful work is performed. The transport of electrons [7,24–29] and redox species [30] will not be considered further except insofar as they relate to the interfacial processes that are the focus of this chapter.

There are three major energy-wasting recombination pathways inside the cell. One involves recombination of e_{TiO2} with the oxidized dye before the dye can be regenerated:

$$e_{TiO_2} + D^+ \rightarrow D \tag{3}$$

This reaction has been studied in some detail [2,4,31,32] and will be considered only briefly here. It is a remarkably slow process (microseconds to milliseconds) at short circuit and, thus, does not limit the short–circuit photocurrent density, J_{sc}. However, the rate of reaction (3) [33] and of the other recombination reactions increases as the potential of the substrate electrode becomes more negative [e.g., as the cell voltage charges from short-circuit (0 V) to its open-circuit photovoltage, V_{oc}, (usually between -0.6 V and -0.8 V versus the counterelectrode)]. At open circuit, no current flows and the rate of charge photogeneration equals the total rate of charge recombination.

Following the initial charge generation/separation [reaction (1)], electrons and holes are confined to their separate chemical phases. However, they are in close proximity to each other throughout the nanoporous film and there is no substantial potential barrier at the interface to prevent them from recombining. Therefore, interfacial charge recombination can be a major energy-loss mechanism in DSSCs, especially if R is anything but I^-. The two remaining recombination reactions are the recombination of e_{TiO2} with the oxidized redox species.

$$e_{TiO_2} + R^+ \rightarrow R \tag{4}$$

and the recombination of an electron in the SnO_2 substrate, e_{SnO2}, with the oxidized redox species,

$$e_{SnO_2} + R^+ \rightarrow R \tag{5}$$

This latter reaction occurs because the nanocrystalline TiO_2 does not completely cover the SnO_2 substrate, leaving spaces between particles where the substrate comes into direct contact with the redox solution [5,12] Reaction (4) is favored by the high surface area of the TiO_2 relative to the SnO_2, but reaction (5) is favored by the high concentration of electrons in the degenerately doped SnO_2 relative to the TiO_2. The relative rates of the two reactions depend also on the chemical nature of R^+, as described in Section VI.

The diffusion length of photogenerated charge carriers is one of the important parameters governing the efficiency of a solar cell. In conventional cells, this is an intrinsic property of the semiconductor and its purity [34]. However, in DSSCs, the diffusion length is a function of the rate of reaction (4) and, thus, varies with different redox couples, surface treatments, and so forth. When the oxidation of R [reaction (2)] is chemically irreversible, the "diffusion length" of electrons is effectively infinite, whereas with kinetically fast, reversible redox couples (see Section VI), it approaches zero with unpassivated interfaces.

III. POTENTIAL DISTRIBUTION IN DSSCs: EFFECTS OF NANOPOROSITY

Knowledge of the potential distribution in a solar cell is crucial to understanding its behavior. However, this problem is complicated in DSSCs because of the semiconductor's nanoporous morphology and the interpenetration of electrolyte throughout its pores. In Section III. A, we ignore the electrolyte ions and consider only the effect of a potential applied across an interpenetrating network of two chemical phases: TiO_2 and solution in our example; however, the analysis is applicable to any two interpenetrating phases. We model this as a distributed resistor network, an analysis which, by itself, is more directly applicable to other types of high-surface-area excitonic solar cells [35–39] in which there is no mobile electrolyte (e.g., OPV cells). However, the additional complication due to the mobile electrolyte in DSSCs causes only a quantitative perturbation, not a qualitative change in behavior.

We then study experimentally the effect of an inert electrolyte solution and show that ion motion forces an applied electrical potential in the dark to drop near the substrate electrode, thus reinforcing the effects of the distributed resistance. Overall, the TiO_2 conduction and valence bands (whose spatial gradients reflect the electric field) remain approximately flat both at equilibrium and under illumination; therefore, charge transfer occurs primarily by diffusion rather than by field-induced drift [4,40–42]. Recent numerical simulations [43,44] and modeling of photogenerated trapped charges [45] show that in an illuminated DSSC there may be, in fact, a very small bulk electric field of about $0.1–3$ mV/μm, but this is not expected to have much influence.

A quantitative analysis of dark currents is a standard tool for characterizing conventional solar cells. However, this is not appropriate for DSSCs or most

OPV cells, as discussed in Section III.D. The necessity of both a nanoporous geometry *and* an electrolyte for screening of photogenerated electric fields is then described. Finally, some comparisons between DSSCs and other high-surface-area solar cells are drawn.

A. Distributed Resistor Network Model of Steady-State Potentials

In this section, we neglect ion motion through the nanoporous film and focus, instead, on the spatial dependence of current flow across a high-surface-area interface between the TiO_2 and the solution containing a redox couple. We employ a simple distributed resistor model that is intuitive and has semiquantitative character. Bisquert et al. have provided a more exact "transmission line" model for analyzing the impedance spectra of nanoporous cells [46,47]. As a negative potential is applied to the substrate electrode in the dark, some electrons are injected into the TiO_2 particles in contact with the substrate. These electrons can proceed along two possible pathways: into the next TiO_2 particle or across the TiO_2–solution interface, reducing the redox species R^+ [reaction (4)]. There are characteristic resistances associated with each pathway. A distributed resistor network model (Fig. 1) can illustrate the potential distribution under steady-state conditions. The first process is modeled by an interparticle resistance, R_{TiO2}, the second by the interfacial charge transfer resistance, R_{ct} [48]. Electrons that proceed into the second TiO_2 particle (second node of the network) have the same choice of two

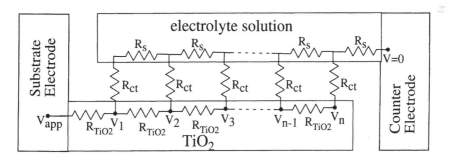

Figure 1 A distributed resistor network models approximately how the applied potential is distributed across a DSSC under steady-state conditions. For various values of the interparticle resistance, R_{TiO2}, and the interfacial charge transfer resistance, R_{ct}, the voltage is calculated for each node of the TiO_2 network, labeled V_1 through V_n. This is purely an electrical model that does not take mobile electrolytes into account and, therefore, potentials at the nodes are electrical potentials, whereas in a DSSC, all internal potentials are electrochemical in nature.

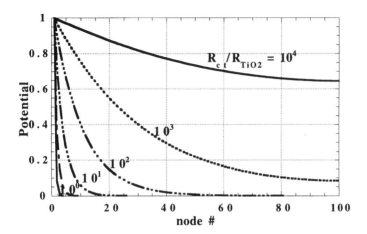

Figure 2 The calculated potential distribution across a DSSC modeled by the resistor network of Fig. 1. The node number corresponds to distance through the cell; in this case, the cell was modeled with 100 circuit elements. See the text for details.

pathways, and so on through the film until the last particle, where the only choice is to transfer into solution. Once electrons are transferred into solution, they are subject to the solution resistance, R_s, which, in this model, we consider negligible compared to the other two, but which could be substantial for a solid-state electrolyte. For simplicity, we neglect charge transfer across the SnO_2–solution interface; if its rate were not negligible, it would cause the applied potential to drop closer to the substrate electrode.

It is possible to solve this model analytically.* The model is useful for calculating the qualitative potential distribution in high-surface-area solar cells and how it is influenced by the relative facility of different charge-transfer pathways. Solutions for a 100-resistor network are shown in Fig. 2, where V_{app} = 1.0 V, R_s = 0.001 Ω, and $R_{TiO2} + R_{ct}$ = $10^4\Omega$.

Figure 2 shows that *only when $R_{ct}/R_{TiO2} \to \infty$ will the potential applied to the substrate electrode, E_{sub}, extend uniformly throughout the TiO$_2$ film.* Otherwise, the (electrochemical) potential of the TiO$_2$ film will be a function of distance from the substrate and vary between E_{sub} and the solution redox potential, E_{redox} (set = 0 V in Fig. 2). Of the three resistances, only R_s is expected to be approximately constant in a standard DSSC. R_{TiO2} is highly variable, depending on light intensity, applied potential, and the concentration of certain "potential-determining" ions such as Li^+ and H^+. In an illuminated DSSC, both the electron concen-

* We thank Mark Hanna and Phil Parilla of NREL for providing the solution to this model.

tration *and* the electron diffusion coefficient increase by three to four orders of magnitude over a cell at equilibrium [28], meaning that R_{TiO2} decreases by six to eight orders of magnitude upon illumination. R_{ct} also decreases somewhat under illumination and with applied negative potential, but much less than R_{TiO2} [32,33]. Only under conditions where $R_{ct} \gg R_{TiO2}$, which may be approached in *thin* cells under conditions of high illumination intensity with a kinetically slow redox couple such as I^-/I_2, is it valid to assume that $E_{sub} \approx E_{TiO2}$, independent of distance. In general, however, the models that neglect the distance dependence of the applied potential are expected to have limited validity.

The current will take the (distributed) path of least resistance through the cell. If R_{TiO2} is the dominant resistance, as it probably is in the dark in conventional DSSCs, much of the TiO$_2$ will remain near solution potential and most current will flow through solution rather than through the TiO$_2$. If R_{ct} is the dominant resistance, as it may be in an illuminated DSSC using the I^-/I_2 redox couple, the TiO$_2$ film will be near the potential of the substrate and current will flow approximately uniformly across the TiO$_2$ – solution interface in thin cells. However, if $R_{ct} \approx R_{TiO2}$ under illumination, only the dyes close to the substrate will contribute to the photocurrent because electrons injected further away will recombine before they reach the substrate. When R_s is the dominant resistance, which might occur in solid-state DSSCs, much of the photocurrent will flow through the TiO$_2$, thereby avoiding the high-resistance hole conductor and also decreasing the efficiency of dyes adsorbed far from the counterelectrode [43,44]. These differences in effective current pathways under different conditions—light versus dark, solid state versus liquid electrolyte, and so forth—are an essential characteristic of DSSCs; however, they make accurate modeling quite difficult. In conventional solar cells, there is only one possible current pathway, so a one-dimensional spatial model is appropriate. However, DSSCs are far more complex.

The distributed resistor model neglects the effect of mobile electrolyte ions. Much of our following discussion of the electrolyte's influence neglects, for simplicity, the distributed resistance. In a real dye cell, both effects operate simultaneously. Both tend toward the same result: An applied potential will be more or less confined near the substrate electrode, depending on the relative rates of charge transport and interfacial charge transfer and on the concentration of electrolyte.

B. Impedance Studies

The energy stored in an electric field is minimized by minimizing the spatial extent of the field. For example, the plates of a charged capacitor are attracted to one another, so moving them closer decreases the energy of the system. In the same way, adding electrolyte to a solution contained between charged electrodes decreases the spatial extent of the field because ions flow to squeeze the electric field into the smallest volume consistent with the decrease in entropy caused by concentrating the ions. Therefore, the potential applied to a DSSC will usually drop over just a fraction of the TiO_2 film. It requires energy to expand the electric field into a greater volume. This can be accomplished, for example, by adding electrons to the TiO_2 film, making it more conductive and increasing the amount of stored charge (capacitance) in the electrochemical double layer (Fig. 3). Such effects of electrolyte motion are investigated experimentally in this and the next section.

The use of inert (electrochemically inactive) electrolyte in our experiments made R_{ct} as high as possible. Zaban et al. measured impedance spectra of DSSCs in the dark between 1 Hz and 10 kHz in dry acetonitrile [42]. The capacitance was independent of both applied potential and surface area (film thickness) of the nanoporous TiO_2 film in the range from $+1.0$ V to about the TiO_2 conduction-band edge potential at -0.5 V versus standard calomel electrode (SCE). At potentials more negative than -0.5 V, the capacitance increased sharply by two to three orders of magnitude and approached a plateau at -1.4 V whereupon it became directly proportional to the surface area (thickness) of the film. We interpreted the capacitance between $+1.0$ and -0.5 V as being dominated by the planar SnO_2 electrode–electrolyte interface, showing that the bulk of the TiO_2

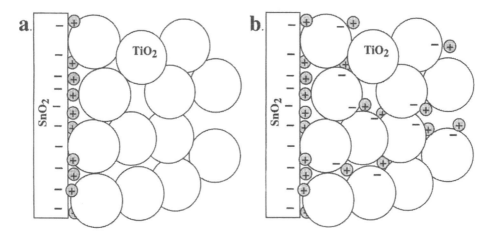

Figure 3 Schematic of a nanoporous TiO_2 film in the dark showing the movement of compensating positive ions (circles with " $+$ ") through the film that screens a negative potential (electrons shown as " $-$ ") applied to the SnO_2 substrate electrode. (a) The electric field is screened close to the substrate when the potential is positive of the conduction band, but (b) extends further into the semiconductor for more negative potentials. The potential distribution also depends on the relative rates of interfacial versus interparticle charge transfer (Fig. 2).

film was not experiencing the potential applied to the substrate. Counterion motion up to the SnO_2 electrode effectively screened the applied potential (Fig. 3a) as it would in a normal electrochemical cell, confining the driving force for charge transfer into solution mostly near the substrate electrode. As the potential is scanned negative of -0.5 V, the TiO_2 becomes progressively more conductive and the electroactive surface area of the film, as measured by its capacitance, expands (Fig. 3b) to eventually include the entire 8-μm-thick film at an applied potential of -1.4 V. If electron–hole pairs were generated in the TiO_2 by illumination with ultraviolet (UV) light, the onset of the capacitance increase was shifted from -0.5 V to $+0.3$ V, reflecting the increased conductivity of the semiconductor.

In these experiments, the potential distribution was measured under conditions where the interfacial current density was minimized by the use of an inert electrolyte. If the electron-transfer rate across the interface had truly been zero $(R_{ct} = \infty)$, the whole TiO_2 film would have eventually charged up to the applied potential; it was the unavoidable leakage current across the interface and the relatively short time scale of our experiments that prevented this from happening. These experiments show that even when R_{ct} is maximized, ion motion through the nanoporous film causes the applied potential to drop near the substrate electrode in nonilluminated DSSCs. As we showed earlier, decreasing R_{ct} causes the applied potential to drop even closer to the substrate electrode.

C. Dye Desorption Experiments

A direct visualization of the potential distribution in the DSSC film was achieved by dye desorption experiments [42]. The common ruthenium-based sensitizing dyes desorb from the TiO_2 surface at potentials negative of a threshold (~ -1.0 V versus SCE), and so the desorption process can provide a visual measure of the potential distribution. Nanoporous TiO_2 films were formed on insulating glass substrates and silver contacts were evaporated on one end. After adsorbing the dye, a film was placed in a stirred solution of dry acetonitrile with an inert electrolyte and held at a negative potential in the dark until no further dye is desorbed. Then, the film was removed and the concentration of dye versus distance from the electrical contact was measured by scanning a laser spot along the film. The results showed that, as expected, the dye desorbed further from the silver contact as the potential was made more negative. However, even at -2.1 V versus SCE, the dye was undisturbed at a distance of more than 0.4 mm from the electrode. In other words, the potential dropped from the applied -2.1 V to somewhat positive of -1.0 V in ~ 0.4 mm, giving visual confirmation of our earlier conclusion. The spatial distribution of the applied potential discussed above plays an important role in all potential dependent measurements and their interpretations.

D. Comparing Dark Currents to Photocurrents

The cathodic dark currents measured under negative bias in a DSSC (Fig. 4) are related to the recombination reactions (4) and (5) in the illuminated cells. However, the relation between the dark current and reaction (4), especially, is difficult to quantify. In conventional solar cells, dark current measurements provide quantitative information about the photorecombination processes [34] because (1) the number of photogenerated charge carriers is only a small perturbation on the dark carrier density and (2) the current flows along the same pathway in the light and in the dark. Neither of these conditions hold for a DSSC.

There are multiple possible current pathways through a DSSC, as shown in Fig. 1, because the nanoporous cell consists of two interpenetrating, bicontinuous chemical phases. The relative conductivity of these two phases and of the connection between them, R_{ct}, depends on the illumination intensity, applied potential, kinetics of the redox couple, and so forth. Therefore, the distribution of current pathways depends also on these variables. In the DSSC, the dark current will take the distributed path of least overall resistance (Sections III.A–III.C), meaning it will flow primarily through solution [50] under the expected conditions of $R_{ct} < R_{TiO2}$. The dark current is thus mainly a measure of reaction (5) in this potential range, even though reaction (4) is expected to be the dominant recombination

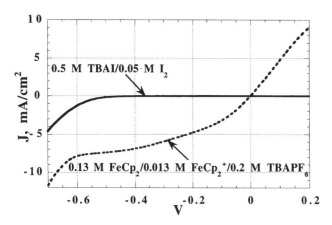

Figure 4   Dark currents in DSSCs with the standard I^-/I_2 redox couple (solid line) and with a kinetically much faster redox couple, ferrocene/ferrocenium, $FeCp_2^{+/0}$. TBA = tetrabutylammonium. The charge-transfer resistance, R_{ct} (see Fig. 1), of the I^-/I_2 couple is ~10^6 times greater than that of the $FeCp_2^{+/0}$ couple, leading to what is sometimes mistaken as diode behavior in the dark for the cell containing the I^-/I_2 couple. (Data from Ref. 49.)

mechanism under illumination, when R_{TiO2} decreases by $> 10^6$ (Section III.A). Only when the voltage is scanned negative of the flat-band potential will an increasing fraction of the TiO_2 become sufficiently conductive that reaction (4) will contribute substantially to the dark current [42]. Upon illumination, however, the density of photoinjected electrons in the TiO_2 becomes much higher than the density of dark carriers, causing reaction (4) to occur over much of the TiO_2 surface. This means that a substantial fraction of the *photo*recombination process [much of reaction (4)] is invisible to measurements of the *dark* current, in contrast to what occurs in conventional cells. Nevertheless, a comparison of photocurrents and dark currents is still informative because, qualitatively, the same criteria hold for DSSCs as for conventional solar cells: The lower the dark current at a given potential, the higher the possible photocurrent.

This analysis is valid for all solar cells that consist of interpenetrating chemical phases—of which there are an increasing number [51]. For those without mobile ions, the distributed resistor model alone leads to the conclusion that dark currents cannot be quantitatively compared to photocurrents; for those with mobile electrolyte, the effect is quantitatively reinforced by the field-induced motion of the electrolyte ions.

E. Screening of Photogenerated Electric Fields

The effect of the *macroscopic* motion of electrolyte ions in response to an applied electric field was treated in Sections III.B–III.D. However, the *nanoscopic* motion of ions in the pores of a DSSC also plays an important role in the charge-separation process. When a photogenerated electron–hole pair is separated across the TiO_2–solution interface, the electrostatic attraction between the opposite charges opposes the separation. In conventional solar cells, this attraction is overcome by the high electric field in the p–n junction: Charge separation causes only a diminution of the built-in field (flattening of the bands; see Section V). The interior of the dye cell, however, contains no significant electric fields, except across the electrochemical double layers, and these are not adequate to screen an injected electron from the "hole" on the oxidized dye [6,8]. Therefore, a transient attractive electric field is created whenever an electron is photoinjected into the TiO_2. This field must be rapidly neutralized in order to avoid charge recombination.

In the "standard" DSSC, mobile electrolyte ions can rapidly rearrange around the photogenerated charge pairs, neutralizing the Coulomb attraction between them [42,52] (with some help from the solvent dielectric properties) and thereby slowing their recombination rate. This is one of the critical functions of the electrolyte in a dye cell: If a solar cell is approximately electroneutral in the dark, like a DSSC, some means of neutralizing the electric field between the photogenerated electrons and holes is essential for efficient charge separation.

For this reason (and others), it is not trivial to make a solid-state version of the dye cell. Initial attempts did not include a mobile electrolyte and thus had no way of neutralizing the Coulomb attraction between the photogenerated charge pairs [42,53]. The best results were achieved by Tennakone et al. [13] in a cell with solid CuI as the hole conductor—in which the ionic mobility of the CuI may have helped neutralize the Coulomb attraction. Later attempts included mobile electrolyte ions, which improved performance [9,54].

The required screening of the Coulomb attraction between photogenerated charge-carrier pairs is only effective when electrolyte ions can practically surround the charge-carrier pair. This cannot occur with *planar* semiconductor electrodes where electrolyte ions are confined to only the half-space. Dye sensitization of planar semiconductor electrodes has been extensively investigated [55–60] and these studies provided much of the conceptual foundation upon which the initial analysis of DSSCs was based. However, the nanostructured geometry of a DSSC makes a fundamental difference: Electrolyte ions can almost surround photoinjected charges in a DSSC, thus no interfacial electric field is required (or even possible in such small colloids) to affect charge separation. However, dye sensitization of *planar* electrodes, where the photogenerated electric field between the injected carriers and their conjugate holes cannot be neutralized by ion motion, *requires* a built-in electric field (Φ_{bi}, or band bending) for efficient charge separation [42,59,61].

This leads to an important historical point: With the exception of a few inefficient organic PV cells (which were generally ignored), all solar cells up to about the year 1990 *required* a built-in electric field to separate photogenerated electrons from holes. Not surprisingly, many researchers came to believe that such a built-in field was an *essential* requirement for solar cells; but this is incorrect, as we discuss further in Section V. What *is* required is a method to separate photogenerated electrons from holes: A built-in electric field is the conventional method; a rapid field neutralization by mobile electrolyte coupled with an interfacial chemical potential gradient as driving force (a lá DSSCs) is another method; a combination of the two should also be effective. All imaginable methods also depend on the slow (relative to charge separation) recombination of the charge pair. It is possible to extrapolate between the two extremes of nanoporous and planar electrodes and predict that the effectiveness of the ionic screening, that is so important for nanoporous films, must diminish as the particle size increases from nanoscopic toward macroscopic (planar) electrodes.

This has implications for the design of high-surface-area solar cells in general: If the bulk of the device is essentially field-free at equilibrium, then mobile electrolyte *and* nanoporosity are required to eliminate the photoinduced electric fields that would otherwise inhibit charge-carrier separation. On the other hand, if the particle size is substantially larger than in the conventional dye cell *or* if there is no mobile electrolyte, then an interfacial or bulk built-in electric field

(Φ_{bi}) is required for efficient separation of the photogenerated charge carriers. There are some size regimes between nanoporous and planar–macroscopic in which a combination of an interfacial electric field and some mobile electrolyte may be the most efficacious.

F.   Comparisons to Other High-Surface-Area Solar Cells

After the success of the DSSC, a number of groups started using the high-surface-area concept in other organic-based solar cells. For example, solar cells have been made with high-surface-area interfaces between conducting polymers and a number of other materials such as C_{60} [35,39], quantum dots [36], nanocrystalline semiconductors [37,62], and other conducting polymers [38]. In these cells, excitons must *diffuse* to an interface in order to dissociate into an electron in one phase and a hole in the other, unlike in DSSCs. Because of limitations in exciton diffusion length and film conductivity in OPV (organic PV) cells, it is often advantageous to structure the interface. The operating principles of these cells have much in common with dye cells but with the exceptions, usually, of containing no electrolyte and of requiring some exciton diffusion. The applied field is not confined by the motion of electrolyte ions in high-surface-area OPV cells, but it is still restricted by the distributed resistance. Photoinduced interfacial charge separation cannot be aided by microscopic electrolyte motion in these cells; therefore, it requires an interfacial electric field and/or very slow interfacial recombination kinetics. The inability of dark current measurements to provide direct quantitative information about the rates of photoinduced recombination reactions (Section III.D) should be common to all solar cells that employ high-surface-area interfaces between bicontinuous phases.

IV.   INTERFACIAL ENERGETICS

Dye-sensitized solar cells are unusual in that the photoactive species (the sensitizing dye) resides in the electrochemical double layer between the TiO_2 and the electrolyte solution [42,52,63]. Therefore, the thermodynamic driving forces for all interfacial reactions—photoinjection [reaction (1)], transfer of the "hole" to solution [reaction (2)], and carrier recombination [reactions (3)–(5)]—are dependent on the semiconductor band edge potential, the solution redox potential, and the magnitude and spatial extent of the potential distribution (double layer) in the interfacial region. The double layer, in turn, is influenced by the species adsorbed to the TiO_2 (including the dye) and by the concentration, chemical nature, and physical size of the electrolyte ions. Thus, changing the composition of the electrolyte can substantially affect the photoconversion process by altering the energetics of the interfacial charge-transfer reactions.

A. Induced pH Sensitivity of the Sensitizing Dye

The oxidation potentials of the common ruthenium-based sensitizing dyes are independent of pH in solution. So naturally, it was assumed at first that this pH independence carried over to the adsorbed sensitizing dye. However, some anomalous results were obtained, such as the apparent independence of electron-transfer rates from their driving forces (altered by varying the pH of solution) [64]. These results led us to investigate the pH dependence of the oxidation potentials of *adsorbed* sensitizing dyes. We discovered that they *became* pH dependent when adsorbed to oxide surfaces [52,63]. This change in behavior upon adsorption was observed also on insulating Al_2O_3 surfaces, showing that it must result from an electrostatic effect rather than from electronic coupling between the dye and the oxide surface. We then broadened the inquiry and tested the response of four ruthenium dyes and two phthalocyanines [52]; all had pH-independent redox potentials in solution and all became pH dependent when adsorbed to TiO_2. The magnitude of the induced pH sensitivity varied between 54 mV/pH unit and 20 mV/pH unit, depending on the size and shape of the dye, whereas the expected change in conduction-band potential of an oxide semiconductor with pH is 59 mV/pH unit [64–66]. The essential parameter appeared to be the *spatial* location of the dye relative to the electrical double layer at the semiconductor–solution interface. Protons and hydroxide ions are so small that they interact with the TiO_2 surface even in the presence of the adsorbed dye, causing the semiconductor bands to move the expected ~59 mV/pH unit. However, the counterions of larger acids, ClO_4^- in our case, are too bulky to fit underneath the specifically adsorbed dye. This locates the dye partially inside the electrochemical double layer (i.e., partially inside the electrostatic field between the adsorbed protons and the counterions in solution). We proposed that the dye's location in a region of high electric field was responsible for altering its effective redox potential and inducing the pH dependence. If the dye were completely inside the field, as the TiO_2 is, its potential would shift the same 59 mV/pH unit as any oxide surface [63,65]. However, a species adsorbed on the surface of TiO_2 cannot be completely inside the double-layer field and thus can experience only some fraction of it, depending on the location of the redox center between the surface and the loci of centers of the counterions.

This work showed that the driving forces for the decisive interfacial electron-transfer reactions in DSSCs are not necessarily fixed by the chemical identity of the participants, but can vary depending on the spatial location of the dye in the electrochemical double layer.

B. Ionic Size Effects

Our proposed explanation for the induced pH sensitivity of the adsorbed sensitizing dye depended only on geometrical and electrostatic factors, not chemical considerations. It depended only on the dye being partially inside the interfacial electric field between the solid and the solution [63]. Thus, we predicted that similar effects would be observed under appropriate conditions in nonaqueous solvents. To test this, we performed, in nonaqueous solvents, both spectroelectrochemical measurements of the reduction of an adsorbed perylene dye and potential-dependent measurements of the photoluminescence of a ruthenium dye in the presence of $LiClO_4$ and $TBAClO_4$ (TBA = tetrabutylammonium) [52]. The lithium ion is small enough to adsorb to TiO_2 and collapse the field in front of the dye; TBA is too large to closely approach the surface, so the field must drop partially across the specifically adsorbed dye (Fig. 5). With these two electrolytes, we were able to vary the spatial extent of the interfacial electric field (the double layer), with the dye located more (TBA^+) or less (Li^+) inside it. Overall, the results were consistent with the mechanism proposed for the induced pH sensitivity and showed clearly that interfacial electrostatics and band edge motion of the semiconductor play crucial roles in determining the relative potentials—the driving forces—in DSSCs.

To further support these conclusions, we studied the potential dependence of the reduction of an adsorbed dye in the dark. Perylene di(hydroximide) is

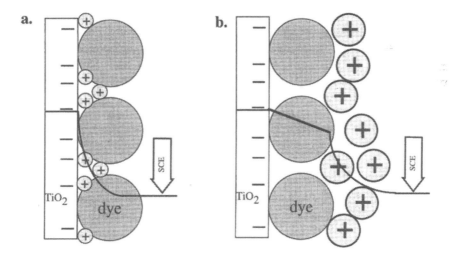

Figure 5 TiO_2–dye–solution interface shown under negative applied or photogenerated bias. The solid line shows the electrical potential drop between the TiO_2 and the solution. (a) In the presence of small cations, the dye oxidation potential is only slightly affected by changes in the TiO_2 potential relative to SCE. (b) In the presence of large cations, the dye potential follows, but always lags, somewhat, the changes in the TiO_2 potential.

electroreduced in solution at ~ -0.6 V versus SCE to the radical anion. This dye adsorbs strongly to TiO$_2$ [67]. When adsorbed, its electroreduction occurs also at ~ -0.6 V when exposed to a Li$^+$-containing solution, but at ~ -0.9 V when exposed to a TBA$^+$-containing solution [52]. In Li$^+$ solution, the fraction of reduced adsorbed dye increased by 10-fold for every 97 mV of applied potential (instead of the 59 mV per decade expected theoretically, showing the effect of the potential distribution): In TBA$^+$ solution, the slope was 230 mV/decade. Again, these results are consistent with the dye being mostly, but not completely, *inside* the double layer with TBA$^+$, and mostly, but not completely, *outside* with Li$^+$ ions [52].

Similar behavior is observed in the potential-dependent luminescence of a ruthenium dye adsorbed to TiO$_2$ [52]. Although the flat-band potential of TiO$_2$ is known to shift positive in the presence of the potential-determining Li$^+$ ion, relative to the TBA$^+$ ion [5,68], this effect cannot explain the observed behavior. For example, in our experiments, the dye injects in both cases, but it takes a much smaller negative potential excursion to turn off the injection process in the presence of Li$^+$ than with TBA$^+$ [52]. This is the opposite of what would be expected if only equilibrium (i.e., dark) band edge motion were responsible for the effect.

The behavior of the dye adsorbed on a negatively charged TiO$_2$ surface for the cases of small and large cations in solution is schematically illustrated in Fig. 5. A similar mechanism also explains the induced pH dependence, although, in this case, the charge on the TiO$_2$ is determined by the adsorbed protons rather than by the applied negative potential. Our experiments and the model of Fig. 5 show that *the redox potential of the sensitizing dye is not fixed relative to either the semiconductor or the solution and that the semiconductor band edge is also not fixed.* This has important implications for the understanding and optimization of the dye cell. For example, during the course of a potential scan from short circuit to open circuit, the redox potential of the dye in an illuminated cell is expected to move relative to either the TiO$_2$ or the solution, or both. In the presence of small cations (Fig. 5a), the driving force for charge injection [reaction (1)] will decrease as the cell approaches its open-circuit photovoltage because the semiconductor band edge is charging negative while the dye potential remains relatively constant. However, in the presence of large cations (Fig. 5b), the dye will tend to follow the semiconductor potential; thus, the driving force for the oxidation of I$^-$ by the oxidized dye [reaction (2)] will decrease with increasing photopotential. Therefore, the photovoltage-limiting kinetic step may be altered merely by changing the *size* of the electrolyte cation [52]. The implications of the variable interfacial potentials and driving forces have not yet been rigorously explored; however, it is clear that they play an important and subtle role in the functioning of a DSSC.

V. INTERFACIAL CHEMICAL POTENTIAL GRADIENTS AND THE PHOTOVOLTAGE-DETERMINING MECHANISM

In this section, we discuss the most fundamental nature of solar cells: how they convert sunlight into electricity or chemical products. Conventional solar cells, almost by definition, all function according to the same photoconversion mechanism epitomized by the photoconversion process in silicon p–n junction solar cells [34,69]. This mechanism is so well known that the assumptions underlying it are sometimes forgotten and it is thought to be the only possible mechanism for photoconversion. Here, we discuss the forces that drive a flux of electrons through a solar cell and use this foundation to describe first the photoconversion mechanism in conventional cells and then in the model that uses a similar mechanism to explain the PV effect in DSSCs. A competing model, based on the interfacial nature of the exciton dissociation process, is then presented and experiments designed to distinguish between the two models are described.

A. Three Potential Energies (U, μ, and E) and the Photoconversion Process

A brief diversion into basic physics and thermodynamics will set the stage for a discussion of the forces and fluxes underlying all solar cells. When a battery is charged, electrical potential energy, U, is converted to chemical potential energy, μ. (Unfortunately, μ is commonly used to denote both the chemical potential energy and the carrier mobility.) There are no macroscopic electric fields inside a charged battery because of the high concentration of electrolyte (as in a DSSC), but if the external electrodes are connected to each other through a load, the stored chemical energy is converted back to electrical energy and can generate a short-circuit current or an open-circuit voltage. Gibbs was the first to clearly formulate the equivalence of these two forms of energy and define the electrochemical potential energy, E, as the sum of the electrical and chemical potential energies, $E = U + \mu$ [70].

The spatial gradient of any potential energy is a force; ∇E is, ultimately, the force that drives the particle fluxes through solar cells and other devices, not ∇U or $\nabla \mu$ individually. (To avoid confusion, ∇E is the electrochemical *force* at a given point and ΔE is the difference in electrochemical *potential energies* between two points, often between one end of the semiconductor and the other: in one dimension, $\int \nabla E \, dx = \Delta E$). We must carefully distinguish between the electrical potential, Φ, and the electrical potential energy, $U = q\Phi$, where q is the electronic charge. Thus, the built-in electrical potential difference across a semiconductor device at equilibrium is usually denoted by $\Phi_{(bi)}$. At equilibrium, the rate of every reaction is exactly counterbalanced by the rate of its reverse reaction (principle of microscopic reversibility); thus, $\nabla E = 0$ and $\nabla U = -\nabla \mu$ at every point. In an electrolyte solution at equilibrium, the electrochemical potential energy is given by the redox potential times the unit charge, $E = q\Phi_{redox}$ [71], whereas in the solid state, it is called the Fermi level, E_F [34,72].

Away from equilibrium, thermodynamics becomes much more complicated and, in marked contrast to equilibrium thermodynamics, it is rarely mechanism independent [73–75]. It becomes, in essence, a generalized form of kinetics. For relatively simple problems, though, like the motion of electrons and holes through electronic devices, it is possible to generalize the relevant forces and their associated fluxes [76,77] to arrive at a generalized kinetic, or "quasithermodynamic," description [55,72,78–80]. In solar cells, the gradients of the quasi-Fermi levels for each relevant particle, say for electrons, ∇E_{Fn}, and for holes, ∇E_{Fp}, are the forces that drive the particle fluxes. This description is *exact* because it is based on generalized kinetics (see below). It is *not* based on the invalid application of equilibrium thermodynamics to nonequilibrium systems [70,73,75–77], although it has often been presented as such.

The general kinetic expression for the one dimensional current density of electrons, $J_n(x)$, through any device is

$$J_n(x) = n(x)\mu_n \nabla U(x) + kT\mu_n \nabla n(x) \tag{6}$$

where $n(x)$ is the concentration of electrons, μ_n is the electron mobility (not to be confused with the chemical potential μ), and k and T are Boltzmann's constant and the absolute temperature, respectively. (We assume an effective one-dimensional geometry for simplicity; in general, and especially in DSSCs, all three dimensions should be taken into account.) There is an exactly analogous equation to describe the flux of holes, so for simplicity, we treat only electrons. The only assumption involved in this *kinetic* equation is that the electron current is only affected by electrical and chemical potential gradients (i.e., that there are no magnetic fields, temperature or pressure gradients, and so forth). Therefore, Eq. (6) is valid both at equilibrium and away from it, both in the dark and in the light.

The quasi-Fermi level (i.e., the nonequilibrium Fermi level) for electrons in a semiconductor is defined as

$$E_{Fn}(x) = E_{cb}(x) + kT \ln\left(\frac{n(x)}{N_c}\right) \tag{7}$$

where $E_{cb}(x)$ is the electrical potential energy of the conduction band edge, $E_{cb}(x) = U(x) + \text{constant}$, and N_c is the density of electronic states at the bottom of the conduction band. Taking the gradient of Eq. (7) and substituting it into Eq. (6) leads to the simplest expression for the electron current through a device:

$$J_n(x) = n(x)\mu_n \nabla E_{Fn}(x) \tag{8}$$

This derivation is based only on kinetic equations and definitions; therefore, it is valid under nonequilibrium conditions. The quasi-Fermi level is, strictly speaking, simply a mathematical substitution that simplifies the final kinetic equation. Although it is clearly related to the equilibrium thermodynamic concept of the electrochemical potential energy, E_F, its validity in the description of nonequilibrium systems is grounded *solely* on the kinetic derivation of Eq. (8). This is crucial because E_{Fn} is often considered to be a "thermodynamic" (i.e., universally valid) property, and it has been derived by the clearly incorrect application of equilibrium thermodynamics to nonequilibrium systems. In this case, the correct result was obtained by an incorrect analysis; unfortunately this "success" has led to further misuses of equilibrium thermodynamics resulting in questionable "thermodynamic" equations for nonequilibrium processes. It is critical to distinguish between true equilibrium thermodynamics, which is universally valid and mechanism independent (at equilibrium!), and the mechanism-dependent, quasi-thermodynamic descriptions of nonequilibrium systems whose credibility depends wholly on their derivation from kinetic equations. Unless unambiguously derived from a kinetic foundation and free of any invocations of equilibrium thermodynamics, it is probably best to ignore any purported "thermodynamic" equations that claim to describe nonequilibrium systems.

In the classical description of nonequilibrium systems, fluxes are driven by forces [73,76,77]. Equation (8) shows that the flux of electrons (J_n) is related to the (photo)electrochemical force (∇E_{Fn}) by a proportionality factor ($n\mu_n$). Equation (8) and the related equation for holes can be employed as a simple and powerful description of solar photoconversion systems. However, it is useful to go beyond this analysis and break ∇E_{Fn} into its component quasithermodynamic constituents, ∇U an $\nabla\mu$, because this helps reveal the fundamental differences between the photoconversion mechanisms of the various types of solar cells. Equation (6) can be separated into two independent electron fluxes, each driven by one of the two generalized forces, ∇U and $\nabla\mu$. Equations (9a) and (9b) are expressed in the form Flux = Proportionality factor × Force:

J_n due to the electrical potential energy gradient is

$$J_n(x) = n(x)\mu_n \nabla U(x) \tag{9a}$$

J_n due to the chemical potential energy gradient is

$$J_n(x) = n(x)\,\mu_n \frac{k\,T}{n(x)\nabla n(x)}$$

or

$$J_n(x) = n(x)\mu_n \nabla\mu(x) \tag{9b}$$

The identity $\nabla\mu(x) = kT/n(x)\nabla n(x)$ can be derived from the equilibrium thermodynamic expression $\mu = kT \ln(n) + \mu^0$ [81]; thus, it is legitimate to call

$\nabla\mu(x)$ the gradient of the chemical potential energy. However, again, it is, strictly speaking, just a mathematical substitution into a kinetic equation—it is a *quasi* thermodynamic potential, not an equilibrium thermodynamic potential. Equations (9a) and (9b) are often referred to as the drift and diffusion components, respectively, of the electron current. One can see immediately from Eqs. (6), (9a), and (9b) that $\nabla U(x)$ and $\nabla\mu(x)$ are *independent* forces in the photoconversion process and, therefore, that *it is possible to drive a solar cell with either one, or both, of these forces*. In fact, the different types of solar cell can be classified according to the relative importance of these two forces in the photoconversion process.

B. Conventional Solar Cells

Conceptually, when a conventional p–n homojunction solar cell is assembled, a highly n-doped and a highly p-doped slab of the same semiconductor material are brought into contact to form a p–n junction. Before contact, each side is electrically neutral, so $\Delta U = q\Delta\Phi = 0$; however, there is a large chemical potential difference, $\Delta\mu$, between them because the n-doped side has a high concentration of free electrons (it is "reduced") and the p-type side has a high concentration of free holes (it is "oxidized"). Upon contact, current flows driven by $\nabla\mu$ according to Eq. (9b), until the electrical potential difference created by the loss of majority carriers is exactly equal and opposite to the chemical potential difference between the two sides (Fig. 6, left). This is equilibrium, where $q\nabla\Phi = -\nabla\mu$ at every point and $\Delta E_{cb} = q\Phi_{bi} = -\mu_{bi}$. Here, $\Phi_{bi} = \Delta\Phi_{equilib}$ is the built-in electrical potential difference or "band bending" across the semiconductor at equilibrium and $\mu_{bi} = \Delta\mu_{equilib}$ is likewise defined as the built-in chemical

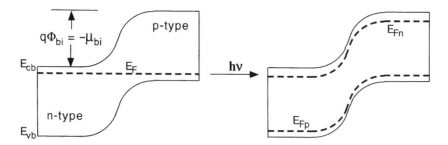

Figure 6 A schematic band diagram (electrical potential energy versus distance) of a conventional p–n homojunction solar cell at equilibrium (left) and at short circuit under *spatially uniform* illumination (right). The energies of the conduction- and valence-band edges are E_{cb} and E_{vb}, respectively. E_F is the Fermi level at equilibrium and E_{Fn} and E_{Fp} are the quasi-Fermi levels of electrons and holes, respectively, under illumination.

potential energy difference across the semiconductor at equilibrium. Under non-equilibrium conditions, we denote these two potentials as $\Phi_{neq}(x)$ and $\mu_{neq}(x)$.

Light absorption (shown on the right side of Fig. 6 as if it were spatially uniform) leads to the production of electron–hole pairs. Because of the large difference in equilibrium carrier concentrations on the two sides, the major effect of light absorption is the increase in the minority carrier density (electrons on the p-type side and holes on the n-type side). This causes a decrease in the chemical potential energy difference between the two sides of the cell, $\Delta\mu_{neq} < \mu_{bi}$. If the electrical potential difference is held constant (short circuit) as shown on the right side of Fig. 6, photogenerated electrons will flow back to the n-type side and holes to the p-type side, driven by the photoinduced difference between the electrical and chemical potentials, $\nabla E_{Fn} = q\nabla\Phi_{neq} + \nabla\mu_{neq}$. If the circuit is opened, the cell will charge up to $V_{oc} = \Phi_{bi} + \Delta\mu_{neq}/q$ (or less if limited by recombination), thereby flattening the bands by the difference between the photoinduced and the equilibrium chemical potentials.

The key mechanistic point in these conventional solar cells is that *both electrons and holes are photogenerated together in the same semiconductor phase*. Because the two carrier types are produced with the same spatial distribution, $\nabla\mu_{neq}$ drives them both in the same direction (although the force on the minority carriers is greater). Therefore, the crucial separation of electrons from holes can occur efficiently *only* by the action of $\nabla\Phi_{neq}$. As the cell charges up toward V_{oc}, the driving force for charge separation, $\nabla\Phi_{neq}$ (band bending), decreases correspondingly. Therefore, the rule for conventional solar cells is that Φ_{bi} sets the absolute upper limit to V_{oc} because when $V_{oc} = \Phi_{bi}$, $\nabla\Phi_{neq} = 0$, the bands are flat, and there is no longer any electrical driving force for charge separation. Excitonic solar cells, however, are fundamentally different because the charge carrier pairs are *already separated across an interface upon photogeneration*; thus, Φ_{bi} is not required for initial charge separation.

C. Junction Model of DSSCs

Before describing the two competing models of DSSCs, the junction model and the interface model, the points upon which both agree are summarized. Both assume that the initial charge separation (photoinjection of electrons from the dye into the nanocrystalline TiO_2 film and transfer of "holes" into solution) is primarily a kinetically controlled process [2,60] and that the subsequent electron transport through the bulk of the film occurs in an essentially electric-field-free regime. Therefore, charge transport occurs by diffusion rather than by field-driven drift [2,17,24,27,40–42,82]. There are no significant electric fields in the individual nanocrystalline TiO_2 particles [4,17] or between the sintered particles in the bulk of the film [42,43,83,84] because, in the first case, the particles are too small (~ 15 nm) and too lightly doped to support a significant space charge and because, in the second case, the sintered particles are surrounded by a concentrated electrolyte solution that screens any existing electric fields within about 1 nm. There can still be an electro*chemical* difference, ΔE_f, between particles (Section III), but it consists mostly of $\Delta\mu$ rather than ΔU. The only significant electrical gradients, ∇U, that can exist in a DSSC are at the various solid–solution interfaces (Section IV).

The equations describing solid-state p–n junction solar cells have sometimes been employed to describe DSSCs. This is equivalent to assuming that DSSCs and p–n junctions function by an identical mechanism; a fairly radical assumption, for which there has often been little attempt at justification. Schwarzburg and Willig, however, proposed a detailed model that rationalized the treatment of DSSCs as p–n junctions [82]. Their "junction model" is in many ways analogous to the model of a conventional solid- state solar cell, but adapted to the physical constraints of a dye cell. It posits the existence of a large equilibrium electrical potential difference, Φ_{bi}, at the TiO_2–SnO_2 substrate interface that drives charge separation across the interface. This junction potential is determined by the difference between the work function of the substrate electrode, Φ_{sub}, and the solution redox potential, Φ_{redox}, in the same fashion as a junction is formed at a conventional semiconductor electrode upon immersion in a redox electrolyte solution [72]. However, the junction potential in the dye cell is restricted to a very narrow region ($\sim$20 nm, about one TiO_2 particle diameter) at the TiO_2–SnO_2 interface because the electrolyte solution that permeates the porous TiO_2 film would screen an electric field within $\sim$ 1 nm or so anywhere else in the cell [42,82]. Therefore, in contrast to conventional cells, the junction potential does not drive charge transport through the bulk of the TiO_2, which remains electric-field-free, but only across the TiO_2–SnO_2 interface. The *key assumption of the junction model* is that the separation of the electrostatically screened electrons in the TiO_2 from their screening charges (usually Li^+) in solution *cannot* occur without this interfacial electrical potential difference, Φ_{bi}, Therefore, Φ_{bi} sets the upper limit to the magnitude of the achievable photopotential, just as it does in a conventional p–n junction cell [82]. Although V_{oc} is limited by Φ_{bi} in this model, too, Φ_{bi} plays a different role than it does in conventional cells: It effects only the final separation of the electron from its screening ion in solution, but it does not separate the photogenerated electron from its conjugate hole nor does it promote the transport of charge carriers. Other groups have deduced the existence of [6,85] or modeled [46] a junction potential occurring at the TiO_2–SnO_2 interface but did not estimate its magnitude or claim it sets an upper limit to V_{oc}.

D.　Interface Model of DSSCs

The more common understanding of DSSCs, that we now refer to as the "interface model" and formerly referred to as the "kinetic model" [12], was developed by a number of groups [2,6,17,18,24,26,40,42,43,52,84,86,87] but usually appears in the literature as a common understanding rather than as a clearly formulated model. We have attempted to formulate this model more precisely [12] and parts of it have been used throughout this chapter. The interface model posits that the photoinjected electrons are driven into the substrate electrode primarily by the chemical potential gradient, $\nabla\mu$, created by the photoinduced increase in the concentration of electrons in the TiO_2 relative to the substrate electrode; that is, E_{Fn} becomes more negative in the TiO_2 when electrons are photoinjected from the excited dye molecules. The counterbalancing electrical potential gradient, ∇U, that would otherwise develop between the photoinjected electrons and the oxidized species is screened by the electrolyte solution that permeates the bulk of the nanocrystalline TiO_2 film [42]. Therefore, ∇E_{Fn} consists almost entirely of $\nabla\mu$.

The photoinduced difference between the quasi-Fermi level for electrons in the TiO_2 and the quasi-Fermi level for "holes" in solution, $\Delta E_{Fn} = E_{Fn,TiO2} - E_{Fp,solution}$, sets an upper limit to the photovoltage, V_{oc}, because it is this potential difference and the fact that electrons and holes are confined to *separate* chemical phases that drives electrons toward the substrate electrode and holes toward the counterelectrode. Although ∇E_{Fn} is mainly comprised of $\nabla\mu$ in DSSCs, there is, nevertheless, a possible role for $q\ \nabla\Phi$ at interfaces where the field cannot be entirely screened by mobile electrolyte.

The interface model predicts that V_{oc} in a dye cell will not be limited by Φ_{bi} because Φ_{bi} does not control the charge-separation process. Rather than having large potential gradients at equilibrium, as in conventional cells, the DSSC has only small and relatively insignificant values of Φ_{bi} and μ_{bi}. Illumination of a DSSC causes the potential gradients to *increase*, whereas in a conventional cell, they *decrease* upon illumination. Because the photoinduced increase in Φ_{neq} is practically eliminated by electrolyte ion redistribution, the photoinduced increase in μ_{neq} can drive an efficient photoconversion process.

E. Comparing the Two Models

Because both models make clear predictions, it is possible to distinguish between them. The difference between the work function of the substrate electrode and the solution redox potential, $\phi_{bi} = \phi_{sub} - \phi_{redox}$, sets the upper limit to V_{oc} in the junction model [82] whereas the interface model suggests that V_{oc} is not strongly correlated to ϕ_{bi} because V_{oc} is controlled mainly by μ_{neq} [12]. Therefore, changing ϕ_{bi} and monitoring its effect on V_{oc} can distinguish between the two models. To do so, we measured V_{oc} in three different redox electrolyte solutions with dye-sensitized TiO_2 films deposited on four different substrates that had vacuum work functions spanning a 1.4-eV range [12].

The conventional dye cell uses I^-/I_2 as the redox couple, and no other known redox couple works nearly as well [49]. The use of most other redox couples, and most substrates besides SnO_2, is expected to accelerate the recombination reactions relative to the conventional dye cell and thus diminish V_{oc}, independent of the mechanistic model (see Section VI). Because this would obscure what we were looking for, the relationship between ϕ_{bi} and V_{oc}, we used only the reduced half of the redox couple in order to minimize recombination rates and maximize V_{oc}. Versus vacuum, the work functions of the clean electrodes are $\phi_{ITO} \approx 4.3$ eV [88], $\phi_{SnO2} \approx 4.8$ eV [88], $\phi_{Au} \approx 5.1$ eV [89], and $\phi_{Pt} \approx 5.7$ eV [89].

There were no substantial differences between the magnitudes of the photovoltages on the different substrate electrodes despite the ~ 1.4-V range in their vacuum work functions (Fig. 7). The slight decrease in V_{oc} on Pt substrates was caused by the enhanced rate of recombination at this highly catalytic electrode [12]. The behavior predicted by the junction model is shown by the theoretical line. Although there are some unavoidable ambiguities in the values of ϕ_{bi} in

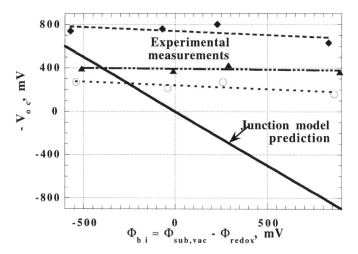

Figure 7   The open-circuit photovoltage plotted versus the difference between the work function of the substrate in vaccum, $\phi_{sub,\ vac}$, and the solution redox potential, ϕ_{redox}. The work function of the substrate in the solution, ϕ_{sub}, the quantity of interest, is difficult to measure directly, but it is related to $\phi_{sub,\ vac}$: see the discussion in Ref. 12. The four types of substrates are, from left to right, ITO, SnO_2, Au and Pt; the filled diamonds are for 0.5 M LiI solution, the open circles are for 0.05 M ferrocene in 0.1 M LiClO$_4$ solution, and the filled triangles are for 0.05 M hydroquinone in 0.1 M LiClO$_4$ solution. The theoretical line shows the behavior predicted by the junction model. (Data from Ref. 12.)

these experiments [12], the overwhelming weight of the evidence points to a simple conclusion: V_{oc} is practically independent of ϕ_{bi}, and, therefore, the junction model is incorrect. In further support of this conclusion, the standard DSSC uses SnO_2 electrodes and a I^-/I_2-containing electrolyte: Given the easily measured solution redox potential and the commonly accepted value of the electrode work function, the junction potential cannot be greater than ~0.3 V [8]. However, typically $V_{oc} = 0.7$–0.8 V. Moreover, the use of Au and Pt substrate electrodes in our experiments creates a ϕ_{bi} that *opposes* electron injection into the SnO_2, yet V_{oc} is almost the same as with the other substrates. For these reasons and others (see Sec. VI.B), we conclude that the photovoltage in dye cells is not determined by the equilibrium junction potential, ϕ_{bi}, in contrast to conventional solar cells. The upper limit to V_{oc} is set by the photoinduced *interfacial* difference between the quasi-Fermi levels of electrons and holes, ΔE_F, not by the equilibrium electric field in the semiconductor alone. We believe that *these results prove unambiguously that the photoconversion mechanism in dye cells is fundamentally distinct from that in conventional cells.*

There are thus two fundamentally different ways to drive a photoconversion process (and combinations should also work), as discussed briefly in Section III.E. In conventional solar cells, which include most cell types made before the year 1990, it is the magnitude of an *equilibrium electrical potential* difference that controls V_{oc}. However, in DSSCs, it is the magnitude of a *photoinduced chemical potential* difference that primarily determines V_{oc}. In OPV cells, *both* contribute [90]. The fact that there are two different ways to drive a photoconversion process is still not universally accepted, probably because conventional solar cells that employ $q\Phi_{bi}$ as the driving force are so well known that the existence of band bending is sometimes mistaken for an absolute requirement. However, Eqs. (6), (9a), and (9b) show that ϕ_{bi} ($= \int \nabla U/q$) is not required for an electron flux. The only universal requirement for a photoconversion process is a photoinduced gradient of the electrochemical potential, ∇E_{Fn} (for electrons). ∇E_{Fn} can consist of a pure electrical potential energy gradient, $q\nabla\phi$, a pure chemical potential energy gradient, $\nabla\mu$, or, more commonly, the sum of the two [12].

VI. INTERFACIAL RECOMBINATION RATES AND PASSIVATION PROCEDURES

The recombination of photogenerated electrons and holes is the bane of all solar cells and a major reason for their less than ideal efficiencies. Excitonic solar cells, in which the electrons and holes exist in separate chemical phases, are subject primarily to interfacial recombination. There is, as yet, no theoretical model to accurately describe interfacial recombination processes, and this is an important area for future research. Wang and Suna [91] have laid a possible foundation for such a model by combining Marcus theory with Onsager theory.

Marcus theory [92,93] was developed for solution-phase electron-transfer (ET) reactions and has been extended to (photo)electrochemical reactions by Marcus, Gerischer, and others [55,72,93]. This theory emphasizes the importance of the solvent reorganization energy as the activation barrier which must be surmounted for electron transfer to occur. Because of its origins in the study of solution-phase ET reactions, it assumes that all electron-transfer processes occur from the distance of closest approach and that any electric field effects are isotropic. These are not viable approximations for electron transfer in solid-state systems. Onsager theory [94–96] was developed decades earlier to describe ET reactions in solids, especially amorphous solids. It has been employed to describe charge-transfer processes in electrophotographic (xerographic) systems [96]. It treats the electrostatic aspects of charge separation and recombination in homogeneous solids, but neglects the reorganization energy. The synthesis of these two theories is an important advance [91], but, unfortunately, no analytical solutions are found for the equations and only numerical results can be obtained.

Despite the lack of theoretical models for interfacial recombination processes in excitonic solar cells, it is obvious empirically that those cells which function efficiently must have a very slow rate of recombination. In DSSCs, this can be explained simply by the slow electron self-exchange rate of the I^-/I_2 redox couple and the absence of field-driven recombination. However, in the case of solid-state, high-surface-area OPV cells, such as the conducting polymer/C_{60}-derivative cells [36,39], the slow rate of interfacial recombination is an important problem that is not yet understood.

In principle, interfacial recombination processes can be inhibited by modifying the interface. The use of t-butylpyridine in the DSSC electrolyte solution to increase its photovoltage is one example [2,97]. We wished to explore general methods for passivating interfacial recombination sites in DSSCs that might allow the use of a variety of redox couples and therefore facilitate making a viable solid-state DSSC.

A. Recombination in DSSCs

Only one redox couple, I^-/I_2, allows conventional dye-sensitized solar cells to function efficiently because of the uniquely slow kinetics for I_2 reduction on SnO_2 and TiO_2 surfaces [reactions (4) and (5)]. The reduction is so slow at both TiO_2 [98] and SnO_2 surfaces that almost every electron photoinjected into the TiO_2 at short circuit survives the transit through the nanoporous film and the SnO_2 substrate and appears in the external circuit. Unlike most OPV cells, interfacial recombination in conventional dye cells is not promoted by the electric field: The "holes" are either an electrically neutral species (I_2) that is unaffected by the field or a species (I_3^-) that has the same negative charge as the electron. Only recently, some other couples with ultraslow kinetics have been discovered, but they are not yet as efficient as I^-/I_2 [99,100]. When faster redox couples such as ferrocene/ferrocenium are employed, the rapid interfacial recombination of photoinjected electrons with the oxidized half of the redox couple (corresponding to a low R_{ct} in Figs 1 and 2) practically eliminates the photovoltaic effect [49].

Several groups [6–8,82] have discussed the reasons for the very slow rates of reactions (4) and (5) when $R/R^+ = I^-/I_2$. This "couple" may be kinetically ideal for current dye cells, but it is far from ideal in many respects [49] and there are ongoing efforts to find a replacement [99,100], especially for solid-state versions of DSSCs. The major effort directed toward solid-state DSSCs [9,13,53,54] is driven primarily by the difficulties associated with the long-term hermetic sealing of a solar cell containing volatile components such as solvent and I_2.

In order to overcome recombination limitations in DSSCs, we sought a general method to inhibit reactions (4) and (5). To gauge our success, we needed a redox couple with a self-exchange rate so fast that the interfacial recombination process would become the photocurrent-limiting reaction. The ferrocene/ferrocenium ($FeCp_2^{+/0}$) couple is kinetically very fast (self-exchange rate, $k_{ex} \approx 10^7$ $M^{-1} s^{-1}$) [101] compared to the I^-/I_2 couple ($k_{ex} \approx 5 \times 10^2 M^{-1} s^{-1}$) [102] and has a redox potential only slightly more positive than the iodine couple (0.31 V versus SCE compared to ~0.15 V). We did not expect $FeCp_2^{+/0}$ to be a viable redox couple for DSSCs, but employed it as a probe of recombination rates. If it were possible to so thoroughly inhibit recombination that a substantial photovoltaic effect could be achieved with $FeCp_2^{+/0}$, then it should be possible to use almost any redox couple in DSSCs.

B. Dark Currents, Photocurrents, and Recombination Rates

The recombination reactions in DSSCs can be qualitatively understood by measuring the dark currents of DSSCs in both I^-/I_2 and $FeCp_2^{+/0}$-containing electrolyte solutions (Fig. 4). A quantitative understanding is not possible yet as discussed in Section III. The dark currents [Fig. 4, corresponding to the sum of reactions (4) and (5)] increase with an overpotential of ~0.4 V for I_2 reduction, but with a vanishingly small overpotential, ~0.0 V, for the much faster redox couple, $FeCp_2^{+/0}$ [49].

The dark $J–V$ curves with the I^-/I_2 couple appear to be "diodelike" (Fig. 4). Because this is similar to the dark $J–V$ curves of conventional solar cells, it has led to suggestions that the current in DSSCs is controlled by a p–n electrical junction. This was used as one justification for modeling DSSCs as p–n junctions (Sec. V.D). However, such $J–V$ data can be fit just as well to the Butler–Volmer equation, a mainstay of electrochemistry [48], as to the diode equation. The

Butler–Volmer equation is more obviously applicable to DSSCs than is the diode equation, which assumes the absence of a mobile electrolyte. Assuming the Butler–Volmer equation, the current onset near 0.45 V would be explained as the result of the very slow kinetics (large electrochemical overpotential) for electron exchange between the electrodes (SnO_2 and TiO_2) and I_2/I_3^-. Assuming the diode equation, on the other hand, the current onset would be explained as a built-in junction voltage of ~0.6–0.8 V resulting from the difference between the solution redox potential, Φ_{redox}, and the work function of the electrical contact, Φ_{sub}. The data in Fig. 4 (as well as the results discussed in Sec. V.E) allow us to distinguish unambiguously between these two explanations. When a kinetically faster redox couple such as $FeCp_2^{+/0}$ is used, the apparent diodelike character disappears (Fig. 4), even though the more positive redox potential of $FeCp_2^{+/0}$ would be expected to *increase* the Φ_{bi} of the putative junction and thus enhance its diodelike behavior. The fact that the opposite occurs shows that the dark currents are controlled by the electrochemical *overpotential*, not by an electrical *junction*. This lends additional support to our earlier conclusion that the photocurrent also is not controlled by an electrical junction (Sec. V.E).

When we derivatize the solvent-exposed oxide interfaces with an electrical insulator (described below), this *cannot create an electrical junction*. Yet, it does impart "diodelike" character to the dark and light currents measured with $FeCp_2^{+/0}$; that is, derivitization *creates an electrochemical overpotential* for the recombination reactions, just as there is naturally an overpotential for the reduction of I_2/I_3^-. This leads to the general conclusion that slowing the recombination reaction(s) relative to the photodriven forward reactions, *by whatever mechanism*, establishes the necessary conditions for a successful photoconversion process.

The photocurrent–voltage curve of a cell made with the I^-/I_2 redox couple (Fig. 8) shows behavior typical of the standard DSSC. The substantial photovoltaic effect is expected from the fact that the dark current (Fig. 4) is negligible positive of about -0.5 V. On the other hand, a cell made with the $FeCp_2^{+/0}$ redox couple shows no measurable photoeffect: Its current under illumination (Fig. 8) is essentially equal to its dark current (Fig. 4). The photovoltaic effect is negligible because practically all photogenerated charge carriers recombine before they can be collected in the external circuit. In general, fast rates of reactions (4) and (5) tend to eliminate the photovoltaic effect in DSSCs.

C. Inhibiting Recombination at the SnO_2 Surface

It is possible to inhibit reaction (5) by employing spray pyrolysis to deposit a compact layer of TiO_2 on the SnO_2 substrate before deposition of the nanocrystalline TiO_2 film [9,10]. It is, however, relatively difficult to achieve a pinhole-free film. We chose to investigate the electropolymerization of an insulating film on the exposed parts of the SnO_2 after the nanocrystalline TiO_2 film has been depos-

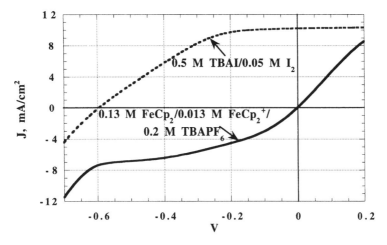

Figure 8 Photocurrent–voltage curves of unpassivated, N3-sensitized TiO_2 films under ~ 1 sun illumination showing the difference between the iodine couple and the ferrocene couple. (Data from Ref. 49.)

ited but before the dye is adsorbed (Fig. 9a). This has the advantage of being a simple technique that will tend to spontaneously plug pinholes because the current density will be highest wherever bare SnO_2 is exposed. We electrodeposited a blocking layer of cross-linked poly(phenyleneoxide-co-2-allylphenyleneoxide), PPO, on the TiO_2-coated SnO_2 surface by adapting a literature procedure [49,103]. The deposition of the PPO film occurs on the SnO_2 surface exposed to the solution rather than on the TiO_2. The sensitizing dye was adsorbed following the electrodeposition of PPO.

The electrodeposition of PPO has a pronounced effect on the $J–V$ characteristics of a DSSC employing fast redox couples. Figure 10a shows the dark currents and photocurrents of an N3-sensitized TiO_2 film in $FeCp_2^{+/0}$ solution before and after treatment with PPO. The dark current is reduced by two to three orders of magnitude in the potential range positive of the TiO_2 flat band, thus allowing a modest photovoltaic effect to appear. The photocurrent is still seriously limited by the rapid recombination of photoinjected electrons in the TiO_2 with $FeCp_2^+$ in solution [reaction (4)], but simply passivating the exposed SnO_2 [reaction (5)] substantially improves cell performance.

D. Inhibiting Recombination at Both the SnO_2 and TiO_2 Surfaces

The sensitizing dye cannot cover all of the TiO_2 surface, and the uncovered areas (Fig. 9) are potential sites for interfacial recombination. A chemical reaction that

(a) SnO$_2$ passivated by electropolymerized PPO

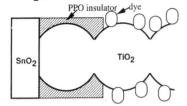

(b) Exposed parts of TiO$_2$ and substrate
passivated by poly(methylsiloxane)

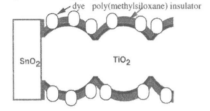

Figure 9   An illustration depicting the two methods of interface passivation used here.
(a) Reaction (5) alone is inhibited by electropolymerizing a film of insulating PPO, poly(p-
henyleneoxide-*co*-2-allylphenyleneoxide), on the SnO$_2$ substrate. (b) Both reactions (4)
and (5) are inhibited by coating exposed oxide surfaces with poly(methylsiloxane).

could cover the exposed oxide surfaces with an ultrathin electrically insulating
film should substantially decrease the rates of both reactions (4) and (5) by physically preventing the close approach of R^+ to e_{TiO2} and e_{SnO2}. As a first attempt,
we employed CH_3SiCl_3 in the vapor phase as a reactant. This reagent should
form covalently bound, cross-linked but stochiometrically imperfect films of po-
ly(methylsiloxane) on the surface of the TiO$_2$ and SnO$_2$ films.

Experimental details are given in Ref. 49. The results were promising but
still far from perfect. Apparently, some of the dye was covered over and/or some
of the pores in the TiO$_2$ film were plugged before all of the exposed TiO$_2$ surface
was covered with the first monolayer of silane. The tendency of the silane, even
at 60°C in the gas phase, to capillary condense in restricted spaces and, therefore,
to deposit thickly in some regions of the film while leaving other regions bare
made the passivation reaction less effective than it could have been. Although
the conditions for this reaction have not been optimized, substantial decreases in
the rates of the recombination reactions could be achieved (Fig. 10b).

The open-circuit photovoltage, short-circuit photocurrent, and fill factor
obtained using the silanization reaction were superior to those obtained with the

PPO reaction (Fig. 10). This is consistent with decreasing the rates of both reactions (4) and (5) with the silanization while only decreasing the rate of reaction 5 with the PPO treatment. There remains substantial room for improving these passivation procedures, and other, as yet unexplored, methods may prove superior to those described here. However, the results achieved so far demonstrate the promise of this approach.

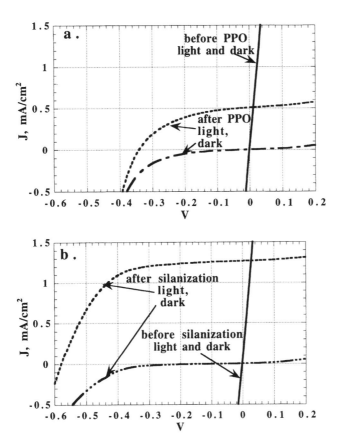

Figure 10   Dark and photocurrents with $FeCp_2^{+/0}$ redox couple before and after interfacial passivation. (a) Effect of blocking recombination at the SnO_2 surface, reaction (5) (Fig. 9a), by electropolymerization of PPO. (b) Effect of blocking recombination at both the SnO_2 and TiO_2 surfaces, reactions (4) and (5) (Fig. 9b), by vapor-phase silane treatment without PPO. (Data from Ref. 49.)

VII. SUMMARY AND CONCLUSIONS

The dye cell is a nanoporous, interface-driven solar cell that seems to defy conventional wisdom. It contains no substantial built-in electrical potential, $\Phi_{bi} \approx 0$, yet it produces a substantial photovoltage; carrier transport occurs primarily by diffusion rather than by drift, yet it is highly efficient. Most fundamentally, electrons and holes are photogenerated on opposite sides of the interface, already separated into their individual phases, leading to the overwhelming importance of interfacial processes over bulk processes. In fundamental ways, dye cells contrast with conventional PV cells. The success of the dye cell highlights the importance of the photoinduced chemical potential energy gradient, $\nabla\mu_{neq}$, as a driving force for photoconversion, whereas decades of experience with conventional solar cells focused attention almost exclusively on the electrical potential gradient, $\nabla U = q\nabla\Phi_{neq}$. The quasithermodynamic force that drives the flux of electrons through all solar cells is the electrochemical potential gradient, $\nabla E = \nabla U + \nabla\mu$. However, in dye cells, ∇U is practically eliminated by the motion of electrolyte ions, leaving $\nabla\mu$ as the major driving force. Thus, the study of dye cells provides an opportunity to understand the "neglected half" of the photoconversion process.

The nanoporous nature of the dye sensitized semiconductor in DSSCs results in a complex spatial distribution of applied or photogenerated potential gradients through the cell. Moreover, the nanoscopic motion of electrolyte ions in response to the Coulomb field between photogenerated electrons and holes plays a crucial role in the charge-separation process. The fact that the sensitizing dyes reside partially inside the electrochemical double layer at the TiO_2–solution interface leads to the unusual and complex situation where the driving forces for the most important electron-transfer reactions vary with potential, light intensity, pH, and even with the size of electrochemically inert counterions and the shape of the dye. Because of the interfacial nature of all important charge-transfer processes in DSSCs, novel methods of passivating the recombination sites are possible and preliminary attempts seem promising.

Organic semiconductor photovoltaic cells share many characteristics with both DSSCs and conventional cells. Charge generation occurs almost exclusively by interfacial exciton dissociation, as in DSSCs, but, in contrast, OPV cells usually contain no mobile electrolyte and thus rely on $\nabla\Phi_{neq}$ *and* $\nabla\mu_{neq}$ to drive charge separation. OPV cells may have planar interfaces, like conventional PV cells, or highly structured interfaces, like DSSCs. They provide a conceptual and experimental bridge between DSSCs and conventional solar cells.

In the future, we see two irreversible trends: (1) The cost of solar energy will continue to decrease, and may drop substantially if organic-based photovoltaics (DSSCs and OPV cells), which are much less expensive than inorganic PV, can be made efficient and reliable and (2) the environmental and societal costs associated with using fossil and nuclear fuels will become ever more difficult to overlook. Although many of these costs are not currently included in the price, being paid instead out of general taxes, the price of nonrenewable energy still will inevitably and irreversibly increase. There is no escape from the laws of thermodynamics: An isolated system like the Earth cannot for long use energy that is not renewable because it will run out, by definition, and because even before that happens, the by-products of using nonrenewable energy may pollute our planet beyond habitability. The future undoubtedly belongs to renewable energy. We do not yet know which specific type of solar cell designs are optimal, but research is leading us toward the most economically and environmentally efficient methods of converting solar energy into electricity and chemical fuels.

ACKNOWLEDGMENTS

I am grateful to all my colleagues and collaborators, most of whom are listed as coauthors in our cited references, to Alison Breeze for a critical review of Section V, and to the U.S. Department of Energy, Office of Science, Division of Basic Energy Sciences, Chemical Sciences Division for supporting this research.

REFERENCES

1. O'Regan, B.; Grätzel, M. *Nature*, **1991**, *353*, 737–740.
2. Kalyanasundaram, K.; Grätzel, M. *Coord. Chem. Rev.*, **1998**, *77*, 347–414.
3. Nazeeruddin, M. K.; Kay, A.; Rodicio, I.; Humphry-Baker, R.; Müller, E.; Liska, P.; Vlachopoulos, N.; Grätzel, M. *J. Am. Chem. Soc.* **1993**, *115*, 6382–6390.
4. O'Regan, B.; Moser, J.; Anderson, M.; Grätzel, M. *J. Phys. Chem.* **1990**, *94*, 8720–8726.
5. Hagfeldt, A.; Grätzel, M. *Acc. Chem. Res.* **2000**, *33*, 269–277.
6. Schlichthörl, G.; Huang, S. Y.; Sprague, I.; Frank, A. J. *J. Phys. Chem. B.* **1997**, *101*, 8141–8155.
7. Huang, S. Y.; Schlichthörl, G.; Nozik, A. J.; Grätzel, M.; Frank, A. J. *J. Phys. Chem. B.* **1997**, *101*, 2576–2582.
8. Cahen, D.; Hodes, G.; Grätzel, M.; Guillemoles, J. F.; Riess, I. *J. Phys. Chem. B.* **2000**, *104*, 2053–2059.
9. Bach, U.; Lupo, D.; Compte, P.; Moser, J. E.; Weissörtel, F.; Salbeck, J.; Spreitzer, H.; Grätzel, M. *Nature*, **1998**, *395*, 583–585.
10. Kavan, L.; Grätzel, M. *Electrochim. Acta* **1995**, *40*, 643–652.
11. Bechinger, C.; Ferrere, S.; Zaban, A.; Sprague, J.; Gregg, B. A. *Nature* **1996**, *383*, 608–610.
12. Pichot, F.; Gregg, B. A. *J. Phys. Chem. B.* **2000**, *104*, 6–10.
13. Tennakone, K.; Kumara, G. R. R. A.; Kumarasinghe, A. R.; Wijayantha, K. G. U.; Sirimanne, A. R. *Semicond. Sci. Technol.* **1995**, *10*, 1689–1693.
14. Tang, C. W. *Appl. Phys. Lett.* **1986**, *48*, 183–185.

15. Gregg, B. A.; Fox, M. A.; Bard, A. J. *J. Phys. Chem.* **1990**, *94*, 1586–1598.
16. Gregg, B. A.; Kim, Y. I. *J. Phys. Chem.* **1994**, *98*, 2412–2417.
17. Hagfeldt, A.; Lindquist, S.-E.; Grätzel, M. *Solar Energy Mater. Solar Cells* **1994**, *32*, 245–257.
18. Kamat, P. V.; Bedja, I.; Hotchandani, S.; Patterson, L. K. *J. Phys. Chem.*, **1996**, *100*, 4900–4908.
19. Tachibana, Y.; Moser, J. E.; Grätzel, M.; Klug, D. R.; Durrant, J. R. *J. Phys. Chem.* **1996**, *100*, 20,056–20,062.
20. Hannappel, T.; Burfeindt, B.; Storck, W.; Willig, F. *J. Phys. Chem. B.* **1997**, *101*, 6799–6802.
21. Ellingson, R. I.; Asbury, J. B.; Ferrere, S.; Ghosh, H. N.; Sprague, J. R.; Lian, T.; Nozik, A. J. *J. Phys. Chem. B.* **1998**, *102*, 6455–6458.
22. Asbury, J. B.; Hao, E.; Wany, Y.; Ghosh, H. N.; Lian, T. *J. Phys. Chem. B.* **2001**, *105*, 4545–4557.
23. Ferrere, S.; Gregg, B. A. *J. Am. Chem. Soc.* **1998**, *120*, 843–844.
24. Södergren, A.; Hagfeldt, A.; Olsson, I.; Lindquist, S.-E. *J. Phys. Chem.* **1994**, *98*, 5552–5556.
25. Schlichthörl, G.; Park, N. G.; Frank, A. J. *J. Phys. Chem. B.* **1999**, *103*, 782–791.
26. Dloczik, L.; Ileperuma, O.; Lauermann, I.; Peter, L. M.; Ponomarev, E. A.; Redmond, G.; Shaw, N. J.; Uhlendorf, I. *J. Phys. Chem. B.* **1997**, *101*, 10,281–10,289.
27. de Jongh, P. E.; Vanmaekelbergh, D. *J. Phys. Chem. B.* **1997**, *101*, 2716–2722.
28. Kopidakis, N.; Schiff, E. A.; Park, N.-G.; van de Lagemaat, J.; Frank, A. J. *J. Phys. Chem. B.*, **2000**, *104*, 3930–3936.
29. Nelson, J. *Phys. Rev. B.* **1999**, *59*, 15,374–15,380.
30. Papageorgiou, N.; Barbe, C.; Grätzel, M. *J. Phys. Chem. B.* **1998**, *102*, 4156–4164.
31. Moser, J. E.; Grätzel, M. *Chem. Phys.* **1993**, *176*, 493.
32. Haque, S. A.; Tachibana, Y.; Willis, R. L.; Moser, J. E.; Grätzel, M.; Klug, D. R.; Durrant, J. R. *J. Phys. Chem. B.* **2000**, *104*, 538–547.
33. Haque, S. A.; Tachibana, Y.; Klug, D. R.; Durrant, J. R. *J. Phys. Chem. B.* **1998**, *102*, 1745–1749.
34. Fahrenbruch, A. L.; Bube, R. H. *Fundamentals of Solar Cells. Photovoltaic Solar Energy Conversion*; Academic Press: New York, 1983.
35. Yu, G.; Gao, J.; Hummelen, J. C.; Wudl, F.; Heeger, A. J. *Science* **1995**, *270*, 1789–1791.
36. Huynh, W. U.; Peng, X.; Alivisatos, A. P. *Adv. Mater.* **1999**, *11*, 923–927.
37. Arango, A. C.; Carter, S. A.; Brock, P. J. *Appl. Phys. Lett.* **1999**, *74*, 1698–1700.
38. Halls, J. J. M.; Walsh, C. A.; Greenham, N. C.; Marseglia, E. A.; Friend, R. H.; Moratti, S. C.; Holmes, A. B. *Nature* **1995**, *376*, 498–500.
39. Shaheen, S. E.; Brabec, C. J.; Sariciftci, N. S.; Padinger, F.; Fromherz, T. *Appl. Phys. Lett.* **2001**, *78*, 841–843.
40. Cao, F.; Oskam, G.; Meyer, G. J.; Searson, P. C. *J. Phys. Chem.* **1996**, *100*, 17,021–17,027.
41. Solbrand, A.; Lindström, H.; Rensmo, H.; Hagfeldt, A.; Lindquist, S.-E.; Södergren, S. *J. Phys. Chem. B.* **1997**, *101*, 2514–2518.
42. Zaban, A.; Meier, A.; Gregg, B. A. *J. Phys. Chem. B.* **1997**, *101*, 7985–7990.

43. Ferber, J.; Stangl, R.; Luther, J. *Solar Energy Mater. Solar Cells* **1998**, *53*, 29–54.
44. Ferber, J.; Luther, J. *J. Phys. Chem. B.* **2001**, *105*, 4895–4903.
45. Vanmaekelbergh, D.; de Jongh, P. E. *J. Phys. Chem. B.* **1999**, *103*, 747–750.
46. Bisquert, J.; Garcia-Belmonte, G.; Fabregat-Santiago, F. *J. Solid State Electrochem.* **1999**, *3*, 337–348.
47. Bisquert, J.; Garcia-Belmonte, G.; Fabregat-Santiago, F.; Ferriols, N. S.; Bogdanoff, P.; Pereira, E. C. *J. Phys. Chem. B.* **2000**, *104*, 2287–2298.
48. Bard, A. J.; Faulkner, L. R. *Electrochemical Methods*; Wiley: New York, 1980.
49. Gregg, B. A.; Pichot, F.; Ferrere, S.; Fields, C. L. *J. Phys. Chem. B.* **2001**, *105*, 1422–1429.
50. Gregg, B. A.; Zaban, A.; Ferrere, S. *Z. Phys. Chem.* **1999**, *212*, 11–22.
51. Gregg, B. A. In *Molecules as Components in Electronic Devices*; Lieberman, M. D. Ed.; American Chemical Society: Washington, DC, 2002.
52. Zaban, A.; Ferrere, S.; Gregg, B. A. *J. Phys. Chem. B.* **1998**, *102*, 452–460.
53. O'Regan, B.; Schwartz, D. T. *J. Appl. Phys.* **1996**, *80*, 4749–4754.
54. Murakoshi, K.; Kogure, R.; Wada, Y.; Yanagida, S. *Solar Energy Mater. Solar Cells* **1998**, *55*, 113–125.
55. Gerischer, H. in *Physical Chemistry. An Advanced Treatise*; Vol. 9A; Jost, W. (Ed); Academic Press: New York, 1970.
56. Gerischer, H. *Photochem. Photobiol.* **1972**, *16*, 243–260.
57. Memming, R. *Photochem. Photobiol.* **1972**, *16*, 325–333.
58. Sonntag, L. P.; Spitler, M. T. *J. Phys. Chem.* **1985**, *89*, 1453–1457.
59. Spitler, M. T. *J. Electroanal. Chem.* **1987**, *228*, 69–76.
60. Parkinson, B. A.; Spitler, M. T. *Electrochim. Acta.* **1992**, *37*, 943–948.
61. Charlé, K. P.; Willig, F. *Chem. Phys. Lett.* **1978**, *57*, 253–258.
62. Arango, A. C.; Johnson, L. R.; Bliznyuk, V. N.; Schlesinger, Z.; Carter, S. A.; Hörhold, H. H. *Adv. Mater.* **2000**, *12*, 1689–1692.
63. Zaban, A.; Ferrere, S.; Sprague, J.; Gregg, B. A. *J. Phys. Chem. B.* **1997**, *101*, 55–57.
64. Yan, S.; Hupp, J. T. *J. Phys. Chem.* **1996**, *100*, 6867–6870.
65. Gerischer, H. *Electrochem. Acta.* **1989**, *34*, 1005–1009.
66. Nozik, A. J. *Annu. Rev. Phys. Chem.* **1978**, *29*, 189–222.
67. Ferrere, S.; Zaban, A.; Gregg, B. A. *J. Phys. Chem. B.* **1997**, *101*, 4490–4493.
68. Rothenberger, G.; Fitzmaurice, D.; Grätzel, M. *J. Phys. Chem.* **1992**, *96*, 5983–5986.
69. Green, M. *Adv. Mater.* **2001**, *13*, 1019–1022.
70. Gibbs, J. W. *The Scientific Papers of J. Willard Gibbs*; Dover: New York, 1961, Vol. 1.
71. Reiss, H. *J. Phys. Chem.* **1985**, *89*, 3783–3791.
72. Nozik, A. J.; Memming, R. *J. Phys. Chem.* **1996**, *100*, 13,061–13,078.
73. Prigogine, I. *Thermodynamics of Irreversible Processes*, 3rd Ed.; Wiley: New York, 1967.
74. Prigogine, I.; George, C. *Proc. Natl. Acad. Sci. USA* **1983**, *80*, 4950–4954.
75. Glansdorff, P.; Prigogine, I. *Thermodynamic Theory of Structure, Stability and Fluctuations*; Wiley: London, 1971.

76. Onsager, L. *Phys. Rev.*, **1931**, *37*, 405–426.
77. Onsager, L. *Phys. Rev.* **1931**, *38*, 2265–2279.
78. Bolton, J. R.; Strickler, S. J.; Connolly, J. S. *Nature* **1985**, *316*, 495–500.
79. Archer, M. D.; Bolton, J. R. *J. Phys. Chem.* **1990**, *94*, 8028–8036.
80. Landsberg, P. T.; Tonge, G. *J. Appl. Phys.* **1980**, *51*, R1–R20.
81. Moore, W. J. *Physical Chemistry*, 4th Ed.; Prentice-Hall: Engelwood Cliffs, NJ, 1972.
82. Schwarzburg, K.; Willig, F. *J. Phys. Chem. B.* **1999**, *103*, 5743–5746.
83. Hodes, G.; Howell, I. D. J.; Peter, L. M. *J. Electrochem. Soc.* **1992**, *139*, 3136–3140.
84. Boschloo, G. K.; Goossens, A.; Schoonman, J. *J. Electroanal. Chem.* **1997**, *428*, 25–32.
85. Levy, B.; Liu, W.; Gilbert, S. E. *J. Phys. Chem. B.* **1997**, *101*, 1810–1816.
86. Enright, B.; Fitzmaurice, D. *J. Phys. Chem.* **1996**, *100*, 1027–1035.
87. Franco, G.; Gehring, J.; Peter, L. M.; Ponomarev, E. A.; Uhlendorf, I. *J. Phys. Chem. B.* **1999**, *103*, 692–698.
88. Grovenor, C. R. M. *Microelectronic Materials*; IOP Publishing: Bristol, 1989.
89. Weast, R. C. *CRC Handbook of Chemistry and Physics*, 61st Ed.; CRC Press: Boca Raton, FL, 1980.
90. Gregg, B. A. *Appl. Phys. Lett.* **1995**, *67*, 1271–1273.
91. Wang, Y.; Suna, A. *J. Phys. Chem. B.* **1997**, *101*, 5627–5638.
92. Marcus, R. A. *J. Chem. Phys.* **1965**, *43*, 679–701.
93. Marcus, R. A.; Sutin, N. *Biochim. Biophys. Acta.* **1985**, *811*, 265–322.
94. Menzel, E. R.; Popovic, E. D. *Chem. Phys. Lett.* **1978**, *55*, 177–181.
95. Popovic, Z. D. *Mol. Cryst. Liq. Cryst.* **1989**, *171*, 103–116.
96. Law, K.-Y. *Chem. Rev.* **1993**, *93*, 449–486.
97. Kay, A.; Humphry-Baker, R.; Grätzel, M. *J. Phys. Chem.* **1994**, *98*, 952–959.
98. Salafsky, J. S.; Lubberhuizen, W. H.; van Faassen, E.; Schropp, R. E. I. *J. Phys. Chem. B.* **1998**, *102*, 766–769.
99. Nusbaumer, H.; Moser, J. E.; Zakeeruddin, S. M.; Nazeeruddin, M. K.; Grätzel, M. *J. Phys. Chem. B.* **2001**, *105*, 10461–10464.
100. Oskam, G.; Bergeron, B. V.; Meyer, G. J.; Searson, P. C. *J. Phys. Chem. B.* **2001**, *105*, 6867–6873.
101. Penner, R. M.; Heben, M. J.; Longin, T. L.; Lewis, N. L. *Science*, **1990**, *250*, 1118–1121.
102. Sun, J.; Stanbury, D. M. *Inorg. Chem.* **1998**, *37*, 1257–1263.
103. Strein, T. G.; Ewing, A. G. *Anal. Chem.* **1992**, *64*, 1368–1373.

3

Photo-Induced Electron Transfer Reactivity at Nanoscale Semiconductor–Solution Interfaces: Case Studies with Dye-Sensitized SnO$_2$–Water Interfaces

Dennis A. Gaal and Joseph T. Hupp
Northwestern University, Evanston, Illinois, U.S.A.

I. INTRODUCTION

One of the most exciting and technologically promising areas of application of contemporary photochemistry is the liquid-junction solar cell. Perhaps the most compelling version of the cell is the "Grätzel cell"—a photovoltaic cell that converts sunlight into electricity with nearly 100% quantum efficiency and greater than 10% overall energy efficiency [1–3]. The cell uses a broadly absorbing coordination compound to sensitize a wide-bandgap semiconductor—a high-area nanocrystalline form of TiO$_2$—to visible light. As shown schematically in Fig. 1, sensitization involves molecular photoexcited-state formation, followed by transfer of an electron ("injection") from the excited molecule into the conduction band of the semiconductor. The oxidized dye is restored to its chromophoric form by reduction with iodide ions present in the surrounding solution. The electrochemical circuit is completed at a dark electrode, which supplies the electrons needed to regenerate iodide from triiodide.

From the diagram, cell operation comprises a series of interfacial electron-transfer (ET) reactions that yields no net photochemistry—only conversion of

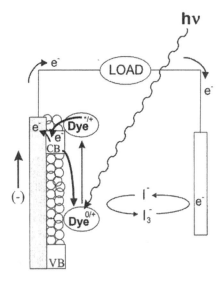

Figure 1 Schematic representation of a Grätzel solar cell. Sub-band-gap light absorption leads to the formation of the sensitizer excited state, followed by electron injection into the conduction band of the high-area nanocrystalline semiconductor. The electrons can be drawn into a circuit to do useful work and returned to the system through the redox mediator, the I/I_3^- couple, at the counterelectrode.

solar energy to heat and electrical energy. In addition, for each of the ET reactions in the light-to-electricity conversion sequence, an energy-wasting back reaction can occur. Clearly, one of the requisites for efficient cell operation is the overall occurrence of comparatively fast forward-ET reactions [4–7] and comparatively slow back-ET processes, such that branching ratios strongly favor the electrical-energy-producing forward sequence. That Grätzel-type cells do this well is evidenced by their near-unity quantum efficiencies [8,9].

We have been especially interested in one particular interfacial ET reaction: back ET from the semiconductor to the dye molecule. Beyond its relevance to solar cell operation [10,11], the reaction is interesting from a fundamental perspective. By monitoring it spectroscopically, one can access interfacial ET kinetics and dynamics at much higher driving forces and on much shorter time scales than generally obtainable by direct electrochemical methods. This, in principle, permits a broader evaluation of factors potentially capable of controlling interfacial redox reactivity—familiar factors such as Marcus-type reorganization energies, surface electronic states, and molecule–surface electronic coupling energies, but also other, less familiar structural and chemical factors. Although the Grätzel cell is based on titanium dioxide, we have focused—for experimental rea-sons—on a closely related electrode material, tin oxide (SnO_2), in much of the work described here. Also, for experimental reasons, we have focused on reactions in aqueous environments. The story is almost certainly different in the nonaqueous environments favored for Grätzel-type solar cells, as evidenced especially by recent reports from Meyer [12,13], Lewis [14,15], and their co-workers, among others [16–18].

In the sections that follow, we first outline and summarize, in the context of contemporary ET theory, the experimental behavior of systems involving only weak, electrostatic interactions between dye molecules and the semiconductor surface. This is followed by (1) a description of the behavior of covalently linked dye–semiconductor combinations, which is remarkably different from that seen with weakly interacting systems, (2) a discussion of the fundamental energetics for the reactions—which again appears to differ significantly for the two reaction subclasses—and (3) a comparative discussion of possible interfacial reaction mechanisms.

II. ELECTRON-TRANSFER THEORY

To place experimental studies in context, it is helpful to recall Marcus' semiclassi-cal formulation of ET reaction rate theory. Briefly, for an ET reaction involving weak electronic interactions, the first-order rate constant can be written as [19,20]

$$k_{ET} = \left(\frac{4\pi^2 H_{ab}^2}{h}\right)(4\pi\lambda RT)^{-1/2} \exp\left(\frac{-\Delta G^*}{RT}\right) \tag{1}$$

where H_{ab} is the initial-state/final-state electronic coupling energy, h is Planck's constant, λ is the reorganization energy, ΔG^* is the activation free energy, R is the gas constant, and T is temperature. In the classical limit and neglecting any barrier round-off due to H_{ab}, Marcus writes the activation free energy as

$$\Delta G^* = \frac{(\Delta G^\circ + \lambda)^2}{4\lambda} \tag{2}$$

where ΔG° is the free-energy driving force for the reaction. As illustrated in Fig. 2, Eq. 2 yields three reactivity regimes: (1) the normal region ($-\Delta G^\circ < \lambda$), where k_{ET} increases with increasing driving force (more negative ΔG°) and ΔG^* decreases, (2) a barrierless point where $-\Delta G^\circ$ equals λ and k_{ET} reaches its maxi-mum value, and (3) the Marcus inverted region ($-\Delta G^\circ > \lambda$), where k_{ET} decreases with increasing driving force and ΔG^* increases.

III. WEAKLY INTERACTING SYSTEMS

A. Model Systems

Nanoparticulate SnO_2 films and colloids generally feature a slight stoichiometric excess of oxygen and, therefore, a net negative surface charge in water at pHs

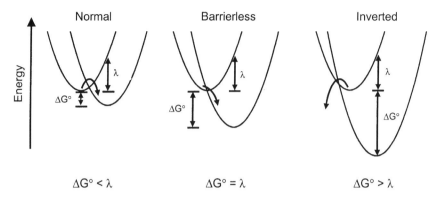

Figure 2 Energy surfaces corresponding to the normal, barrierless, and inverted electron-transfer reactivity regions of Marcus theory.

where enough surface oxygen atoms are deprotonated (i.e., pHs greater than 7) [21]. The negative charge can be used to advantage to bind cationic dye molecules via weak electrostatic interactions. The binding turns out to be easier to control if the SnO_2 photoelectrode in Fig. 1 is replaced by colloidal SnO_2 and the reaction sequence is limited to dye injection and back electron transfer. Under these conditions, as Rogers and Ford [22–24] as well as Kamat and co-workers [25–27] have shown, interfacial ET kinetics can be conveniently followed by simple pump/probe-type transient absorbance measurements.

The dyes most often used in the Grätzel cell are polypyridyl complexes of ruthenium, featuring pendant carboxylates for covalent surface attachment [28–30]. The ubiquitous N3 dye (**1**), for example, has the formula Ru^{II} (4,4′-carboxylate-2,2′- bipyridine)$_2$(NCS)$_2^{4-}$ [31–34] Sensitization is based on visible-region metal-to-ligand charge-transfer (MLCT) absorption, followed by electron injection from one of the two coordinated bipyridines into the photoelectrode's conduction band. The back reaction entails ET from the semiconductor to the oxidized ruthenium center. With this in mind, we examined a set of about a dozen derivatized dicationic *tris*-2,2′-bipyridine (bpy) (**2**) and 1,10-phenanthroline (phen) (**3**) complexes of ruthenium(II) and osmium(II) as colloidal tin oxide sensitizers [35,36]. The idea was to span a very wide range of back-ET driving forces with systems anticipated to be homologous in terms of reorganization energy. To expand the range of driving forces still further, we also examined a few complexes featuring a pair of 4-pyrrolidinopyridine (4pypy) (**4**) or 1-methyl-imidazole (m-im) (**5**) ligands in place of one of the chelating ligands.

1

Complex 1

2   3

4   5

Complexes 2–5

B. Preliminary Observations: Dye Sensitivity, Kinetic Heterogeneity, and Reaction Orders

Dye adsorption onto colloidal SnO_2 is accompanied in most cases by essentially complete quenching of dye emission. The simplest interpretation is excited-state consumption via rapid electron injection—a suggestion supported by comparative studies with SiO_2 (an insulator incapable of accepting electrons) in place of SnO_2 and confirmed, for several dyes, by measuring product spectra.

Figure 3 shows representative single-wavelength absorbance transients for three dyes electrostatically bound to colloidal SnO_2. The transients correspond to photoinitiated bleaching and recovery of the respective MLCT absorbances. From Fig. 3, it is clear that (1) injection is rapid in comparison to back ET, (2) back ET is complex kinetically, but (3) the complex recovery rates depend strongly on the identity of the dye, at least in the first few hundred nanoseconds of the recovery. In order to isolate the shorter-time recovery kinetics, transients were fit, somewhat pragmatically, to a bi-exponential decay function:

$$\Delta A = \Delta A_0 [d \exp(-k_{1,\text{app}}t) + b \exp(-k_{2,\text{app}}t)] + \text{constant} \qquad (3)$$

where A is absorbance and ΔA_0 is the change in absorbance just after photoexcita-

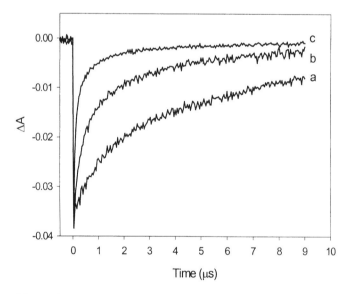

Figure 3 Transient absorbance spectra obtained at the respective MLCT maximum for aqueous SnO_2 colloid sensitized with (a) $Ru(5\text{-Cl-phen})_3^{2+}$, (b) $Ru(phen)_3^{2+}$, and (c) Os $(phen)_3^{2+}$. (Adapted from Ref. 36.)

tion. Similar complexity is routinely encountered for back-ET reactions involving TiO_2. There it has typically been ascribed to surface-state mediation of the recovery kinetics, where the rate-limiting step can either be interfacial ET from any of several states of differing energy or site-to-site electron hopping on the semiconductor surface prior to interfacial ET, or some combination of the two processes [37–39].

Focusing on the shorter time-scale component, the characteristic recovery time shows a strong dependence on the pump-laser power or, equivalently, the number of electrons injected: The higher the power, the shorter the recovery time. Similar behavior has been noted by Ford et al. [40]. If $k_{1,app}$ is plotted versus the number of electrons injected per particle (Fig. 4), a linear correlation is obtained. In other words, the reaction appears to be first order in electrons (and first order in the oxidized dye). What does this mean mechanistically? The simplest interpretation—sketched in Scheme 1—is that the injected electrons are free to return to any available dye molecule, not just the molecule from which they originated. This would be the case if injected electrons avoided surface states (at least at these shorter times) and remained in the conduction band. (Notably, the power-dependent kinetic behavior persists in a rigid glass matrix. Consequently, possible

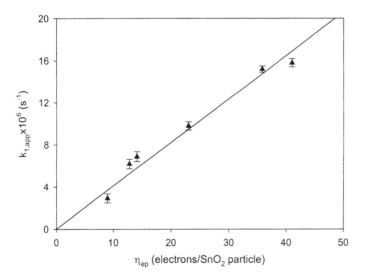

Figure 4   Plot of $k_{1,app}$ for back ET to $Os(phen)_3^{3+}$ versus η_{ep}, the number of electrons injected per colloidal SnO_2 particle. Note that Eq. (4) contains η_e, the number of electrons injected per unit volume. The estimated average volume for one particle is $1800 \, nm^3$. (Adapted from Ref. 36.)

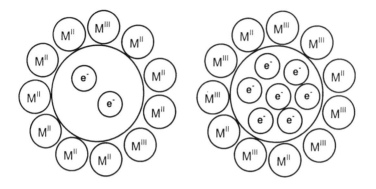

Long ET Lifetimes   Short ET Lifetimes

Scheme 1

alternative explanations centering on injection-induced dye desorption and subsequent diffusive recovery can be excluded [35].)

In contrast, geminate recombination, caused, for example, by proximal trapping of the injected electron, would yield power-independent recovery times; see Scheme 2. (Of course, if the trapped electrons were not immobilized, but instead were able to migrate rapidly from surface state to surface state, overall second-order behavior would be recovered [37].)

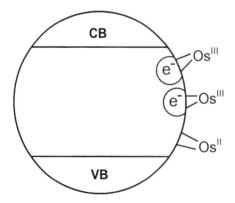

Scheme 2

The second-order kinetics problem for semiconductor–solution interfaces has been considered in some detail by Lewis and co-workers [41,42]. For molecules immobilized on a semiconductor surface and assuming that electrons are transferred from the conduction band, not surface states, their rate law can be written as

$$\text{Rate} = \frac{d[M^{II}L_3^{2+}]}{dt} = k_{1,app}\Gamma(M^{III}L_3^{3+}) = k_{bET}\eta_e\Gamma(M^{III}L_3^{3+}) \qquad (4)$$

where $\Gamma(M^{III}L_3^{3+})$ is the surface concentration of oxidized dye (mol/cm^2) and η_e is the concentration of electrons in the semiconductor. In this formulation, the units of k_{bET} are cm^3/s, as expected for a second-order interfacial process. The rate constant can be obtained in an approximate way from the slopes of best fit lines of plots of $k_{1,app}$ versus η_e, constrained to pass through the origin (Fig. 4). A better approach, however, is to refit absorbance transients directly to second-order decays. This procedure yields k_{bET} values that are typically slightly smaller—at most, a factor of 2.

C. Driving-Force Effects

Marcus theory promises that electron-transfer rates will systematically change as $\Delta G°$ is changed. For back-ET reactions at the semiconductor–solution interface, $\Delta G°$ equals the difference quantity, $E_{cb} - E_f$, where E_{cb} is the electrochemical potential of the conduction-band edge and E_f is the dye's ground-state formal potential [here, the Ru(III/II) or Os(III/II) couple; see Table 1 for values]. Because the colloidal SnO$_2$ particles are large enough ($\sim$ 15 nm in diameter) for quantum confinement effects to be neglected, E_{cb} for the particles has been equated with E_{cb} for macroscopic tin oxide electrodes (-0.88 versus saturated calomel electrode (SCE) at pH 9) [22].

Figure 5 shows that back-ET rates indeed are sensitive to the driving force, with the log k_{bET} versus $\Delta G°$ mapping out a classical Marcus curve that extends significantly into the inverted region. The slight deviations at high driving force have been tentatively ascribed to the participation of a high-frequency vibrational or librational mode. The maximum in the rate plot corresponds to the driving force where the reaction is barrierless and $-\Delta G°$ equals λ. From the plot, λ is $\sim$ 1.4 eV. This value is roughly twice as large as found by Lewis and co-workers in a study of N3-type dyes on titanium dioxide electrodes in acetonitrile as solvent [14] and considerably larger than expected from available estimates of internal and solvent reorganization energies. For example, Brown and Sutin [43] report a kinetically derived reorganization energy of $\sim$ 0.8 eV for the RuL_3^{3+}/RuL_3^{2+} self-exchange in homogeneous solution. Also, x-ray crystallographic studies of a related redox pair, Fe(phen)$_3^{3+}$ Fe(phen)$_3^{2+}$, indicate no detectable difference in the M(III)—N versus M(II)—N bond lengths [44] and, therefore, little internal reorganization energy. The solvent contribution (λ_s) to the reorganization energy

Table 1 Electrochemical Data, Driving-Force Information, and Electron-Transfer Values for All Ru and Os Sensitizers

Compound	E_f (V, vs. SCE)	ΔG_{bET} (eV) (pH = 9)	$k_{bET} \times 10^{13}$ (cm^3/s)	ΔH^* (kJ/mol)
Ru(5-Cl-phen)$_3^{2+}$	1.39[a]	−2.27	0.8	20
Ru(phen)$_3^{2+}$	1.27[a]	−2.15	1.7	14
Ru(bpy)$_3^{2+}$	1.29[a]	−2.17	1.7	16
Ru(5-CH$_3$-phen)$_3^{2+}$	1.26[a]	−2.14	1.8	13
Ru(5,6-CH$_3$-phen)$_3^{2+}$	1.23[a]	−2.11	1.7	13
Ru(4,7-CH$_3$-phen)$_3^{2+}$	1.12[a]	−2.00	3.0	12
Ru(3,4,7,8-CH$_3$-phen)$_3^{2+}$	1.05[a]	−1.93	6.8	10
Os(5-Cl-phen)$_3^{2+}$	0.90[b]	−1.78	5.1	8.6
Os(phen)$_3^{2+}$	0.80[c]	−1.68	10	7.2
Os(4-CH$_3$-phen)$_3^{2+}$	0.79[c]	−1.67	12	6.4
Os(4,7-CH$_3$-phen)$_3^{2+}$	0.62[c]	−1.50	14	7.2
Os(3,4,7,8-CH$_3$-phen)$_3^{2+}$	0.57[b]	−1.45	15	4.0
Os(phen)$_2$(4pypy)$_2^{2+}$	0.52[c]	−1.40	17	5.7
Os(phen)$_2$(m-im)$_2^{2+}$	0.46[c]	−1.34	13	6.4
Os(4,7-CH$_3$-phen)$_2$(4pypy)$_2^{2+}$	0.41[c]	−1.29	16	5.7
Os(4,7-CH$_3$-phen)$_2$(m-im)$_2^{2+}$	0.38[c]	−1.26	12	11
Os(3,4,7,8-CH$_3$-phen)$_2$(4pypy)$_2^{2+}$	0.35[c]	−1.23	13	6.9
Os(3,4,7,8-Ch$_3$-phen)$_2$(m-im)$_2^{2+}$	0.31[c]	−1.19	11	12

[a] Juris, A.; Balzani, V.; Barigelletti, F.; Campagna, S.; Belser, P.; Von Zelewsky, A. *Coord. Chem. Rev.* **1998**, *84*, 85−277.
[b] Leidner, C. R.; Murray, R. V. *J. Am. Chem. Soc.* **1984**, *106*, 1606−1614.
[c] Experimentally measured and corrected from Ag/AgCl to SCE.

can be estimated using continuum theory for a sensitizer at a semiconductor interface as [14,45]

$$\lambda_s = \frac{(\Delta e)^2}{8\pi\varepsilon_0}\left[\frac{2}{d}\left(\frac{1}{n^2} - \frac{1}{\varepsilon}\right) - \frac{1}{2r}\left(\frac{n_{sc}^2 - n^2}{n_{sc}^2 + n^2}\frac{1}{n^2} - \frac{\varepsilon_{sc} - \varepsilon}{\varepsilon_{sc} + \varepsilon}\frac{1}{\varepsilon}\right)\right] \quad (5)$$

where Δe is the amount of charge transferred, n_{sc} and ϵ_{sc} are the refractive index and dielectric constant of the semiconductor, respectively, n and ϵ are the corresponding parameters for the solvent, d is the diameter of the molecule, and r is the separation distance between the sensitizer and the surface. Application of the equation to the tin oxide reactions yields a solvent reorganizational contribution of just 0.3–0.6 eV, leaving roughly 1 eV of reorganization unaccounted for. One possible explanation is that significant reorganizational demands exist on the semiconductor side of the interface—and, indeed, resonance Raman studies of

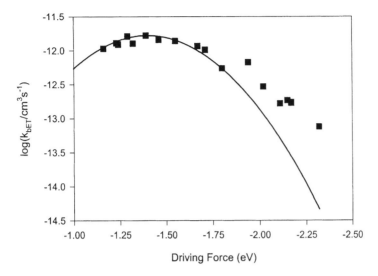

Figure 5 Marcus behavior for the back-ET process at electrostatically sensitized SnO_2 interface for a series of complexes spanning a broad range of redox potentials. Line drawn is a classical Marcus curve [Eq. (1)] based on $\lambda = 1.4$ eV. (Adapted from Ref. 36.)

electron transfer between $Fe(CN)_6^{4-}$ and titanium dioxide offer one example of such an effect (albeit, with contributions amounting to much less than 1 eV) [46].

D. Photoacoustic Assessment of Reaction Energetics

There is an intriguing alternative explanation for the seemingly exceptionally large reorganization energies: Back ET conceivably could be occurring from trap states having much lower energies than E_{cb}, as sketched in Fig. 6. If so, the true driving force would be less than the difference in energy between the SnO_2 conduction band and the sensitizer redox potential, the ΔG° values in Fig. 5 would be gross overestimates, and the true maximum in the Marcus curve (corresponding to λ) would occur at a much smaller ΔG°.

This idea has been explored via time-resolved interfacial photoacoustic spectroscopy. The method reports on the heat evolved during a reaction [47,48]. Table 2 summarizes results from five back-ET reactions on tin oxide and compares ΔH° values with ΔG° values calculated assuming that electrons are transferred from the conduction band. At short times (corresponding to times used to evaluate $k_{1,app}$), the agreement is good, validating the Marcus-curve assessment of λ. At longer times (corresponding roughly to times used to evaluate $k_{2,app}$), ΔH° is about 1 eV *less* than the conduction-band edge/formal-potential energy

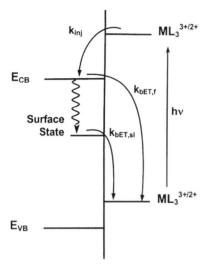

Figure 6 Diagram illustrating qualitative energetics for back ET via two pathways: direct transfer from the bottom of the conduction band and transfer mediated by interfacial surface states (trap states). Note that the rate constants for the two processes may differ.

Table 2 Photoacoustic Energies for Back-ET Reactions at the SnO_2–Water Interfaces

Compound	$-\Delta G_{bET}$ (eV)[a]	$-\Delta H'_{bET}$ (eV)[b]	$-\Delta H^{sl}_{bET}$ (eV)[b]
$Ru(phen)_3^{2+}$	2.15	2.4(2)	1.0(2)
$Os(5\text{-}Cl\text{-}phen)_3^{2+}$	1.78	1.8(2)	0.6(2)
$Os(phen)_3^{2+}$	1.68	1.8(2)	0.5(2)
$Os(4,7\text{-}dimethyl\text{-}phen)_3^{2+}$	1.50	1.1(3)	<0.3
$Os(phen)_2(ethylenediamine)^{2+}$	1.41	1.1(3)	ca. 0.2

[a] ΔG_{bET} is the back-ET driving force determined from the conduction-band energy and the redox potential of the sensitizer in solution.
[b] $\Delta H'_{bET}$ and ΔH^{sl}_{bET} are the thermodynamic reaction enthalpies associated with the fast and slow kinetic components, respectively, determined by photoacoustic spectroscopy.

difference—indicating that under these conditions, back ET occurs from deep traps.

E. Another Look at Driving-Force Effects: pH Effects

Band edges for metal oxide semiconductors, including tin oxide, shift by about -60 mV/pH unit [49,50]. By modulating E_{cb}, pH variations can be used to change $\Delta G°$ for back ET in a systematic fashion. Shown in Fig. 7 are pH-dependent rate data reported for two dye couples, $Os(3,4,7,8,-CH_3\text{-phen})_2(m\text{-im})_2^{3+/2+}$ and Os $(5\text{-Cl-phen})_3^{3+/2+}$. The back ET rate for the first increases with increasing pH, consistent with its identification as a normal-region reactant in Fig. 5. The second decreases with increasing pH (and increasing driving force), as expected for Marcus inverted-region reactivity. Evidently, pH shifts and formal potential shifts are fully equivalent ways of changing the back-reaction driving force.

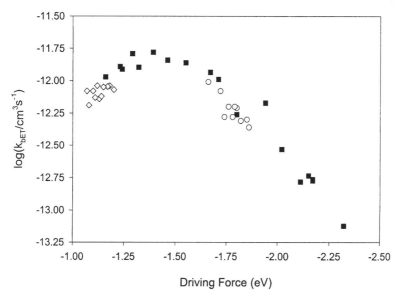

Figure 7 Rate constant versus driving force for the back-ET process at electrostatically sensitized SnO_2 interface for a series of complexes spanning a range of redox potentials. The filled square symbols (■) correspond to data collected at pH = 9. Also included in the figure are the results of variable pH studies for $Os(5\text{-Cl-phen})_3^{3+}$ (◇) and $Os(3,4,7,8\text{-}CH_3\text{-phen})_2(m\text{-im})_2^{3+}$ (○). The pH variations serve to shift the potential of the conduction band edge. (Adapted from Ref. 36.)

F. Temperature Effects and the Nature of Reorganization Energies

As illustrated by the modified Arrhenius plots in Fig. 8, the back-ET process is thermally activated in both the normal and inverted region, but nearly activationless for reactions occurring at or near the maximum in the rate versus driving-force plot. The findings—especially activation in the inverted region—point to a largely classical reorganizational barrier. Substantive involvement of high-frequency modes would lead to the largely temperature-independent rate behavior usually encountered for inverted kinetics [51,52]. Figure 9 illustrates the mechanistic distinction: Strongly activated behavior implies thermal barrier crossing; temperature-independent or nearly independent behavior implies nuclear tunneling.

Table 1 summarizes the behavior, in the form of activation enthalpies (ΔH^*), for each of 18 reactions. The values listed are somewhat larger than published values [36], reflecting corrections for unrecognized thermal control errors in the original investigation. As expected from classical Marcus theory, decreases in rate are accompanied by increases in ΔH^*. Curiously, however, as the reaction is pushed progressively further into the inverted region, ΔH^* increases by

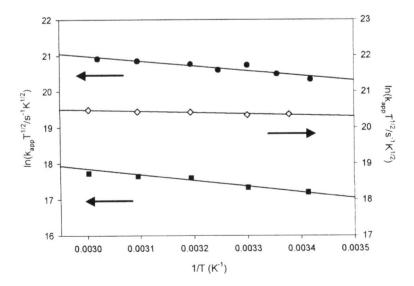

Figure 8 Modified Arrhenius plots for back electron transfer from colloidal SnO_2 to adsorbed $Ru(phen)_3^{3+}$ ($\blacksquare$), $Os(3,4,7,8-CH_3-phen)_3^{3+}$ ($\Diamond$), and $Os(3,4,7,8-CH_3-phen)_2$ $(m-im)_2^{3+}$ ($\bullet$). (Adapted from Ref. 36.)

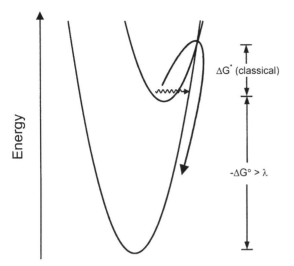

Energy

$\Delta G^{\ast}$ (classical)

$-\Delta G^{\circ} > \lambda$

Figure 9 Activated versus tunneling behavior in the Marcus inverted region.

more than the amount needed to accommodate the decreases in rate. Following Marcus and Sutin [53], it has been suggested that the existence of a significant entropic driving force for back ET could account for this peculiar finding [36].

G. Dynamics, Electronic Coupling, and Nonadiabaticity

In addition to providing information about the reorganization energy, rate plots such as Fig. 5 contain quantitative information about dynamics. At the top of the plot, $\Delta G^{\ast}$ goes to zero and the rate constant can be equated with the pre-exponential part (i.e., the dynamics part) of the applicable classical or semiclassical expression. Under nonadiabatic conditions, Eq.(1) is applicable and, in principle, the initial-state/final-state electronic coupling energy, H_{ab}, can be determined. With an assumption about the effective electronic coupling length (l_{sc}) between the semiconductor and the adsorbed complexes, application of a modified version of Eq. (1) appropriate for second-order kinetics yields for the three reactions closest to the top of the plot H_{ab} values of 20–50 cm^{-1} for the coupling constant [36]. The values are smaller than the frequencies of potentially Franck–Condon active metal–ligand modes, semiconductor phonon modes, and solvent (water) vibrational modes, as well as the frequency of solvent longitudinal relaxation, suggesting that the reactions, indeed, are weakly nonadiabatic.

Because the magnitude of H_{ab} is highly distance sensitive [54], a further diagnostic for nonadiabaticity is a falloff in reaction rate with increasing semiconductor/redox-center separation distance. The separation distance can be altered by incorporating spacers as substituents on the periphery of a ligand. Experiments of this kind using alkyl groups have yielded the expected rate decreases for back-ET reactions. The findings are summarized in Fig. 10, where k_{bET} is plotted versus the number of carbons comprising the alkyl chain. The observed rate decreases point to decreases in the frequency factor. Variable temperature rate measurements confirm that the rate effects come from changes in pre-exponential factors, not activation barriers, and corroborate the conclusion (above) that back ET is nonadiabatic; see Table 3.

Returning to Fig. 10, because the majority of the alkyl groups are flexible and because the interfacial electrostatic binding geometries are unknown, the rate of falloff of k_{bET} or H_{ab}^2 with semiconductor/molecule separation distance cannot be evaluated quantitatively. A curious finding that remains unexplained is that the falloff with osmium complexes is considerably weaker than with ruthenium species. Finally, although the injection reaction was not the focus of the study, spacers clearly do decrease its rate, as shown, for example, by an increase in emission quantum yield (decrease in injection efficiency) with the largest spacers.

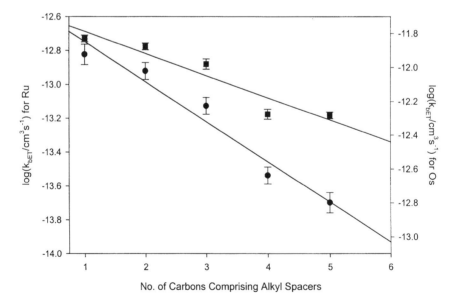

Figure 10 Plot of rate constants for back electron transfer from SnO_2 to electrostatically bound ruthenium (●) and osmium (■) complexes as a function of the number of carbon atoms comprising alkyl spacers. Within experimental error, the driving force for each series of reactions is unaffected by changing the size of the alkyl spacer.

Table 3 Kinetic Parameters for Back Electron Transfer from Colloidal SnO_2 to Several Dyes Featuring Peripheral Alkyl "Spacer" Groups

Compound	$k_{bET} \times 10^{13}$ (cm^3/s)	ΔH^*(kJ/mol)
Ru(4,4'-dimethyl-bpy)$_3^{2+}$	1.5(2)	14
Ru(4,4'-diethyl-bpy)$_3^{2+}$	1.2(1)	14
Ru(4,4'-dipropyl-bpy)$_3^{2+}$	0.7(1)	14
Ru(4,4'-dibutyl-bpy)$_3^{2+}$	0.3(1)	15
Ru(4,4'-dipentyl-bpy)$_3^{2+}$	0.2(1)	14
Ru(4,4'-di-t-butyl-bpy)$_3^{2+}$	0.4(1)	15
Os(4,4'-dimethyl-bpy)$_3^{2+}$	14(1)	7.7
Os(4,4'-diethyl-bpy)$_3^{2+}$	13(1)	8.7
Os(4,4'-dipropyl-bpy)$_3^{2+}$	10(1)	7.8
Os(4,4'-dibutyl-bpy)$_3^{2+}$	5.3(3)	7.2
Os(4,4'-dipentyl-bpy)$_3^{2+}$	5.2(2)	9.2
Os(4,4'-di-t-butyl-bpy)$_3^{2+}$	6.0(3)	7.4

IV. STRONGLY INTERACTING SYSTEMS

A. Surface Attachment Chemistry

Simple electrostatic binding is usually insufficient for long-term dye use in liquid-junction solar cells. Better suited are dyes that are chemically appended to the semiconductor. One approach that works well is carboxylate binding (see N3 dye structure, **1**), either based on esterlike linkages to Ti, Sn, Zr, and so forth or based on interfacial chelation of these atoms [12,55–57]. Even better for certain mechanistic studies, such as rate measurements at elevated temperature or at extreme pHs, are phosphonate linkages. Binding via phosphonate functionalities can be exceptionally robust and extraordinarily persistent [58–62]. By analogy to carboxylates, binding presumably involves either phosphoester linkages to surface metal atoms or phosphonate chelation of the atoms. The available crystallographic literature on Zr(IV) phosphonate compounds supports the phosphoester description [63–65]. In any case, coordination complexes featuring one or more 4,4'-methylphosphonate-2,2'-bipyridine ligands (phosbpy; **6**) have been used to achieve persistent dye attachment to high-area tin oxide photoelectrodes and colloidal particles. As outlined below, the change in mode of binding, although structurally unexceptional, has striking mechanistic consequences.

B. Rate Behavior

Phosphonate attachment of Ru(bpy)$_3^{2+}$ analog to SnO_2 in water as solvent is accompanied by efficient luminescence quenching. Transient absorbance mea-

6

Complex 6

surements show that the quenching is due to efficient electron injection. In contrast to the behavior of electrostatically bound dyes, the corresponding back-ET reactions exhibit power-independent decay times—in other words, first-order rather than second-order recovery kinetics. (The decays again are complicated, and only the initial portions have been examined in detail.) Similar behavior has been recorded for carboxylated and phosphonated dyes at titanium dioxide–water interfaces [66,67]. The findings have been interpreted in terms of geminate recombination of the injected electron with the oxidized dye, as shown in Scheme 2.

Driving-force studies, using mixed-ligand coordination to alter formal potentials, show only normal-region behavior (Fig. 11), not the inverted Marcus curve seen for electrostatically bound compounds. Similar behavior has been reported for phosphonate-bound dyes on TiO_2 in water [67]. Other studies on TiO_2 in nonaqueous environments have yielded Marcus inverted rate behavior or else no sensitivity to driving force, suggesting that water may induce mechanistically distinct behavior [14,37,68].

As illustrated in Fig. 12, rate measurements as a function of pH or H_0 show almost no change over an extraordinarily wide range—18 pH units (H_0 is the Hammett acidity parameter; it is useful for characterizing proton activities in extremely acidic solutions [69]). The pH range examined corresponds to a change of more than 1.1 eV in E_{cb} and, therefore, $\Delta G°$! Variations in the back-ET rate by several orders of magnitude would be expected if the behavior paralleled the driving-force dependence summarized in Fig. 11. Again, similar rate behavior has been recorded for phosphonate-bound dyes on titanium dioxide [66]. The origin of the insensitivity to pH-modulated changes in reaction driving force is considered in some detail below.

Variable-temperature studies show that the back reaction is thermally activated. The dependence of ΔH^* on driving force (variations in dye formal potential) is illustrated in Fig. 13 and is consistent with normal-region reactivity [67]. If ΔS^* can be neglected, application of Eq. (1) yields H_{ab} values of approximately

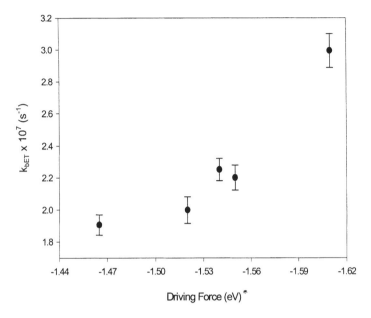

Figure 11 Dependence on driving force of first-order rate constant for back electron transfer from colloidal SnO_2 films to covalently attached complexes. The variations indicate that the reactions occur in the Marcus normal region. The identities of the molecular redox couples, listed from highest driving force to lowest, are, $Ru^{III/II}$ (5-Cl-phen)$_2$ (phosbpy)$^{1-/2-}$, $Ru^{III/II}$(phen)$_2$(phosbpy)$^{1-/2-}$, $Ru^{III/II}$(4,7-CH$_3$-phen)$_2$(phosbpy)$^{1-/2-}$, and $Ru^{III/II}$(3,4,7,8-CH$_3$-phen)$_2$(phosbpy)$^{1-/2-}$

1 cm^{-1}. These place the reactions in the weakly nonadiabatic regime, implying, therefore, that electronic coupling controls the reaction dynamics. Electronic-coupling-limited rate behavior is consistent with the need to traverse a total of eight bonds in the back reaction, as sketched in Fig. 14. Note that only four bonds are traversed in the injection reaction, suggesting that electronic coupling could be considerably more favorable for this process. Related work has been described by Lian and co-workers for covalently attached dyes on TiO_2 electrodes [70,71].

C. Semiconductor Energetics

The remarkable insensitivity of back-ET rates to pH-induced shifts in the conduction-band edge raises the question of why E_{cb} responds to pH in the first place. Experimentally, conduction-band edges for metal oxide semiconductor electrodes, including SnO_2 electrodes, are observed to shift in the negative direction

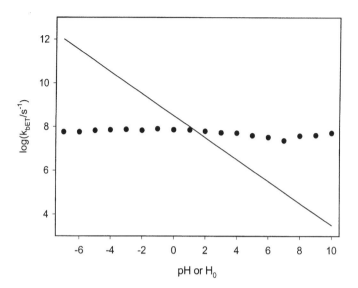

Figure 12 Plot of the first-order rate constant for back ET from high-area SnO_2 electrode to covalently bound $Ru^{III}(bpy)_2(phosbpy)^{1-}$ as a function of pH or H_0. Notably, k_{bET} is nearly pH independent (ca. factor of 3 variations; see Fig. 18). The solid line shows the behavior expected for an inverted reaction based on the 1.1-eV variation in driving force introduced by the pH variations.

by about 60 mV per pH unit [72–78]. The behavior can extend over an enormous pH range: 31 pH units in the case of titanium dioxide, the most extensively examined system [78]. Typical textbook explanations for these "Nernstian shifts" focus on the well-known protonation/deprotonation equilibria of surface oxygen atoms (terminal and bridging oxo and hydroxo groups) [79,80]. The accompanying changes in surface charge induce changes in interfacial potential. The expected form of the interfacial potential versus pH plot, however, is an acid/base titration curve centered at the pK_a of the surface hydroxo or aquo functionality, not an extended Nernstian shift [81–83]. The surface-protonation explanation actually instead describes the origin of the zeta potential—which does show the expected titration-curve behavior for E versus pH.

   Another explanation is that metal oxide electrodes behave like glass pH electrodes. The analogy, however, really does not work: Glass membranes in pH electrodes connect a sample solution with a second, reference solution of known pH. The semiconductor electrode, on the other hand, is in contact with only a single solution.

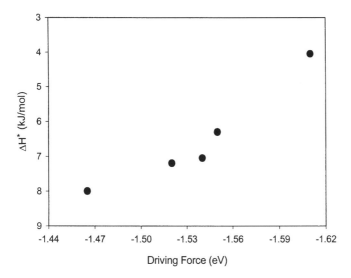

Figure 13 Activation enthalpy for back ET from colloidal SnO_2 films to a series of covalently attached dyes as a function of reaction driving force. See caption to Fig. 11 for identification of the dyes.

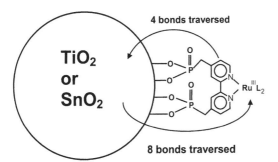

Figure 14 Proposed phosphonate anchoring scheme illustrating the number of bonds traversed in the ligand-based forward reaction (injection reaction) and the metal-ion-based back reaction.

The third explanation, supported by experiment, centers on coupling of proton uptake to electron addition. When the potential of a semiconductor electrode is poised, under dark conditions, at a potential negative of E_{cb}, electrons accumulate near the interface—either in the conduction band or, more typically for nanocrystalline electrodes, in trap states close in energy to E_{cb} [84,85]. Electrochemical quartz-crystal microgravimetry measurements in water with high-area tin oxide, zinc oxide, titanium dioxide, and zirconium dioxide electrodes show that charge-compensating intercalation of cations occurs [49,78,86–90]. With the exception of zinc oxide, which is stable in microgravimetry experiments only at high pHs, the uptake has been shown to occur over a wide range of pHs. Studies with D_2O establish that the species taken up is the proton. Mass comparisons show that even at high pHs, the intercalating species is the proton—indicating that H_2O can serve as the proton source when the concentration of hydronium ions is too low to provide protons. (In nonhydroxylic solvents, electrolyte cations, such as Na^+ and Li^+ are taken up [90,91]—with E_{cb} shifting progressively more negative as the cation size increases and uptake becomes more difficult sterically [92]. In other words, the semiconductor behaves essentially identically to a conventional Li^+/metal oxide battery electrode) [49,90,93].)

From the Nernst equation, proton-coupled electron addition leads to a -59 mV shift in potential per pH unit. Figure 15 shows the behavior of titanium dioxide and Figure 16 shows the behavior of tin oxide. The plots comprise Pourbaix diagrams for these materials. The breaks observed at extreme pHs with TiO_2 define pK_a's for $Ti^{IV}O(OH)$ and $Ti^{III}O(OH)$, with the relevant electrochemical equilibrium at less extreme pH or H_0 values $[-8 < \text{pH} (H_0) < 23]$ described by [78]

$$Ti^{IV} O_2 + e^- + H^+ \rightleftharpoons Ti^{III} O(OH) \tag{6}$$

The analogous process for tin oxide can be written as

$$Sn^{IV} O_2 + e^- + H^+ \rightleftharpoons Sn^{III}O(OH) \tag{7}$$

Inorganic chemists generally are unhappy with formulations of tin in oxidation state III. The formulation here ought not to be taken too seriously. Disproportionation into Sn(IV) and Sn(II) conceivably may occur. Alternatively, $Sn^{III}O(OH)$ may well correspond to trapping of an electron by a cluster of tin ions.

D. Driving Forces and Interfacial Thermodynamics

Perhaps the simplest explanation for the pH independence of back-ET rates at metal oxide semiconductor–solution interfaces is that the formal potential of the dye moves in registry with the conduction-band edge. The energy *difference*, $E_{cb} - E_f$ (dye), is then unchanged with respect to pH and the back-ET reaction experiences a pH-independent driving force. To amplify briefly, the idea is that

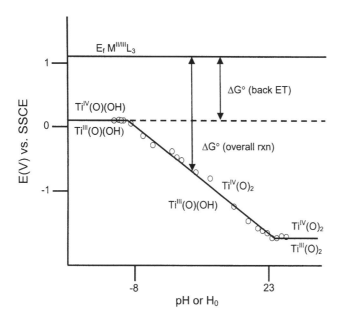

Figure 15 Modified Pourbaix diagram for TiO_2 illustrating the origins of pH-dependent band energetics and the pH-independent back-ET kinetics for covalently anchored dye species. The open circles are experimentally determined values of E_{cb} (combined electrochemical quartz microbalance and reflectance measurements). The driving force for the overall back reaction [coupled electron and proton transfer; cf. Eqs. (10) and (11) for analogous reactions at SnO_2] is pH dependent, but the driving force for the back ET in isolation [cf. Eq. (10)] is pH independent. (Data from Ref. 78.)

a surface-confined redox couple (the dye) will sense the pH-modulated electric field at the interface and the formal potential will be altered accordingly [94,95]. There is good experimental evidence, based on mediator-coupled spectroelectrochemical measurements with TiO_2, that E_f for some dyes, indeed, can vary with pH—at least in the vicinity of the pH of zero charge (see Chapter 2) [96,97]. As the dye potential shifts, however, matching with the mediator potential becomes less satisfactory and E_f (dye) becomes difficult to determine with good reliability at pHs more than two or three pH units away from the optimal pH for such measurements.

A more broadly applicable approach to evaluating the formal potential of an adsorbed dye is cyclic voltammetry (CV). Under dark conditions, nanocrystalline SnO_2, TiO_2, and ZrO_2 electrodes behave as insulators at the potentials needed to oxidize the dyes discussed here. If the dye loading is high, however, the percola-

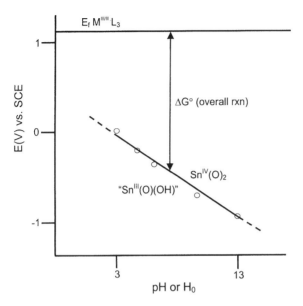

Figure 16 Modified Pourbaix diagram for SnO_2 illustrating the origin of the pH dependence of E_{cb} (see Ref. 73) and showing how the overall back-reaction driving force (E_{cb} − E_f) changes with pH. Insufficient data are available to estimate the driving force for the ET step in isolation (cf. Fig. 15).

tion threshold for charge transport by dye-to-dye electron hopping can be exceeded and CV responses can readily be observed at an underlying conductive platform [98–100]. Because the measurement does not rely upon the conductivity of the photoelectrode, the approach can even be used to measure dye potentials on insulating materials such as alumina.

What do the voltammetry measurements show? Figure 17 is a plot of the formal potential of the $Fe^{III/II}(phosbpy)_3^{9-/10-}$ couple on TiO_2 as a function of pH or H_0 between − 3 and + 10. The potentials are reported versus Ag/AgCl at pH 7, but they are corrected for two effects: (1) liquid-junction potentials, which can be significant under conditions of extreme acidity, and (2) protonation of solution-exposed phosphonate substituents. The corrections were done by using a solution-phase redox couple as an internal reference. With the assumption that no more than four of the six available phosphonate groups of the iron complex can bind to the nanocrystalline semiconductor surface, $Ru^{III/II}(bpy)_2(phosbpy)^{1-/2-}$ was employed as the internal reference. Figure 17 reveals a sigmoidal variation of E_f with solution pH, but with the variations in E_f confined to pHs close to the pH of zero charge (approximately pH 6) [83,101,102].

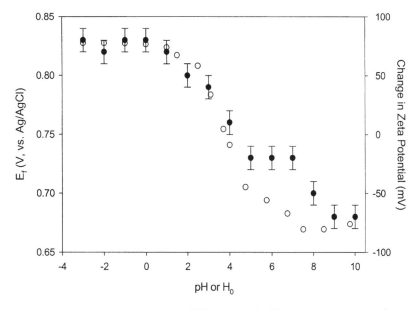

Figure 17 Formal potential of $Fe^{III/II}(phosbpy)_3^{9-/10-}$ on nanocrystalline TiO_2 as a function of pH (●), and zeta potentials on TiO_2 function of pH (○). Zeta-potential variations were calculated from electrophoretic transport data contained in Ref. 81. Note the strong correlation between formal potential variations and zeta-potential variations.

If probed over a limited pH range, the sigmoidal dependence could be mistaken for a Nernstian or sub-Nernstian variation of the dye potential with proton activity and a coupling of changes in E_f to changes in E_{cb}. If evaluated over a broader pH range, however, the striking result is not how much E_f changes, but how little. The ~ 180-mV variation is only a small fraction of the roughly 720 mV change expected if E_f and E_{cb} shifted in tandem. Furthermore, the variations that do occur have the wrong functional form to be explained by a conduction-band coupling effect. Instead, as shown in Fig. 17, the changes in E_f map with very good fidelity onto the changes expected from a simple zeta-potential effect. (This effect is formally analogous to an ion-adsorption-based "effect" upon the apparent formal potential of a surface-confined redox couple at a metal electrode–solution interface [103]. Although not included in the figure, similar behavior is seen for surface-bound $Ru^{III/II}(phosbpy)_3^{9-/10-}$, $Os^{III/II}(phosbpy)_3^{9-/10-}$, $Ru^{III/II}(bpy)_2(phosbpy)^{1-/2-}$, and $Os^{III/II}(bpy)_2(phosbpy)^{1-/2-}$ [where the internal reference used for the latter two was the solution-phase $Ru^{III/II}(phen)_3^{3+/2+}$ couple].

What about other metal oxide surfaces? The behaviors of the phosphonate-anchored dyes on semiconducting SnO_2 and ZrO_2 and insulating Al_2O_3 surfaces are qualitatively similar (i.e., only weak, non-Nernstian variations of formal potentials with pH are seen). Nevertheless, there are differences. For example, over the investigated range, dye potentials on alumina, in contrast to TiO_2, do not show the sigmoidal variation with pH. This is consistent with the substantially higher pH of zero charge for alumina (pH $\sim$ 9) [101,104]; shifts in zeta potential, which are responsible for the shifts in E_f on TiO_2, are unimportant here. For the dyes and surfaces investigated, no evidence for coupling of dye potentials to band-edge potentials is found, and the "coupling" notion can be discarded as an explanation for the approximate pH independence of back-ET rates.

Beyond the extensive studies in aqueous environments, work by Qu and Meyer in acetonitrile as solvent should be noted [12]. Briefly, for dyes immobilized on TiO_2 and ZrO_2 electrodes that had been pretreated in aqueous solutions at pH 1 versus pH 11, they found only minor differences in formal potential—roughly 80 mV—but large differences in E_{cb}.

E. Residual Kinetic Effects

Although back-ET rates for phosphonate-anchored dyes are, for the most part, unaffected by changes in pH, residual effects do exist. Figure 18 presents a more detailed plot of a portion of the rate data contained in Fig. 12. As shown in the figure, modest residual variations in the back-ET rate constant (factor of 3) are paralleled by small variations in the dye potential (-60 mV), where the shapes of both plots are reasonably well described by zeta-potential effects. Assuming that the shifts in E_f translate directly into changes in the free-energy driving force, the combined results are consistent with Marcus normal region behavior (cf. Fig. 11).

F. Mechanistic Interpretations

Any mechanism for back ET for phosphonate-attached dyes must reconcile the seemingly contradictory observations that rates respond to changes in driving force when the changes are introduced by changing the dye's formal potential, but not when they are introduced by changing the conduction-band-edge energy. The mechanism also needs to account for first-order kinetics and for the persistence of normal-region behavior even at very high driving forces. The following sequence [Eqs. (8)–(11)], which parallels a mechanism proposed for back-ET reactions on TiO_2 [67], is consistent with the available data:

$$Sn^{IV}O_2 + ML_3^{2+*} \rightarrow SnO_2(e_{cb}^-) + ML_3^{3+} \rightarrow Sn^{III}O_2$$

$$+ ML_3^{3+} \qquad \text{(injection/trapping)} \qquad (8)$$

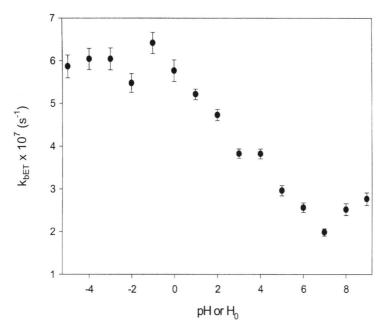

Figure 18 Dependence of back-electron rate constant on pH for $Ru^{III}(bpy)_2(phosbpy)^{1-}$ covalently bound to SnO_2. The observed small reactivity variations (ca. factor of 3) are consistent with a residual zeta-potential-based driving-force effect.

$$Sn^{III}O_2 + H^+ + ML_3^{3+} \rightarrow Sn^{III}O(OH)$$
$$+ ML_3^{3+} \qquad \text{(proton intercalation)} \quad (9)$$

$$Sn^{III}O(OH) + ML_3^{3+} \rightarrow Sn^{IV}O(OH) + ML_3^{2+} \qquad \text{(back ET)} \quad (10)$$

$$Sn^{IV}O(OH) + ML_3^{2+} \rightarrow Sn^{IV}O_2 + H^+$$
$$+ ML_3^{2+} \qquad \text{(proton expulsion)} \quad (11)$$

First, injection occurs from the photoexcited dye into the tin oxide conduction band, but is followed by very rapid trapping at a site that is energetically close to the conduction band and physically close to the dye. Trapping is accompanied by rapid, charge-compensating uptake of a proton—either from a hydronium ion or from a water molecule. Perhaps because of the proton uptake, the trapped electron remains proximal to the dye for at least a few hundred nanoseconds. The proximity enables each electron to return precisely to the dye that initially injected it. In other words, the recombination is geminate and the process is first order. Finally, following back ET, the charge-compensating proton is rapidly released by the trap site. The rate-limiting step in this sequence is Eq. (10). Because proton release occurs as a separate, following step, the energy of the electron in this step is governed by the pH-independent $Sn^{III}O(OH)/Sn^{IV}O(OH)$ couple. The band-edge energy, on the other hand, is controlled by the pH-dependent $Sn^{III}O(OH)/Sn^{IV}O_2$ couple. In this scenario, the driving force for the overall back reaction is approximately the difference in energy between E_{cb} and the dye formal potential. This difference increases with increasing pH. The driving force for the rate-determining step, however, is the difference between the pH-*independent* $Sn^{III}O(OH)/Sn^{IV}O(OH)$ potential and the dye formal potential (also pH independent, apart from small zeta-potential-related perturbations). It is worth noting that similar mechanisms are common in solution-phase photochemistry. For example, the rate of oxidative quenching of $Ru(bpy)_3^{2+*}$ by anthraquinone disulfonate is pH independent, despite the Nernstian pH dependence of the quinone/hydroquinone formal potential [105]. The rate-limiting step is an isolated electron-transfer step, with proton uptake by the semiquinone occurring in a following step.

    Returning to interfacial reactions, the energy diagram in Fig. 15 illustrates, for TiO_2, how ΔG for the isolated back-ET step differs from ΔG for the overall back reaction (ET + proton expulsion). At typical pHs, the driving force for the isolated back-ET step is considerably smaller than for the overall back reaction. Recognizing the distinction, it is clear, for example, that the kinetically relevant driving forces for back ET from SnO_2 in Fig. 11 are smaller than implied by the figure—meaning that the reorganization energy is similarly smaller.

V. COMPARISONS AND CONCLUSIONS

At least for short to intermediate time scales, the kinetics of back electron transfer at the tin oxide – aqueous solution interface can be described by semiclassical Marcus theory. It follows that interfacial ET, rather than a process such as trap-to-trap electron hopping, must be rate determining for the family of dyes described here. The back-ET reaction is borderline nonadiabatic and is characterized by a surprisingly large reorganization energy. The origin of the large reorganization energy is unclear, but the persistence of thermally activated rate behavior in the Marcus inverted region points to largely classical (i.e., low frequency) contributions.

    The detailed mechanism of the back-ET reaction shows a remarkable sensitivity to the mode of binding of the dye to the semiconductor surface. For electrostatically bound dyes, the overall reaction sequence is well described by Eqs. (12)–(14), with back ET evidently occurring directly from the tin oxide conduction band:

$$Ru^{II}L_3^{2+} + h\nu \rightarrow RuL_3^{2+*} \qquad \text{MLCT excitation} \qquad (12)$$

$$RuL_3^{2+*} + SnO_2 \rightarrow Ru^{III}L_3^{3+}$$

$$+ SnO_2(e_{cb}^-) \qquad \text{electron injection} \qquad (13)$$

$$Ru^{III}L_3^{3+} + SnO_2(e_{cb}^-) \rightarrow Ru^{II}L_3^{2+}$$

$$+ SnO_2 \qquad \text{back electron transfer} \qquad (14)$$

For covalently attached dyes, the mechanism is more complex, involving trap states as intermediates and entailing coupled proton transfer. Why do the mechanisms differ? It appears likely that phosphonate- and carboxylate-binding perturb semiconductor surfaces sufficiently to create new trap states that can be rapidly populated following injection. The states are necessarily spatially proximal to the attached dyes, but apparently sufficiently separated from each other to preclude fast trap-to-trap hopping.

If the back ET rate is fast enough, it can attenuate a cell's quantum efficiency. Under certain circumstances (albeit, probably not achieved here), the rate can also influence a cell's photovoltage [10,106,107]. The differing mechanisms for back ET suggest differing criteria for optimization of cell performance under conditions where back ET rates do play a significant role. For example, higher driving forces yield faster rates for normal region reactions (covalently attached reactants), but slower rates for inverted-region reactions (selected electrostatically bound reactants). For polypyridyl-based MLCT chromophores, lower driving forces often go hand in hand with broader spectral coverage. For inverted-region reactions, this implies a trade-off between kinetic optimization (i.e., minimization of back-ET rates) and light collection. For normal-region reactions, on the other hand, kinetic and spectral optimization may well be achievable without a trade-off.

Are the mechanisms described here applicable to cells operating in nonaqueous environments? It is conceivable that the sequence described by Eqs. (12)–(14) occurs under certain conditions. The more complex sequence involving coupled electron and cation transfer probably does not. Although Li^+ (the electrolyte cation most often used in Grätzel-type cells) is known to intercalate into high-area metal oxide semiconductors [49,90,108–111], the rate is probably too slow to be coupled to injection and back ET in the same way that aqueous proton uptake and release are coupled to these processes. The ability to use water itself as a proton source means that solution-phase diffusional limitations on proton uptake are absent. Alkali metal ion uptake from nonaqueous solutions, on the other hand, clearly is subject to diffusional limitations.

The available nonaqueous reactivity data from Meyer and co-workers [13] and from Schmehl and co-workers [60] on injection and back ET to and from TiO_2 appear to point, instead, to a special role for equilibrium cation adsorption. In the absence of initial alkali metal ion adsorption, injection from MLCT-type chromophores does not occur. In their presence, injection does occur, but in proportion to the amount of cation present. Curiously, the cation concentration appears not to affect rate constants for injection or back ET on TiO_2—only the yields for these reactions. Presumably, similar behavior would be encountered with SnO_2 electrodes. A speculative interpretation is that alkali metal cation adsorption serves to poise the conduction-band edge of an individual nanoparticle at a potential compatible with exoergic injection. At low cation concentrations, only a fraction of the nanoparticles comprising a high-area photoelectrode may have adsorbed enough cations to permit injection. Obviously, implied in this scheme are local modulation of E_{cb} and a spatially heterogeneous distribution of E_{cb} values under certain conditions.

In summary, at nanostructured tin-oxide semiconductor–aqueous solution interfaces, back ET to molecular dyes is well described by conventional Marcus-type electron-transfer theory. The mechanistic details of the reaction, however, are remarkably sensitive to the nature of the semiconductor–dye binding interactions. The mechanistic differences point, potentially, to differing design strategies for kinetic optimization of the corresponding liquid-junction solar cells.

ACKNOWLEDGMENTS

We gratefully acknowledge the Office of Naval Research for support of this work at the earliest stages and the Basic Energy Science Program of the U.S. Dept. of Energy (grant No. DE-FG02-87ER13808) for subsequent support.

REFERENCES

1. O'Regan, B.; Grützel, M. *Nature* **1991**, *353*, 737.
2. Bach, U.; Lupo, D.; Comte, P.; Moser, J. E.; Weissörtel, F.; Salbeck, J.; Spreitzer, H.; Grätzel, M. *Nature* **1998**, *398*, 583.
3. Hagfeldt, A.; Grätzel, M. *Acc. Chem. Res.* **2000**, *33*, 269.
4. Hannappel, T.; Burfeindt, B.; Storck, W.; Willig, F. *J. Phys. Chem. B.* **1997**, *101*, 6799.
5. Ellingson, R. J.; Asbury, J. B.; Ferrere, S.; Ghosh, H. N.; Sprague, J. R.; Lian, T. Q.; Nozik, A. J. *J. Phys. Chem. B.* **1998**, *102*, 6455.
6. Tachibana, Y.; Moser, J. E.; Grätzel, M.; Klug, D. R.; Durrant, J. R. *J. Phys. Chem.* **1996**, *100*, 20,056.
7. Benko, G.; Kallioinen, J.; Korppi-Tommola, J. E. I.; Yartsev, A. P.; Sundstrom, V. *J. Am. Chem. Soc.* **2002**, *124*, 489.
8. Grätzel, M. *Prog. Photovolt.* **2000**, *8*, 171.
9. Parkinson, B. A.; Spitler, M. T. *Electrochim. Acta.* **1992**, *37*, 943.
10. Argazzi, R.; Bignozzi, C. A.; Heimer, T. A.; Castellano, F. N.; Meyer, G. J. *J. Phys. Chem. B.* **1997**, *101*, 2591.
11. Bonhôte, P.; Moser, J.; Humphrey-Baker, R.; Vlachopoulos, N.; Zakeeruddin, S. M.; Walder, L.; Grätzel, M. *J. Am. Chem. Soc.* **1999**, *121*, 1324.

12. Qu, P.; Meyer, G. J. *Langmuir* **2001**, *17*, 6720.
13. Kelly, C. A.; Farzad, F.; Thompson, D. A.; Stipkala, J. M.; Meyer, G. J. *Langmuir* **1999**, *15*, 7047.
14. Kuciauskas, D.; Freund, M. S.; Gray, H. B.; Winkler, J. R.; Lewis, N. S. *J. Phys. Chem. B.* **2001**, *105*, 392.
15. Sauve, G.; Cass, M. E.; Coia, G.; Doig, S. J.; Lauermann, I.; Pomykal, K. E.; Lewis, N. S. *J. Phys. Chem. B.* **2000**, *104*, 6821.
16. Haque, S. A.; Tachibana, Y.; Willis, R. L.; Moser, J. E.; Grätzel, M.; Klug, D. A.; Durrant, J. R. *J. Phys. Chem. B.* **2000**, *104*, 538.
17. Kamat, P. V.; Bedja, I.; Hotchandani, S.; Patterson, L. K. *J. Phys. Chem.* **1996**, *100*, 4900.
18. Bauer, C.; Boschloo, G.; Mukhtar, E.; Hagfeldt, A. *J. Phys. Chem. B.* **2001**, *105*, 5585.
19. Marcus, R. A. *J. Chem. Phys.* **1965**, *43*, 679.
20. Sutin, N. *Adv. Chem. Phys.* **1999**, *106*, 7.
21. Mulvaney, P.; Grieser, F.; Meisel, D. *Langmuir* **1990**, *6*, 567.
22. Ford, W. E.; Rodgers, M. A. J. *J. Phys. Chem.* **1994**, *98*, 3822.
23. Ford, W. E.; Rodgers, M. A. J. *J. Phys. Chem.* **1994**, *98*, 7415.
24. Ford, W. E.; Rodgers, M. A. J. *J. Phys. Chem. B.* **1997**, *101*, 930.
25. Barazzouk, S.; Lee, H.; Hotchandani, S.; Kamat, P. V. *J. Phys. Chem. B.* **2000**, *104*, 3616.
26. Martini, I.; Hartland, G. V.; Kamat, P. V. *J. Phys. Chem. B.* **1997**, *101*, 4826.
27. Nasr, C.; Liu, D.; Hotchandani, S.; Kamat, P. V. *J. Phys. Chem.* **1996**, *100*, 11,054.
28. Hara, K.; Sugihara, H.; Tachibana, Y.; Ashraful, I.; Yanagida, M.; Sayama, K.; Arakawa, H. *Langmuir* **2001**, *17*, 7280.
29. Zakeeruddin, S. M.; Nazeeruddin, M. K.; Humphry-Baker, R.; Pechy, P.; Quagliotto, P.; Barolo, C.; Viscardi, G.; Grätzel, M. *Langmuir* **2002**, *18*, 952.
30. Kleverlaan, C. J.; Indelli, M. T.; Bignozzi, C. A.; Pavanin, L.; Scandola, F.; Hasselmann, G. M.; Meyer, G. J. *J. Am. Chem. Soc.* **2000**, *122*, 2840.
31. Nazeeruddin, M. K.; Kay, A.; Rodicio, I.; Humphry-Baker, R.; Muller, E.; Liska, P.; Vlachopoulos, N.; Grätzel, M. *J. Am. Chem. Soc.* **1993**, *115*, 6382.
32. Oskam, G.; Bergeron, B. V.; Meyer, G. J.; Searson, P. C. *J. Phys. Chem. B.* **2001**, *105*, 6867.
33. Fillinger, A.; Parkinson, B. A. *J. Electrochem. Soc.* **1999**, *146*, 4559.
34. Kallioinen, J.; Lehtovuori, V.; Myllyperkio, P.; Korppi-Tommola, J. E. I. *Chem. Phys. Lett.* **2001**, *340*, 217.
35. Dang, X. J.; Hupp, J. T. *J. Am. Chem. Soc.* **1999**, *121*, 8399.
36. Gaal, D. A.; Hupp, J. T. *J. Am. Chem. Soc.* **2000**, *122*, 10956.
37. Hasselmann G. M.; Meyer, G. J. *J. Phys. Chem. B.* **1999**, *103*, 7671.
38. Nelson, J.; Haque, S. A.; Klug, D. R.; Durrant, J. R. *Phys. Rev. B.* **2001**, *30*, 5321.
39. Tachibana, Y.; Haque, S. A.; Mercer, I. P.; Durrant, J. R.; Klug, D. A. *J. Phys. Chem. B.* **2000**, *104*, 1198.
40. Ford, W. E.; Wessels, J. M.; Rodgers, M. A. J. *J. Phys. Chem. B.* **1997**, *101*, 7435.
41. Royea, W. J.; Fajardo, A. M.; Lewis, N. S. *J. Phys. Chem. B.* **1997**, *101*, 11,152.
42. Lewis, N. S. *Annu. Rev. Phys. Chem.* **1991**, *42*, 543.

43. Brown, G. M.; Sutin, N. *J. Am. Chem. Soc.* **1979**, *101*, 883.
44. Brunschwig, B. S.; Creutz, C.; MaCartney, D. H.; Sham, T. K.; Sutin, N. *Faraday Discuss. Chem. Soc.* **1982**, *74*, 113.
45. Marcus, R. A. *J. Phys. Chem. B.* **1991**, *95*, 2010.
46. Blackbourn, R. L.; Johnson, C. S.; Hupp, J. T. *J. Am. Chem. Soc.* **1991**, *113*, 1060.
47. Leytner, S.; Hupp, J. T. *Chem. Phys. Lett.* **2000**, *330* 3–4, 231.
48. Braslavsky, S. E.; Heibel, G. E. *Chem. Rev.* **1992**, *92*, 1381.
49. Lemon, B. I.; Lyon, L. A.; Hupp, J. T. In *Nanoparticles and Nanostructured Films: Preparation, Characterization, and Applications*; Fendler, J. H., Ed., Wiley–VCH: New York, 1998, p. 335.
50. Finklea, H. O., *Semiconducting Electrodes*; Elsevier: New York, 1988.
51. Bixon, M.; Jortner, J. *J. Phys. Chem.* **1991**, *95*, 1941.
52. Moser, J. E.; Grätzel, *M. Chem. Phys.* **1993**, *176*, 493.
53. Marcus, R. A.; Sutin, N. *Biochim. Biophys. Acta* **1985**, *811*, 265.
54. Closs, G. L.; Miller, J. R. *Science,* **1988**, *240*, 440.
55. Lees, A. C.; Kleverlaan, C. J.; Bignozzi, C. A.; Vos, J. G. *Inorg. Chem.* **2001**, *40*, 5343.
56. Finnie, K. S.; Bartlett, J. R.; Woolfrey, J. L. *Langmuir* **1998**, *14*, 2744.
57. Argazzi, R.; Bignozzi, C. A.; Heimer, T. A.; Castellano, F. N.; Meyer, G. J. *Inorg. Chem.* **1994**, *33*, 5741.
58. Gillaizeau-Gauthier, I.; Odobel, F.; Alebbi, M.; Argazzi, R.; Costa, E.; Bignozzi, C. A.; Qu, P.; Meyer, G. J. *Inorg. Chem.* **2001**, *40*, 6073.
59. Yan, S. G.; Lyon, L. A.; Lemon, B. I.; Prieskorn, J. S.; Hupp, J. T. *J. Chem. Educ.* **1997**, *74*, 657.
60. Andersson, A. M.; Isovitsch, R.; Miranda, D.; Wadhwa, S.; Schmehl, R. H. *Chem. Commun.* **2000**, *6*, 505.
61. Ferrere, S. *Chem. Mater.* **2000**, *12*, 1083.
62. Pechy, P.; Rotzinger, F. P.; Nazeeruddin, M. K.; Kohle, O.; Zakeeruddin, S. M.; Humphry-Baker, R.; Grätzel, M. *J. Chem. Soc. Chem. Commun.* **1995**, *1*, 65.
63. Alberti, G. *Acc. Chem. Res.* **1978**, *11*, 163.
64. Clearfield, A.; Smith, G. D. *Inorg. Chem.* **1969**, *8*, 431.
65. Chakraborty, D.; Chandrasekhar, V.; Bhattacharjee, M.; Kratzner, R.; Roesky, H. W.; Noltemeyer, M.; Schmidt, H. G. *Inorg. Chem.* **2000**, *39*, 23.
66. Yan, S. G.; Hupp, J. T. *J. Phys. Chem.* **1996**, *100*, 6867.
67. Yan, S. G.; Prieskorn, J. S.; Kim, Y.; Hupp, J. T. *J. Phys. Chem. B.* **2000**, *104*, 10,871.
68. Martini, I.; Hodak, J. H.; Hartland, G. V. *J. Phys. Chem. B.* **1998**, *102*, 607.
69. Rochester, C. H. *Acidity Functions*; Academic Press: New York, 1970.
70. Asbury, J. B.; Hao, E.; Wang, Y.; Lian, T. Q. *J. Phys. Chem. B.* **2000**, *104*, 11,957.
71. Asbury, J. B.; Hao, E.; Wang, Y.; Ghosh, H. N.; Lian, T. Q. *J. Phys. Chem. B.* **2001**, *105*, 4545.
72. Bedja, I.; Kamat, P. V.; Hua, X.; Lappin, A. G.; Hotchandani, S. *Langmuir* **1997**, *13*, 2398.
73. Bolts, J. M.; Wrighton, M. S. *J. Phys. Chem.* **1976**, *80*, 2641.
74. Watanabe, T.; Fujishima, A.; Tatsuoki, O.; Honda, K. *Bull. Chem. Soc. Japan*, **1976**, *49*, 8.

75. Redmond, G.; O'Keefe, A.; Burgess, C.; MacHale, C.; Fitzmaurice, D. *J. Phys. Chem.* **1993**, *97*, 11,081.

76. Gottesfeld, S.; McIntyre, J. D. E. *J. Electrochem. Soc.* **1979**, *126*, 742.

77. Natan, M. J.; Mallouk, T. E.; Wrighton, M. S. *J. Phys. Chem.* **1987**, *91*, 648.

78. Lyon, L. A.; Hupp, J. T. *J. Phys. Chem. B.* **1999**, *103*, 4623.

79. Kalaysundarum, K.; Grätzel, M.; Pelizetti, E. *Coord. Chem. Rev.* **1986**, *67*, 57.

80. Healy, T. W.; White, L. R. *Adv. Colloid Interf. Sci.* **1978**, *9*, 303.

81. Boxall, C. *Chem. Soc. Rev.* **1994**, *23*, 137.

82. Boxall, C.; Kelsall, G. H. *J. Electroanal. Chem.* **1992**, *328*, 75.

83. Nelson, B. P.; Candal, R.; Corn, R. M.; Anderson, M. A. *Langmuir* **2000**, *1*, 6094.

84. Boschloo, G.; Fitzmaurice, D. *J. Phys. Chem. B.* **1999**, *103*, 7860.

85. Boschloo, G.; Fitzmaurice, D. *J. Electrochem. Soc.* **2000**, *147*, 1117.

86. Lemon, B. I.; Hupp, J. T. *J. Phys. Chem. B.* **1997**, *101*, 2426.

87. Lemon, B. I.; Hupp, J. T. *J. Phys. Chem.* **1996**, *100*, 14,578.

88. Lemon, B. I. Ph.D. dissertation, Northwestern University, Evanston, IL, 1999.

89. Lyon, L. A. Ph. D. dissertation, Northwestern University, Evanston, IL, 1996.

90. Lyon, L. A.; Hupp, J. T. *J. Phys. Chem.* **1995**, *99*, 15,718.

91. Krtil, P.; Kavan, L.; Fattakhova, D. *J. Solid State Electron.* **1997**, *1*, 83.

92. Redmond, G.; Fitzmaurice, D. *J. Phys. Chem.* **1993**, *97*, 1426.

93. Exnar, I.; Kavan, L.; Huang, S. Y.; Grätzel, M. *J. Power Sources* **1997**, *68*, 720.

94. Creager, S. E.; Weber, K. *Langmuir* **1993**, *9*, 844.

95. Smith, C. P.; White, H. S. *Anal. Chem.* **1992**, *64*, 2398.

96. Zaban, A.; Ferrere, S.; Sprague, J.; Gregg, B. A. *J. Phys. Chem. B.* **1997**, *101*, 55.

97. Zaban, A.; Ferrere, S.; Gregg, B. A., *J. Phys. Chem. B.* **1998**, *102*, 452.

98. Bonhôte, P.; Gogniat, E.; Tingry, S.; Barbe, C.; Vlachopoulos, N.; Lenzmann, F.; Comte, P.; Grätzel, M. *J. Phys. Chem. B.* **1998**, *102*, 1498.

99. Heimer, T. A.; D'Arcangelis, S. T.; Farzad, F.; Stipkala, J. M.; Meyer, G. J. *Inorg. Chem.*, **1996**, *35*, 5319.

100. Trammell, S. A.; Meyer, T. J. *J. Phys. Chem. B.* **1999**, *103*, 104.

101. Deo, G.; Wachs, I. E. *J. Phys. Chem.* **1991**, *95*, 5889.

102. Moser, J.; Punchihewa, S.; Infelta, P. P.; Grätzel, M. *Langmuir* **1991**, *7*, 3012.

103. Hupp, J. T.; Weaver, M. J. *J. Electroanal. Chem.* **1983**, *145*, 43.

104. Veeramasuneni, S.; Yalamanchili, M. R.; Miller, J. D. *J. Colloid Interf. Sci.* **1996**, *184*, 594.

105. Darwent, J. R.; Kalyanasundaram, K. *J. Chem. Soc. Faraday Trans.* **1981**, *77*, 373.

106. Cahen, D.; Hodes, G.; Grätzel, M.; Guillemoles, J. F.; Riess, I. *J. Phys. Chem. B.* **2000**, *104*, 2053.

107. van de Lagemaat, J.; Park, N. G.; Frank, A. J. *J. Phys. Chem. B.* **2000**, *104*, 2044.

108. Shouji, E.; Buttry, D. A. *Electrochim. Acta*, **2000**, *45*(22–23), 3757.

109. Guerfi, A.; Paynter, R. W.; Dao, L. H. *J. Electrochem. Soc.* **1995**, *142*, 3457.

110. Chemseddine, A.; Morineau, R.; Livage, J. *J. Solid State Ionics* **1983**, *9–10*, 357.

111. Yu, A. S.; Frech, R. *J. Power Sources*, **2002**, *104*, 97.

4

Current Status of Dye-Sensitized Solar Cells

Hironori Arakawa and Kohjiro Hara

National Institute of Advanced Industrial Science and
Technology, Tsukuba, Japan

I. INTRODUCTION

It is known that the photoelectrochemical cell (PEC), which is composed of a photoelectrode, a redox electrolyte, and a counter electrode, shows a solar light-to-current conversion efficiency of more than 10%. However, photoelectrodes such as n- and p-Si, n-and p-GaAs, n- and p-InP, and n-CdS frequently cause photocorrosion in the electrolyte solution under irradiation. This results in a poor cell stability; therefore, many efforts have been made worldwide to develop a more stable PEC.

On the other hand, oxide semiconductor materials such as ZnO and TiO_2 have good stabilities under irradiation in solution. However, such stable oxide semiconductors cannot absorb visible light because of their wide band-gap character. Sensitization of wide-band-gap oxide semiconductor materials by photosensitizers, such as organic dyes which can absorb visible light, has been extensively studied in relation to the development of photography technology since the middle of the nineteenth century. In the sensitization process, dyes adsorbed onto the semiconductor surface absorb visible light and excited electrons of dyes are injected into the conduction band of the semiconductor. Dye-sensitized oxide semiconductor photoelectrodes have been used for PECs.

In the late 1960s, Gerischer and Tributsch researched a ZnO photoelectrode sensitized by organic dyes, including rose bengal, fluorescein, and rhodamine B

[1,2]. Single-crystal and polycrystalline materials, which could not adsorb a large amount of dye, were used for the photoelectrode. Therefore, they had low light-harvesting efficiencies and, consequently, low photon-to-current conversion efficiencies. Additionally, the organic dyes have a narrow absorption range in visible light, which also contributed to low solar cell performance. For example, Matsumura and colleagues reported a solar light-to-current conversion efficiency (η) of about 1% using a PEC composed of a sintered ZnO photoelectrode, a Pt counterelectrode, and an aqueous solution of I^-/I_3^- redox couple and rose bengal dye [3,4]. Therefore, in order to improve light-harvesting efficiencies and solar cell efficiency, two approaches were conducted by researchers in this field. One is the development of photoelectrodes with larger surface areas that could adsorb a large amount of dye. The other is synthesis of new dye photosensitizers with a wider absorption range. Significant improvements in the performance of dye-sensitized solar cells (DSSCs) have been established both by the development of high-performance nanoporous TiO_2 thin-film electrodes adsorbing a large amount of dye photosensitizer and the synthesis of new ruthenium (Ru) dye photosensitizers having the capacity to absorb visible and near-infrared (IR) light from 400 to 900 nm.

Ruthenium bipyridyl complexes are suitable photosensitizers because their excited states have a long lifetime and the oxidized Ru(III) center has a long-term chemical stability. Therefore, Ru bipyridyl complexes have been studied intensively as photosensitizers for homogeneous photocatalytic reactions and dye-sensitization systems. The Ru bipyridyl complex, bis(2,2'-bipyridine)(2,2'-bipyridine-4, 4'-dicarboxylate)ruthenium(II), having carboxyl groups as anchors to the semiconductor surface was synthesized and single-crystal TiO_2 photoelectrodes sensitized by this Ru complex were studied in 1979 and 1980 [5,6].

Recently, significant improvements in the performance of DSSCs have been made by Grätzel and co-workers at the Swiss Federal Institute of Technology (EPFL). They achieved the solar energy conversion efficiency (η) of 7–10% under AM 1.5 irradiation using a DSSC composed of a nanoporous and nanocrystalline TiO_2 thin-film electrode having a large surface area, a novel Ru bipyridyl dye, and an iodine I^-/I_3^- redox couple in organic solvent [7,8]. They also developed another new Ru terpyridine complex, which absorbs light in the near-IR region up to 900 nm, as a photosensitizer for the nanocrystalline TiO_2 photoelectrode. This DSSC showed a η of 10.4% under AM 1.5 with a short-circuit photocurrent (J_{sc}), of 20.5 mA/cm^2, an open-circuit photovoltage (V_{oc}) of 0.72 V, and a fill factor (ff) of 0.70 [9,10].

The new DSSC developed by Grätzel and co-workers, the so-called "Grätzel cell," has very attractive characteristics as follows:

1. High solar-energy conversion efficiency. A high efficiency equal to that of the amorphous Si solar cell has been obtained as a laboratory development and efficiencies greater than 10% might be possible.

2. Low production cost. The production procedure of DSSC is relatively simple and the materials for DSSC are relatively inexpensive. Fabrication costs will, therefore, be less than that of conventional solar cells. For example, the estimated production cost, 0.60 US$/Wp for a DSSC with 10% efficiency, is quite competitive with those of conventional solar cells [11,12].

3. Low limitation of material resources for DSSCs. Oxide semiconductors such as TiO_2, dyes, and iodine complexes are abundantly available. Although Ru resources are limited, the amount of Ru complex utilized in the DSSC is only 1×10^{-7} mol/cm^2. Organic dye photosensitizers could be used rather than Ru complexes if Ru sources are limited.

4. Various types of DSSC for colorful, adaptable consumer products. Colorful and transparent solar cells can be made using various kinds of dyes according to requests. For example, transparent solar cells could be used in place of window panes. Additionally, plastic substrates might be utilized rather than glass and it would expand the use of DSSCs, although it is necessary to a establish the low-temperature processing technology for TiO_2 film.

5. Less emission against the environmental pollution. The TiO_2, dyes, and iodine used in the DSSCs are nontoxic. The only component that could potentially cause harm is the organic solvents used as the electrolyte solution. Future research should be directed toward developing a solid-state electrolyte.

6. Potential to recycle use. The organic dye photosensitizers adsorbed on the electrode can be removed or detached by washing the electrode with alkali solutions or combustion, allowing recycling photoelectrodes for DSSCs.

As mentioned earlier, the DSSC is a very attractive and promising device for solar cell applications that has been intensively investigated worldwide, and its photovoltaic mechanism has also intensively investigated [11–20]. Moreover, commercial applications of the DSSC have been under investigation. In this chapter, we describe the DSSC, including its component materials, structure, working mechanism, efficient preparation procedure, current researches, and long-term stabilities. We also introduce the subjects for improvement of its performance and commercial applications.

II. MATERIALS AND MECHANISM

The Grätzel cell is composed of a Ru dye attached to a porous TiO_2 film of about 10 μm thickness on conducting glass (photoelectrode), an electrolyte solution, a spacer in between two electrodes, a Pt-sputtered conducting glass (counter elec-

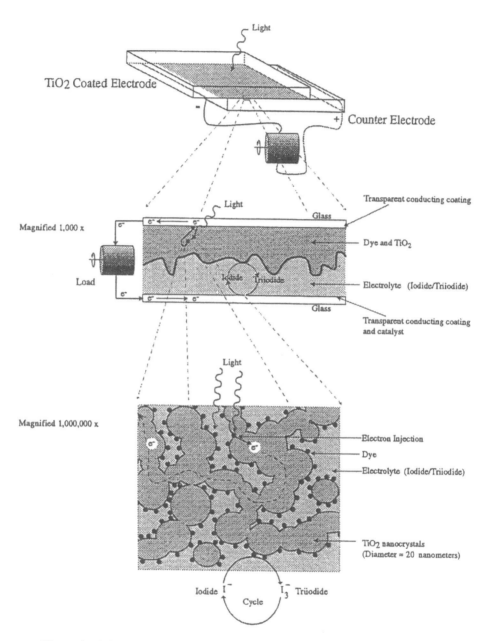

Figure 1   Schematic structure of dye-sensitized nanocrystalline TiO_2 solar cell, Grätzel cell. (From Ref. 12.)

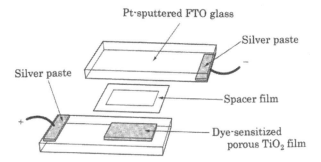

Figure 2 Assembly of dye-sensitized solar cell.

trode), and a cell sealant. Figure 1 shows the schematic structure of the DSSC. The complete DSSC assembly is shown in Fig. 2. The materials, structure, working mechanism, and electron-transfer kinetcs of DSSC are introduced in this section.

A. Materials and Structure

1. Conducting Glass

Transparent conducting oxide (TCO)-coated glass is used as the substrate for the TiO_2 photoelectrode. In order to obtain a high solar cell efficiency, the substrate must have a low sheet resistance and high transparency. In addition, the sheet resistance of conducting glass must be almost independent of calcination temperature of the TiO_2 film up to 500°C. Indium-tin oxide (ITO) is the most commonly used TCO material. However, ITO has a low thermal stability of resistance; however, it, nevertheless, has low resistance at ambient temperature. Usually, fluorine-doped SnO_2 (FTO) glass is used as the TCO substrate for DSSC. In our case, the FTO conducting glass from Nippon Sheet Glass Co. (R = 8–10 Ω/ square) is normally used.

2. Porous and Thin TiO_2 Film on FTO Glass

Titanium oxide (TiO_2) has a very good chemical stability under irradiation in solution and it is nontoxic and inexpensive, in addition to good photoconductivity. The porous and nanocrystalline TiO_2 thin-film photoelectrode is prepared by a simple process. TiO_2 colloidal paste is coated on an FTO glass substrate and then it is calcined at 450–500°C, producing a porous TiO_2 film with about 10 μm thickness. Because this film is composed of TiO_2 nanoparticles (10–30 nm), the material has a nanoporous structure. The factor of the internal surface area of TiO_2 film compared to its external surface area, the roughness factor (rf), is more than 1000; that is, a TiO_2 film with 1 cm^2 and 10 μm thickness has an actual surface area of more than 1000 cm^2. The dye is fixed onto the TiO_2 surface in a monolayer. Thus, if the nanoporous TiO_2 film has a high roughness factor, the amount of dye fixed is significantly increased (on the order of 10^{-7} mol/cm^2). This results in a significant increase of light-harvesting efficiency (i.e., up to nearly 100% at the peak absorption wavelength of the dye). In comparison, the amount of adsorbed dyes on the surface of single-crystal and polycrystalline materials is quite small, resulting in materials with only 1% light-harvesting efficiency, even at the peak wavelength.

A TiO_2 film containing a small proportion of large TiO_2 particles (250–300 nm), which can scatter incident photons effectively, is usually used to improve the light-harvesting efficiency. The porosity of the TiO_2 film is also very important, because the electrolyte solution containing the redox ions as an electron carrier must be able to penetrate the film effectively in order to maintain good contact between redox ions and dyes on the TiO_2 surface. Appropriate porosity of the TiO_2 film, 50–70%, is achieved by the addition of a proper amount of polymer substrate such as polyethylene glycol (PEG) and ethyl cellulose (EC) into the TiO_2 colloidal solution or paste following the calcination process. Figure 3 shows a scanning electron microscope (SEM) photograph of a typical porous and nano-

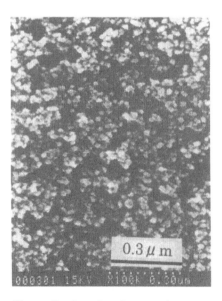

Figure 3 Scanning electron microscope photograph of a typical nanocrystalline TiO_2 film.

crystalline TiO_2 film. A detailed explanation of the procedure for preparing TiO_2 film is described in Section III.

3. Ru Dye Photosensitizer

The Ru dye photosensitizer, which contributes the primary steps of photon absorption and the consequent electron injection, is fixed onto the TiO_2 surface. The chemical structures of typical Ru dye photosensitizers developed by Grätzel and co-workers are shown in Fig. 4. TBA denotes the tetrabutylammonium cation

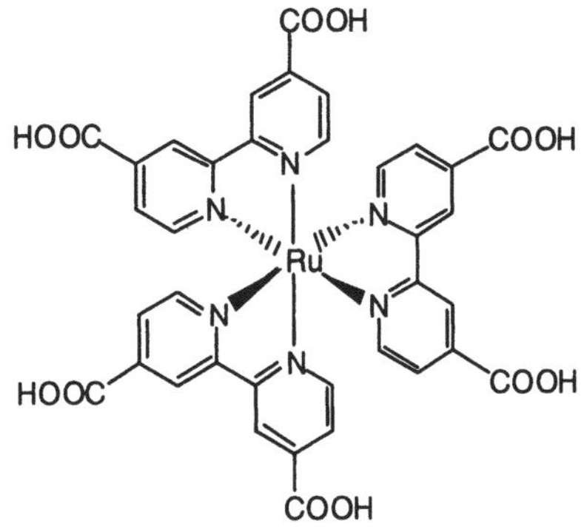

RuL_3

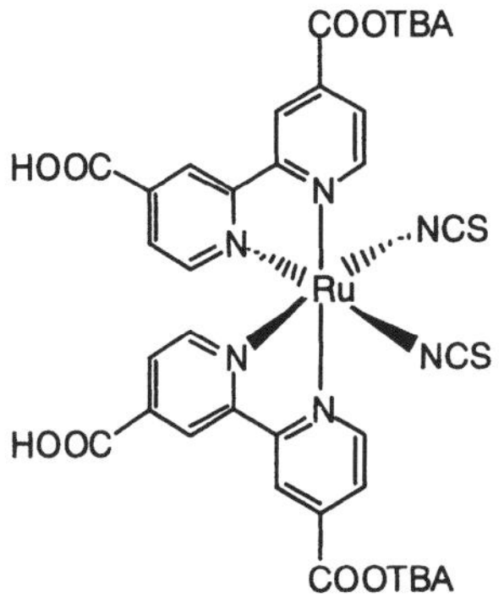

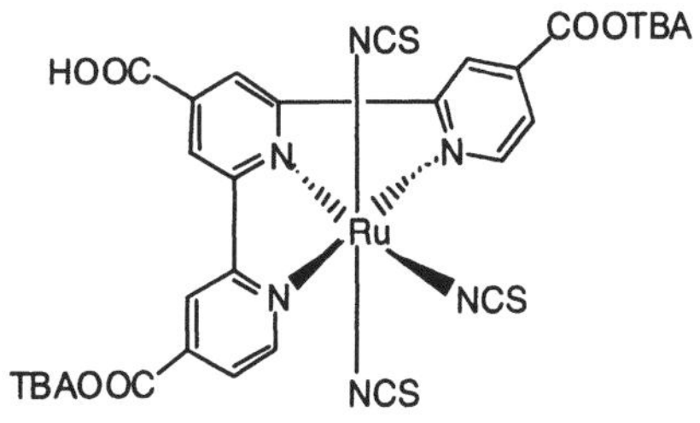

(N719 dye)
(TBA adduct of N3 dye)

$RuL'(NCS)_3$

(Black dye)

Figure 4 Molecular structures of Ru dye photosensitizers developed by Grätzel and co-workers.

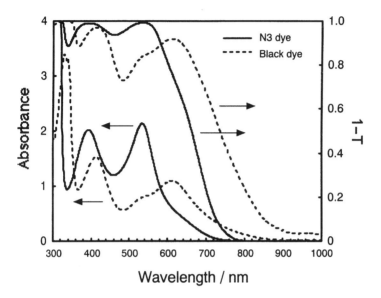

Figure 5 Light absorption properties of N3 dye, black dye, and their TiO_2 photoelectrodes represented by absorbance and light-harvesting efficiency.

$(C_4H_9)_4N^+$. Figure 5 shows the absorption properties of these dyes in solution and dye-attached TiO_2 film. The axes show the absorbance (A) of the dye and $1\text{-}T\ (= 1-10^{-A})$ of the dye-attached TiO_2 film (T denotes transmittance). The $RuL_2(NCS)_2$ complex, cis-bis(4,4'-dicarboxy-2,2'- bipyridine)dithiocyanato ruthenium(II), which is termed N3 dye (or red dye), absorbs light widely in the range from 400 to 800 nm. The $RuL'(NCS)_3$ complex, trithiocyanato 4,4'4''-tricarboxy-2,2':6',2''-terpyridine ruthenium(II) (black dye), absorbs light in the near-IR region up to 900 nm. Absorptions of these dyes in the visible and near-IR regions are attributed to the metal-to-ligand charge-transfer (MLCT) transition. The highest occupied molecular orbital (HOMO) and the lowest unoccupied molecular orbital (LUMO) are derived from the d-orbital of Ru metal and the π^*-orbital of the ligand, respectively. The NCS ligand shifts the HOMO level to higher energy, leading to a red shift in the absorption property of the dye, and it contributes as an electron acceptor from the iodide ion, I^- These Ru dyes have carboxyl functional groups which allows them to chemically bond to the TiO_2 surface. Dye fixation causes a large electronic interaction between the ligands of dye and the conduction band of TiO_2, resulting in effective electron injection from the Ru complex into the TiO_2. It is clarified by a Fourier transform infrared (FTIR) spectroscopic analysis that Ru dye is bonded to the TiO_2 surface using a carboxylate bidentate

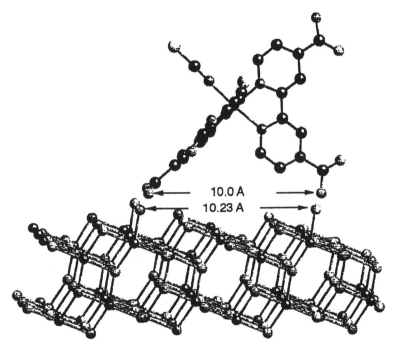

Figure 6 N3 dye structure sitting on the (101) surface of TiO_2: Top: N3 dye structure; bottom: TiO_2 lattice. (From Ref. 20.)

coordination or an ester bonding [21–25]. Figure 6 shows the anchored structure of the N3 dye on the (101) surface of TiO_2. The coverage of the TiO_2 surface with the N3 dye is estimated to be nearly 100% from the calculation of the surface area of TiO_2 and the number of dyes on TiO_2 surface.

4. Electrolyte Solution

The electrolyte solution used in the DSSC contains I^-/I_3^- redox ions, which carry electrons between the TiO_2 photoelectrode and the counterelectrode. Mixtures of 0.05–0.1 M of I_2 and 0.1–0.5 M of an iodide salt such as LiI, NaI, KI, tetraalkylammonium iodides (R_4NI), and imidazolium iodides are dissolved in organic solvents. As an organic solvent, acetonitrile, propionitrile, methoxyaceto-nitrile, propylene carbonate, and their mixtures are used. The performance of the DSSC depends on the counter cations of the applied iodide salt such as Li^+, Na^+, K^+, and R_4N^+. The cell performance varies because the cations influence the ion conductivity of the electrolyte solution and they adsorb on the TiO_2 surface differently, resulting in a shift of the conduction band level of the TiO_2 electrode [26,27]. The viscosity of the solvent also affects ion conductivity of the electrolyte directly and, consequently, the cell performance. In order to improve cell performance, low viscosity solvents are preferable. The diffusion coefficient of I_3^- in methoxyacetonitrile is estimated as $(5.4–6.2) \times 10^{-6}$ cm^2 sec^1 [26]. Usually, basic additives such as *tert*-butylpyridine (TBP) are added to the electrolyte solution in order to improve cell efficiency [8]. As an electron carrier for DSSC, the Br^-/Br_2 redox pair and hydroquinone were also tested [27,28]; however, the I^-/I_3^- pair gives the best efficiency.

5. Counter Electrode

The triiodide ion, I_3^-, formed by the reaction of oxidized Ru (III) dye with the I^- ion, is reduced back to the I^- ion at the surface of the counterelectrode. In order to reduce the triiodide ion effectively, the counterelectrode must have a high electrocatalytic activity. Pt sputtered (5–10 $\mu g/cm^2$, or approximately 200 nm thickness) FTO glass or carbon material is usually used as the counterelectrode.

6. Cell-Sealing Materials

A sealing material shown in Fig 2 is essential for keeping the electrolyte solution in between the two electrodes and preventing leakage and evaporation of the electrolyte solution. The chemical and photochemical stability of the sealing material against the electrolyte components as well as the solvent is very important. For example, thermal adhesive polymer sheets such as Surlyn (Du Pont Co. Ltd.) and the copolymer of polyethylene and polyacrylic acid are good sealing materials. In addition, glues such as epoxy resin are usually utilized as outer sealing materials to achieve long-term stability of DSSCs.

B. Working Mechanism and Electron-Transfer Kinetics

1. Working Mechanism

Figure 7 shows the schematic energy diagram of a DSSC. The following steps contribute the photon to current conversion:

1. Ruthenium dye photosensitizers attached on the TiO_2 surface absorb incident photon flux.
2. The dye photosensitizers are excited from the ground state (S) to the excited state (S*) owing to the MLCT transition. The excited electrons are injected into the conduction band of the TiO_2 film electrode, resulting in the oxidation of the dye photosensitizers.
3. Injected electrons in the conduction band of TiO_2 film are transported though the surface or the bulk of interconnected TiO_2 nanoparticles

with diffusion toward the back contact (FTO) and consequently, they reach the counterelectrode through the external load and wiring.

4. The oxidized Ru dye photosensitizers (S^+) accept electrons from the redox mediator, I^- ion, regenerating the ground state of dye photosensitizer (S), and the I^- ion is oxidized to the I_3^- ion.

5. The oxidized redox mediator, the I_3^- ion, diffuses to the counterelectrode and is re-reduced to the I^- ion over the Pt surface.

Overall, electric power is generated without any permanent chemical transformation. The performance of the DSSC is predominantly based on four energy levels of the components: the excited state (approximately LUMO) and the ground state (HOMO) of the Ru dye photosensitizer, the Fermi level of the TiO_2 photoelectrode, which is located near the conduction-band level, and the redox potential of the redox mediator (I^-/I_3^-) in the electrolyte solution. The photocurrent obtained from the DSSC is determined by the energy difference between the HOMO and LUMO of the Ru dye photosensitizer corresponding to the band gap, E_g, for inorganic semiconductor materials. The smaller the HOMO–LUMO energy gap, the larger will be the photocurrent because of utilization of the long-wavelength region in the solar spectrum. The energy gap between the LUMO level and the conduction-band level of TiO_2 film, ΔE_1, is important, and the energy level of the LUMO must be sufficiently negative with respect to the conduction band of TiO_2 film in order to inject electrons effectively. In addition, substantial electronic coupling between the LUMO and the conduction band of TiO_2 film also leads to effective electron injection. The HOMO level of the complex must be sufficiently more positive than the redox potential of the I^-/I_3^- redox mediator to accept electrons effectively (ΔE_2). The energy gaps, ΔE_1 and ΔE_2, must be larger than approximately 200 mV the driving force for the electron-transfer reactions to take place with an optimal efficiency [18].

In the case of solid–liquid–junction solar cells, PEC, the photovoltage is attributed to the energy gap between the Fermi level (near the conduction-band level for the n-type semiconductor) of the semiconductor photoelectrode and the redox potential of the mediator in the electrolyte. As shown in Fig. 7, the photovoltage of the DSSC is developed by the energy gap between the Fermi level of a TiO_2 electrode and the redox potential of the I^-/I_3^- in the electrolyte solution. The conduction-band level of the TiO_2 photoelectrode and the redox potential of I^-/I_3^- are estimated to be -0.5 V versus saturated calomel electrode (SCE) and 0.4 V versus SCE, respectively, shown in Fig. 7 [20] (or -0.7 V versus SCE and 0.2 V versus SCE, respectively [15,18]). Thus, in the case of a DSSC using a TiO_2 photolectrode and I^-/I_3^- redox mediator, the maximum photovoltage is expected to be approximately 0.9 V, depending on the components of electrolyte solution because the Fermi level of the TiO_2 photoelectrode depends on the species and concentrations of electrolyte components.

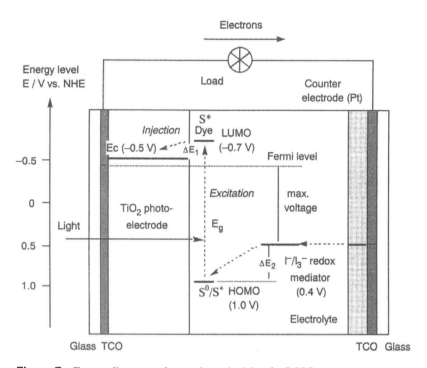

Figure 7   Energy diagram and operating principle of a DSSCs.

In contrast to a conventional p–n-junction-type solar cell, the mechanism of the DSSC does not involve a charge-recombination process between electrons and holes because electrons are injected from the dye photosensitizers into the semiconductor, and holes are not formed in the valence band of the semiconductor. In addition, electron transport takes place in the TiO_2 film, which is separated from the photon absorption sites (i.e., the photosensitizers); thus, effective charge separation is expected. This photon-to-current conversion mechanism of the DSSC is similar to that for photosynthesis in nature, where chlorophyll functions as the photosensitizer and electron transport occurs in the membrane.

In conventional p–n-junction-type solar cells and classical PECs using poly-crystalline or single–crystal photoelectrodes, the electronic contact between components forms the photovoltaically active junction. In addition, the equilibrium between electronic charge carriers in the materials leads to space charge formation. Photogenerated charges are separated by the electric field in the space charge layer. In the DSSC, however, the individual particle size is too small to form a space charge layer. Charge separation in the DSSC has been discussed relative to an electrical field at the electrolyte–semiconductor interface, although it is not due to a space charge in the semiconductor [29] (see Chapter 3). Small cations, such as Li^+ in the electrolyte and H^+ released from the dyes upon binding, can adsorb (or intercalate) on the semiconductor surface. A dipole is formed across the Helmholtz layer between these cations and negatively charged species (iodide ions and the dye). The electrical potential drop across the Helmholtz layer will help to separate the charges and reduce recombination with the dye cations or the redox mediator. Under illumination, this potential will decrease, as the electrons injected in the semiconductor will neutralize some of the positive charge at the surface.

Photovoltaic performance of the DSSC is described as follows: Figure 8 shows the external spectral response curve of the photocurrent for nanocrystalline TiO_2 solar cells sensitized by N3 and black dyes with the I^-/I_3^- redox mediator, where the incident photon-to-current conversion efficiency (IPCE) is represented as a function of wavelength. IPCE is obtained by the following equation:

$$IPCE\ (\%) = \frac{1240\ (eV\ nm) \times J_{sc}\ (\mu A/cm^2)}{\lambda\ (nm) \times \Phi\ (\mu W/cm^2)} \times 100 \tag{1}$$

where J_{sc} is the short-circuit photocurrent density for monochromatic irradiation, λ is the wavelength, and Φ is the monochromatic light intensity. As shown in Fig. 8, solar cells sensitized by the Ru dye photosensitizers can efficiently convert

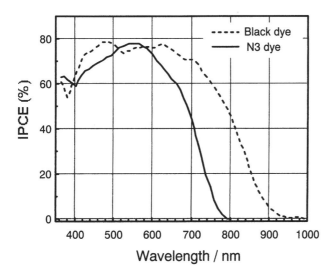

Figure 8 Spectral responses (IPCE) of N3- and black dye-sensitized TiO_2 solar cells. IPCE is plotted as a function of wavelength. (Data from Ref. 20.)

visible light to current. N3 dye [RuL$_2$(NCS)$_2$] responds to light from 400 to 800 nm and black dye [RuL'(NCS)$_3$] responds to the near-IR region up to 950 nm. The IPCE of the N3-dye-sensitized solar cell reaches 80% at 550 nm and exceeds 70% in the region from 450 to 650 nm. Taking into consideration of light losses due to light reflection and absorption by the FTO glass substrate, the incident photon-to-current conversion efficiency is almost 90–100%, indicating a high performance of the DSSC. IPCE is also given by

$$\text{IPCE} = \text{LHE}\phi_{inj}\eta_c \qquad (2)$$

$$\text{LHE} = 1 - T = 1 - 10^{-A} \qquad (3)$$

where LHE is the light-harvesting efficiency, ϕ_{inj} is the quantum yield of electron injection, and η_c is the efficiency of collecting injected electrons at the back contact. According to Eq. (2), if ϕ_{inj} and η_c are almost equal to unity, IPCE is determined by the LHE (i.e., $1 - T$) of the dye adsorbed on the film, as shown in Fig. 5.

Solar energy-to-electricity conversion efficiency (η) under white-light irradiation (e.g., AM 1.5) can be obtained from

$$\eta = \frac{J_{sc} V_{oc} \text{ ff}}{I_0 \; 100} \qquad (4)$$

where I_0 is the photon flux (approximately 100 mW/cm^2 for AM 1.5). A current versus voltage curve obtained for a nanocrystalline TiO$_2$ solar cell sensitized by the black dye is shown in Fig. 9 [20]. The evaluation of this cell was carried out at the National Renewable Energy Laboratory (NREL) in the United States. An efficiency of 10.4% was obtained (cell size = 0.186 cm^2, J_{sc} = 20.53 mA/cm^2, V_{oc} = 0.721 V, and ff = 0.704) [10,20].

2. Electron-Transfer Kinetics

Recently, the electron-transfer kinetics in the DSSC, shown as a schematic diagram in Fig. 10, have been under intensive investigation. Time-resolved laser spectroscopy measurements are used to study one of the most important primary processes—electron injection from dye photosensitizers into the conduction band of semiconductors [30–47]. The electron-transfer rate from the dye photosensitizer into the semiconductor depends on the configuration of the adsorbed dye photosensitizers on the semiconductor surface and the energy gap between the LUMO level of the dye photosensitizers and the conduction-band level of the semiconductor. For example, the rate constant for electron injection, k_{inj}, is given by Fermi's golden rule expression:

$$k_{inj} = \left(\frac{4\pi^2}{h}\right) \mid V \mid^2 \rho(E) \qquad (5)$$

where V is the electronic coupling between the photosensitizer and the semicon-

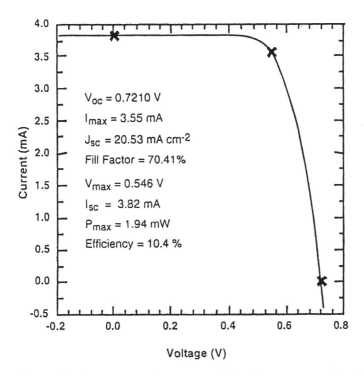

Voc = 0.7210 V

Imax = 3.55 mA

Jsc = 20.53 mA cm⁻²

Fill Factor = 70.41%

Vmax = 0.546 V

Isc = 3.82 mA

Pmax = 1.94 mW

Efficiency = 10.4 %

Figure 9 Photocurrent–voltage curve obtained for a nanocrystalline TiO$_2$ solar cell sensitized by black dye. The results plotted were obtained at the NREL calibration laboratory. (From Ref. 20.)

ductor, ρ (E) is the density of states of the conduction band, and h is the Planck constant. The value of V is attributed to overlap between the wave function of the excited states of the dye photosensitizer and the conduction band, and it depends strongly on the distance between the adsorbed dye photosensitizer and the semiconductor surface. In a DSSC, the dye photosensitizers are strongly absorbed onto the semiconductor surface, with carboxyl groups as anchors, resulting in a very large V between the π^*-orbital of the excited state of the dye photosensitizer and the conduction band of the TiO$_2$ film, which consists of the unoccupied 3d-orbital of Ti^{4+}. In addition, the conduction band of the semiconductor has a continuous and relatively large density of states. Thus, electron injection from the dye phtosensitizer to the semiconductor occurs at a higher rate than does relaxation from the excited state to the ground state. For example, it has been observed by time-resolved laser spectroscopy that electron injection from N3 dye into the TiO$_2$ film occurs on the order of femtoseconds [30, 34]. This ultrafast

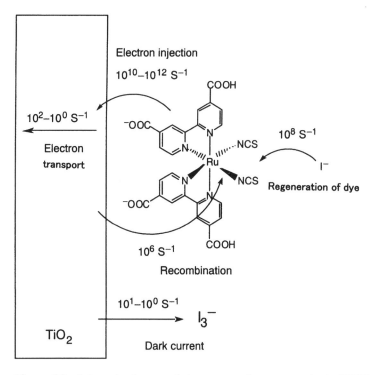

Figure 10   Schematic diagram of electron-transfer processes in the DSSC.

rate of electron injection contributes to the high energy conversion efficiencies of the DSSC.

In addition, the rate constant for electron injection depends strongly on the semiconductor materials employed. A slower electron injection rate was observed in the ZnO system with coumarin dyes and N3 dye compared to the TiO_2 system [37,39,47]. The different rate constant, may be caused by the difference in the electronic coupling between the π^*-orbital of the dye and the accepting orbitals in ZnO and TiO_2 and/or their density of states. The states near the conduction-band edge of ZnO consist of the $4s$-orbitals of Zn^{2+}, whereas those of TiO_2 consist of the $3d$-orbitals of Ti^{4+}, resulting in an expected difference in electronic coupling with the π^*-orbital of the dye.

The charge-recombination process between injected electrons and oxidized dyes must be much slower than the process of electron injection and electron transfer from the I^- ion into oxidized dyes (i.e., regeneration of dyes) to accomplish effective charge separation. It was reported that charge recombination between injected electrons of TiO_2 and cations of N3 dye occurs on the order of microseconds to milliseconds, in contrast with ultrafast electron injection [30,41,42,48–50]. The much slower charge recombination compared to electron injection leads to effective charge separation and, consequently, a high cell efficiency. Charge recombination in the N3 dye/TiO_2 system is caused by the direct electron transfer from TiO_2 film to Ru(III), the center of dye, whereas electron injection occurs from the bipyridyl ligands of dye to TiO_2 film. Thus, it is considered that long-distance electron transfer from TiO_2 film to the Ru(III) metal center and leads to a much smaller electron-transfer rate.

Electron transfer from I^- into the oxidized Ru photosensitizer (cation), or regeneration of the Ru photosensitizer, is one of the primary processes needed to achieve effective charge separation. The kinetics of this reaction has also been investigated by time-resolved laser spectroscopy [48,51]. The electron-transfer rate from I^- into the Ru(III) cation of the N3 dye was estimated to be 100 nsec [48]. This reaction rate is much faster than that of charge recombination between injected electrons and Ru dye cations. Thus, fast regeneration of the oxidized Ru dye photosensitizer also contributes to the accomplishment of effective charge separation.

Recombination of injected electrons in the semiconductor with the triiodide ion (I_3^-) at the interface, corresponding to the dark current, is one of the unwanted processes in DSSCs. This reaction could also occur on the SnO_2 surface because the nanocrystalline TiO_2 does not completely cover the TCO substrate, but predominantly occurs on the TiO_2 surface because of the high surface area of the TiO_2 relative to the SnO_2. This reaction contributes to loss of photovoltaic performance in a DSSC, which is analogous to the forward bias injection of holes and electrons in a p–n junction. The V_{oc} in the DSSC is given using the injection current, I_{inj}, as represented by the following equation as well as p–n-junction solar cells:

$$V_{oc} = \frac{kT}{q} \ln\left(\frac{I_{inj}}{I_0} + 1\right) \tag{6}$$

where k is the Boltzmann constant, q is the magnitude of the electron charge, T is the absolute temperature, and I_0 is the dark current. I_{inj} and I_0 are represented by the following equations:

$$I_{inj} = q\eta\,\Phi_0 \tag{7}$$

$$I_0 = q\,n_0\,k_{et}\,[I]_3^- \tag{8}$$

where η is the quantum yield for photogenerated electrons, Φ_0 is the incident photon flux, n_0 is the electron density on the-conduction band of the semiconductor in the dark, k_{et} is the rate constant for recombination [reaction (5)], and

$[I_3^-]$ is the concentration of the oxidized redox mediator, I_3^-, in the solution. From Eqs. (6)–(8), we obtain the following equation:

$$V_{oc} = \frac{kT}{q} \ln\left(\frac{\eta \, \Phi_0}{n_0 \, k_{et} \, [I_3^-]} + 1\right) \qquad (9)$$

Usually, $\eta \, \Phi_0 >> n_0 \, k_{et} \, [I_3^-]$ and Eq. (9) is simplified as follows [6,12,50,51]:

$$V_{oc} = \frac{kT}{q} \ln\left(\frac{\eta \, \Phi_0}{n_0 \, k_{et} \, [I_3^-]}\right) \qquad (10)$$

Dark current is considered to occur at the TiO_2–electrolyte interface where the dye photosensitizers are not adsorbed. To suppress dark current, pyridine derivatives such as TBP have been employed as coadsorbates on the TiO_2 surface, resulting in the improvement of the photovoltage [8,52]. TBP is considered to adsorb on the uncovered TiO_2 surface. Figure 11 shows the current–voltage characteristics for N3-dye-sensitized TiO_2 solar cell under illumination and dark using the electrolyte with and without TBP. This clearly indicates that TBP suppresses dark current, resulting in improvement of the V_{oc}. The decrease of J_{sc} by the

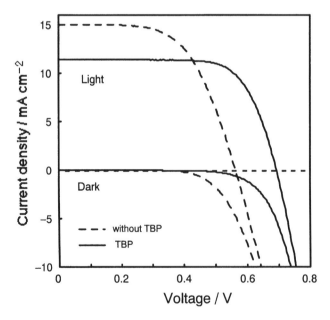

Figure 11   Current–voltage characteristics for N3-dye-sensitized TiO_2 solar cells under illumination and dark using the electrolyte with and without TBP.

addition of TBP is considered to arise due to a negative shift of the conduction band level of TiO_2 by adsorption of TBP.

The kinetics of recombination has been investigated and discussed intensively [8,52–59]. If this reaction occurs with a large reaction rate, the DSSC does not work. Taking into consideration of this and a slow transport of the photoinjected electrons through the nanocrystalline TiO_2 film, the recombination must be extremely slow. In fact, the time scale for recombination has been estimated to be on the order of 0.1 to several seconds [56]. This slow recombination would be due to a low electrocatalytic activity of TiO_2 for the reduction of triiodide ions.

Electron transport in TiO_2 film is an important process related to the photovoltaic performance in the DSSC and has been studied by many researchers [60–73]. It has been discussed with respect to different mechanisms: a diffusion model [60–63], a mechanism that involves tunneling through potential barriers between the particles [63], a trapping/detrapping mechanism [65–68], and an insulator–metal (Mott) transition mechanism [69]. The electronic conductivity in TiO_2 is very low, resulting in a slow response of the photocurrent. For example, diffusion coefficients for the electrons in nanocrystalline TiO_2 film have been estimated to be 1×10^{-7} [60] and 1.5×10^{-5} cm^2/sec [61]. In the DSSC, electronic conductivity of the TiO_2 film is significantly increased due to electron injection from dye photosensitizers under irradiation [62,63]. In addition, the conductivity and response of the photocurrent increases with increasing incident light intensity [62,63]. It has been suggested that when injected electrons fill trap sites and/or surface levels in the TiO_2 film, the diffusion coefficient of the electron increases drastically, leading to a enhanced electronic conductivity and good photocurrent response.

III. PROCEDURE FOR PREPARATION OF DSSC WITH GREATER THAN 8.0% EFFICIENCY

A. Preparation of TiO₂ Paste [18,74]

Commercial TiO_2 powder, such as P25 (Degussa or Nippon Aerosil [75]), are usually used to prepare the TiO_2 photoelectrode. However, colloidal TiO_2 solution prepared by hydrolysis of Ti(IV) alkoxides, such as isopropoxide and butoxide, has also been used to produce high-performance solar cells. Generally, anatase rather than rutile is more suitable for electrodes [76]. Preparation involves the following steps:

1. Hydrolysis of Ti alkoxides by 0.1 M HNO$_3$
2. Peptization by heating at 80°C for 8 hr with strong agitation, followed by filtration
3. Hydrothermal treatment by autoclaving at 200–250°C for 12 hr

4.   Sonication with an ultrasonic bath
5.   Concentration by an evaporator

TiO_2 sol formation is conducted by controlled hydrolysis of a Ti-alkoxide, such as Ti-isopropoxide. To obtain monodispersed particles of the desired size, the hydrolysis and condensation kinetics must be controlled. Suitably modified Ti-alkoxides with acetic acid or acetyl acetonate yield colloids having a large surface area (>200 m^2/g) and smaller particle diameter (5–7 nm) [18,74]. The peptization process results in segregation of the agglomerates of TiO_2 particles to the primary particles and then the large agglomerates are removed by filtration. Autoclaving of the colloidal TiO_2 solution leads to growth of the primary particles to 10–25 nm as well as an increase of the crystallinity of anatase particles. At a higher autoclaving temperature, the particle size increases and more rutile formation occurs, particularly at temperatures above 240°C. Electrodes prepared using colloids autoclaved at 230°C are transparent, whereas those made from colloids autoclaved at higher temperatures are translucent or opaque. This effect is due to light scattering by larger particles. After autoclaving, the precipitates are redispersed using an ultrasonic processor equipped with a Tihorn (e.g., Sonics & Materials Inc.; 400–600 W [77]). The colloidal solution is then concentrated at 45°C using a rotary evaporator and an appropriate concentration of TiO_2 is approximately 11 wt%.

B.   Preparation of Porous TiO$_2$ Film Electrode

Two methods are applicable for preparing TiO_2 thin films: One is the doctor blade method and the other is the screen-printing method.

1.   Doctor Blade Method. In order to increase the porosity of the TiO_2 film, 0.02–0.07 g of polyethylene glycol (PEG, molecular weight 20,000) is added into 1 mL of the concentrated colloidal TiO_2 solution (TiO_2, 11 wt%). If a commercial powder such as P25 is used, the powder is dispersed by grinding with water, a particle stabilizer such as acetylacetone, and a nonionic surfactant such as Triton X. The colloidal TiO_2 solution is spread on a TCO glass substrate and then calcined at 450°C for 30 min under air. The resulting TiO_2 film is transparent.

2.   Screen-Printing Method. The TiO_2 colloid is separated from acidic water after hydrolysis, washed carefully, and then mixed with ethyl cellulose (EC) as a binder and α-terpineol in ethanol. α-Terpineol is used to increase the viscosity of the solution. Finally, TiO_2 paste is produced after evaporating ethanol solvent. The paste is printed on a FTO glass substrate using a screen-printing machine and then calcined at 500°C for 1 hr under air. The film thicknesses are easily controlled in screen printing by the selection of paste composition (i.e., ratio of TiO_2 included), screen mesh size, and printing times.

The film prepared by above-mentioned two methods has a film thicknesses of 5–15 μm and the film mass is 1–2 mg/cm^2. The optimum film thickness is 13–14 μm. The film has a porosity of 60–70% [74]. High porosity produces effective diffusion of the redox mediator into the film. The roughness factor (described in Sec II.A.2) of a 10-μm film reaches approximately 1000, allowing absorption of a large amount of Ru dye photosensitizer and, consequently, increased light-harvesting efficiency. The TiO$_2$ film prepared from the paste including 10–20-nm TiO$_2$ particles is transparent. The scattering property of the film is important for improvement of the light-harvesting efficiency of the dye-coated film, resulting in improved IPCE performance of solar cell. This effect of the scattering in the TiO$_2$ film has been investigated in detail [18,74,78–80]. The path length of the incident light and the light absorption by adsorbed dyes are increased by light scattering in the TiO$_2$ film. This can be achieved by mixing of some amounts of larger TiO$_2$ particles with small TiO$_2$ particles during film preparation, whereas larger particles have a small surface area and, consequently, they cannot absorb a large amount of the dye. A simulation of light scattering in the TiO$_2$ electrode of the DSSC predicts that a suitable mixture of small TiO$_2$ particles (e.g., 20 nm in diameter) and larger particles (250–300 nm in diameter) as effective light-scattering centers has the potential to enhance solar light absorption significantly [79]. Actually, the photocurrent of the DSSC increased using a scattering film, compared to that for a transparent film [80]. This improvement in the photoresponse of the DSSC due to the scattering effect is observed especially in the low-energy region (e.g., 650–900 nm). As shown in Figs. 5 and 8, the IPCE values obtained in the red region in Fig. 8 are higher than light absorption ratios of dye-fixed TiO$_2$ film (1−T in Fig. 5). On the low-energy side, a significant part of the incident radiation penetrates the layer due to the low absorption coefficient of the dye, whereas photons of 500–650 nm are mainly absorbed near the TCO–TiO$_2$ interface because of the large absorption coefficient. Multiple reflections of the low-energy light in highly scattering films result in the increase of light absorption and hence the increase of photoresponse compared with the absorption spectra in solution.

It has also been reported that TiCl$_4$ treatment of the film significantly improves cell performance, especially the photocurrent [8]. After printing, the TiO$_2$ films are immersed in 0.1–0.5 M TiCl$_4$ solutions at room temperature and then calcined at 450°C for 30 min. It is speculated that TiCl$_4$ treatment improves the photocurrent by improving the connections between TiO$_2$ particles.

C. Ruthenium Dye Fixation

After preparation of the porous nanocrystalline TiO$_2$ film, the N3 dye is adsorbed onto the TiO$_2$ film surface. The film is immersed into the dye solution (0.2–0.3 mmol in ethanol or *tert*-butanol/acetonitrile, 1:1 mixed solution) followed by storage at room temperature for 12–18 hr. This treatment produces intense coloration of the film. Before use, the film is washed with alcohol or acetonitrile to remove excess nonadsorbed dyes in the nanoporous TiO_2 film.

D. Preparation of Electrolyte Solution

As described in Section I.B.4, electrolyte solution is prepared using an iodine redox electrolyte, additives, and an organic solvent. Typical organic solvents are nitrile derivatives having relative low viscosity, such as acetonitrile, propionitrile, methoxyacetonitrile, and methoxypropionitrile, which produce a high degree of ion conductivity. It has been reported that imidazolium derivatives, such as 1,2-dimethyl-3-hexylimidazolium iodide (DMHImI) and 1,2-dimethyl-3-propylimidazolium iodide (DMPImI), decrease the resistance of the electrolyte solution and improve photovoltaic performance [81,82]. A typical electrolyte solution producing a high solar cell performance for the Ru dye/TiO_2 solar cell reported by Grätzel and co-workers is a mixture of 0.02 M I_2, 0.04 M LiI, 0.5 M DMHImI, and 0.5 M TBP in acetonitrile [74]. TBP shifts the conduction-band level of the TiO_2 electrode to the negative direction and suppresses the dark current that corresponds to a reduction of I_3^- ions by injected electrons, leading to improvement of the photovoltage [8,52].

E. Counterelectrode

Sputtered Pt (5–10 μg/cm^2, or 200 nm thickness) on as FTO glass substrate has been usually employed as a counterelectrode. When Pt is sputtered, the photocurrent is slightly increased due to the light-reflection effect by a mirrorlike effect. In addition, the activity of reducing triiodide ions (I_3^-) over the Pt-sputtered FTO glass is improved by the formation of nanosize Pt particles over Pt-sputtered FTO glass [83]. Therefore, a small amount of an alcoholic solution of H_2PtCl_6 is dropped on the surface of the Pt-sputtered FTO glass substrate, followed by drying and heating at 385°C for 10 min. This results in formation of small Pt particles on the surface. The properties of the Pt counterelectrode directly affect the fill factor of the solar cell. A desirable exchange-current density over the Pt-sputtered FTO electrode for the reduction of triiodide ions is 0.01–0.2 A/cm^2 [18,83].

F. Cell Assembly and Measurement of Performance

We usually prepare an unsealed DSSC and measure its photovoltaic performance. A spacer film, such as polyethylene (15–30 μm thickness), is placed on the dye-coated TiO_2 photoelectrode and then the electrolyte solution is dropped on the surface of the TiO_2 photoelectrode using a pipette (one or two drops). The counterelectrode is placed over the TiO_2 photoelectrode and then the two electrodes are clipped together with two binder clips. If a low-melting-point polymer film such as Surlyn is used instead of the spacer film, we can produce a sealed cell. For obtaining long-term stability of the cell, the cell edge is sealed again carefully using glue such as ethylene vinyl acetate and epoxy resin. The overall preparation procedure of efficient DSSC is shown in Fig. 12.

Since Grätzel and co-workers reported the high performance of DSSC in 1991, many researchers worldwide have attempted to reproduce their result. Some reported performances are shown in Table 1. These cells were prepared using N3 dye, as shown in Fig. 4, and a nanocrystalline TiO_2 photoelectrode. In many cases, the light condition is AM 1.5 (approximately 100 mW/cm^2) produced with a solar simulator in the laboratory. Grätzel and co-workers reported $\eta = 9.6\%$ in 1993, and they achieved 10.0% at NREL in 1997. Other high efficiencies were reported by Lindquist and co-workers at Uppsala University (6.9%, 1994), NREL (7.5%, 1997), and Ishihara Sangyo Co. Ltd. in Osaka (6.3%, 1994) and by Yanagida and co-workers at Osaka University (6.1%, using 21 mW/cm^2, 1995). Our group achieved 7.2% in 1998, 8.4% in collaboration with EPFL in 1999, and 8.3% in cooperation with Sumitomo Osaka Cement Co. Ltd. in 2000. A venture company, Institut für Angewandte Photovoltaik GmbH (INAP), that has been investigating commercialization of the DSSC in cooperation with EFPL achieved 7.0% in 1997 using a 144-cm^2 cell (12 × 12 cm^2). Although these groups have reproduced efficiency values greater than 7%, additional studies are necessary to reproduce 10% efficiency. We are still investigating the improvement of cell

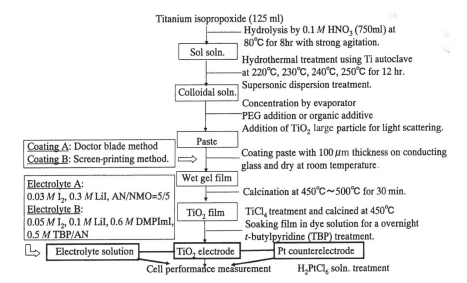

Figure 12   Preparation procedure for the DSSC.

Table 1   Photovoltaic Performance of N3 or N719 Dye-Sensitized TiO_2 Solar Cells

Institute	Cell size/cm^2	J_{sc}/mA cm^{-2}	V_{oc}/mV	Fill factor	η/%	Light source	Year
EPFL	0.31	18.2	720	0.73	9.6	AM 1.5	1993
EPFL-NREL	0.17	18.6	740	0.73	10.0	AM 1.5	1997
Uppsala Univ.	1.0	30.4	610	0.37	6.9	ELH lamp	1994
ISC	0.5	14.2	630	0.71	6.3	AM 1.5	1994
Osaka Univ.	0.5	3.9	570	0.67	6.1	21 mW cm^{-2}	1995
NREL	0.44	14.5	730	0.71	7.5	AM 1.5	1997
NIMC	0.13	14.5	698	0.71	7.2	AM 1.5	1998
EPFL-NIMC	0.21	15.2	780	0.71	8.4	AM 1.5	1999
INAP	144				7.0	AM 1.5	1997

EPFL: Swiss Federal Institute of Technology, Switzerland; NREL: National Renewable Energy Laboratory, U.S.A.; ISC: Ishihara Sangyo Co. Ltd., Japan; NIMC: National Institute of Materials and Chemical Research, Japan (the former name of PCRC/AIST); INAP: Institut fur Angewandte Photovoltaik Gmbh, Germany.

efficiencies by changing the preparation conditions of TiO_2 films and the component of electrolyte solutions. Table 2 shows the current data of our experiment. The best efficiency of the N719 dye (TBA adducts of N3 dye)/TiO_2 solar cell is 8.7% so far. Table 2 shows the performances of DSSCs using various TiO_2 photoelectrodes made of a different mixing ratio of TiO_2 colloidal pastes.

Table 2   Photovoltaic Performance of Dye-Sensitized TiO_2 Solar Cells Prepared at PCRC/AIST

Name	Cell area (cm^2)	Thickness (μ)	Jsc (mA/cm^2)	Voc (V)	Fill factor	Efficiency η (%)
SP-1	0.26	18.0	17.2	0.71	0.72	8.71
SP-2	0.26	14.0	14.9	0.76	0.74	8.33
SP-3	0.26	15.0	15.6	0.74	0.71	8.19
SP-4	0.26	13.0	14.5	0.77	0.73	8.11
SP-5	0.26	15.0	15.1	0.74	0.73	8.11

TiO_2 film: TiO_2 paste for printing on FTO glass: mixture of two kinds of TiO_2 pastes. Mixing ratio of two pastes was changed in each sample. TiO_2 particle sizes in each TiO_2 paste: 14 ~ 20 nm and 50 ~ 250 nm. Film was prepared by screen printing method.
Dye: $Ru(Hdcbpy)_2(NCS)_2(TBA)_2$[N719dye], 3×10^{-4} M in AN/t-BuOH.
Electrolyte: 0.05M-I_2, 0.1M-LiI, 0.62M-DMPImI, 0.5M-TBP/Acetonitrile(AN).
Measurement: AM1.5 (100 mW/cm^2) using a solar simulator (Wacom, WXS-80C).

The standardization of measurement conditions is essential for obtaining correct cell performance because cell efficiency depends on measurement conditions, such as light intensity and spectrum. Generally, under the low light intensity, the fill factor of DSSC is improved due to a low photocurrent (i.e., low series resistance), resulting in improved cell efficiency. The light under AM 1.5 irradiation should be used as the light source for measurements. Spectral response (IPCE) performance of DSSC also depends on light conditions [63]. The DSSC shows a relatively slow response due to the low electron mobility of TiO_2 film, as described in Section II.B.2. Under a high intensity of irradiation, the response of the DSSC increases with increasing electron injection as well as electron trap filling. Therefore, IPCE performance should be measured by a DC method under a monochromatic light of high intensity or the AC method using a white bias irradiation and low chopper frequencies (e.g., <50 Hz [63]).

IV. CURRENT RESEARCH TRENDS

Since Grätzel and co-workers developed the DSSC with a high efficiency of 7–10% in 1991 and 1993, much attention has been paid to the development of this new type of DSSC. In this section, current research efforts in the DSSC with new oxide semiconductor photoelectrodes, new metallic dyes, pure organic dyes, and new electrolyte systems are introduced.

A. New Oxide Semiconductor Photoelectrodes

Up to now, porous and nanocrystalline TiO_2 films have been predominantly used as the photoelectrodes of the DSSC. However, other oxide semiconductor materials, such as ZnO [25,84–90], SnO_2 [25,51,89–91], Nb_2O_5 [25,45,89,90], In_2O_3 [25,89,90], $SiTiO_3$ [92], and NiO [93], have been also investigated. Table 3 shows the photovoltaic performances of DSSCs using various oxide semiconductor photoelectrodes. Nevertheless, the TiO_2 photoelectrode has the best performance, and oxide semiconductor photoelectrodes exceeding the performance of the TiO_2 photoelectrode have not been found yet. The physical properties of oxide semiconductor materials, such as the band gap, energy level of the conduction band, and electric conductivity, influence the performance of the oxide semiconductor photoelectrode significantly, finally affecting the cell performance.

Mixed-oxide photoelectrodes composed of two different kinds of single-oxide semiconductor materials are also investigated. Tennakone and co-workers reported that the DSSC composed of a nanocrystalline SnO_2/ZnO mixed photoelectrode with N3 dye revealed a high efficiency, which was almost equal to that of the TiO_2 solar cell [94,95]; that is, $\eta = 7.94\%$ ($J_{sc} = 22.8$ mA/cm^2, $V_{oc} = 0.67$ V, and ff $= 0.50$) under 80 mW/cm^2 and $\eta = 15\%$ under 10 mW/cm^2 [94]. They employed a mixed-oxide film consisting of small particles of SnO_2 (15 nm)

Table 3 Photovoltaic Performance of DSSC Using Different Oxide Semiconductor Photoelectrode

Reference	Photoelectrode	Dye	Conditions	Performance
[85]	ZnO	N3	56 m W cm^{-2}	$\eta = 2\%$
[89]	ZnO	Mercurochrome	AM 1.5 (99 mW cm^{-2}), 0.09 cm^2	$\eta = 2.5\%$ ($J_{sc} = 7.4$ mA cm^{-2}, $V_{oc} = 0.52$, $ff = 0.64$)
[89]	SnO$_2$	Mercurochrome	AM 1.5 (100 mW cm^{-2}), 0.25 cm^2	$\eta = 0.65\%$ ($J_{sc} = 2.0$ mA cm^{-2}, $V_{oc} = 0.58$, $ff = 0.56$)
[89]	In$_2$O$_3$	Mercurochrome	AM 1.5 (100 mW cm^{-2}), 0.25 cm^2	$\eta = 0.38\%$ ($J_{sc} = 5.4$ mA cm^{-2}, $V_{oc} = 0.24$, $ff = 0.29$)
[25]	Nb$_2$O$_5$	N3	520 nm (4 mW cm^{-2}), 1 cm^2	$\eta = 2.6\%$ ($J_{sc} = 0.29$ mA cm^{-2}, $V_{oc} = 0.61$, $ff = 0.58$)
[90]	Nb$_2$O$_5$	N3	Xe lamp (100 mW cm^{-2}), UV and IR cut off	$\eta = 1.2\%$ ($J_{sc} = 3.3$ mA cm^{-2}, $V_{oc} = 0.67$, $ff = 0.54$)
[92]	SrTiO$_3$	N3	AM 1.5 (1 sun)	$\eta = 1.8\%$ ($J_{sc} = 3$ mA cm^{-2}, $V_{oc} = 0.79$, $ff = 0.70$)
[94]	SnO$_2$/ZnO	N3	900 W cm^{-2}	$\eta = 8\%$ ($J_{sc} = 22.8$ mA cm^{-2}, $V_{oc} = 0.67$ V, $ff = 0.5$)
[96]	Nb$_2$O$_5$/TiO$_2$	N3	Xe lamp	$J_{sc} = 11.4$ mA cm^{-2}, $V_{oc} = 0.73$ V, $ff = 0.564$
[97]	TiO$_2$/ZnO	N3	Xe lamp (81 mW cm^{-2}), UV and IR cut off	$\eta = 9.8\%$ ($J_{sc} = 21.3$ mA cm^{-2}, $V_{oc} = 0.71$, $ff = 0.52$)
[90]	Nb$_2$O$_5$/TiO$_2$	N3	Xe lamp (100 mW cm^{-2}), UV and IR cut off	$\eta = 2.0\%$ ($J_{sc} = 7.1$ mA cm^{-2}, $V_{oc} = 0.68$, $ff = 0.42$)

and large particles of ZnO (2 μm, 53 wt%). The efficiency of the DSSC using the combined photoelectrode was improved dramatically compared to that using single ZnO or SnO$_2$ photoelectrodes, independently. It is speculated that the Ru dyes are fixed onto SnO$_2$ nanoparticles, and ZnO contributes to the charge-separation process as well as the electron-transfer process. This result indicates the possibility of the development of a DSSC composed of a non-TiO$_2$ photoelectrode with a high efficiency, although the detailed mechanism of the new mixed-oxide photoelectrode has not been elucidated up to now.

The effect of surface modification of the TiO$_2$ photoelectrode by other oxide materials has been also investigated by several research groups. For example, Zaban et al. prepared a nanocrystalline TiO$_2$ photoelectrode coated with a

Nb_2O_5 layer, whose conduction-band level was more negative than that of TiO_2. They measured the cell performance of a DSSC using this photoelectrode with N3 dye [96]. Both the J_{sc} and V_{oc} were improved compared to those of the TiO_2 photoelectrode. Wang et al. have studied the cell performance of the DSSC using a nanocrystalline ZnO-modified TiO_2 photoelectrode with N3 dye [97]. TiO_2 nanoparticles were covered by ZnO layers. In this study, both the J_{sc} and V_{oc} were also improved compared to those of the TiO_2 photoelectrode. They speculated that the improvement was caused by a positive shift of the conduction-band level of the photoelectrode and suppression of dark current due to the ZnO coating [97].

B. New Dye Photosensitizers

1. Metal Complex Dyes

As described in Section II.A.3, *cis*-bis(4,4′-dicarboxy-2,2′-bipyridine)dithiocyanato ruthenium(II) (N3 dye) and trithiocyanato 4,4′4″-tricarboxy-2,2′:6′,2″-terpyridine ruthenium(II) (black dye) exhibit quite good performance; therefore, the systems using these two Ru dyes have been intensively investigated. However, other Ru dye photosensitizers have also been synthesized and characterized, and their performances as photosensitizers in DSSCs have been reported by many researchers [98–116]. Several kinds of electron withdrawing ligand for Ru dye photosensitizers are shown in Fig. 13. Figure 14 shows the structures of newly synthesized Ru dye photosensitizers and their absorption properties. The axis shows the molar absorption coefficient, ε (i.e., absorption coefficient per M; unit: $M^{-1}\,cm^{-1}$).

We synthesized the Ru phenanthroline complex, *cis*-bis [4,7-dicarboxy-1,10-phenanthroline)dithiocyanato ruthenium(II), which is abbreviated as (Ru(dc-phen)$_2$(NCS)$_2$)]. The complex has the absorption maximum due to the MLCT transition at around 520 nm, which is similar to that of N3 dye. [108,112,114] By using this dye, a nanocrystalline TiO_2 solar cell showed an efficiency (η) of 6.1–6.6% under AM 1.5 irradiation. We also synthesized the new Ru bipyridyl complex having one acetylacetonato ligand (β-diketonato ligand), which is an electron-donating ligand, instead of two thioisocyanato ligands (—NCS), and this dye also showed a high performance for nanocrystalline TiO_2 solar cells [115]. The Ru biquinoline complex, whose absorption is red-shifted compared to that for N3 dye, has also been synthesized [111]. Because the LUMO level of this complex is not sufficiently negative to inject electrons effectively, a high solar cell efficiency was not obtained in the case of nanocrystalline TiO_2 electrode. However, electron injection from this complex into the nanocrystalline SnO_2 electrode occurred effectively, because its conduction-band level was more positive than that of TiO_2 [111]. In order not only to inject electrons effectively into the conduction band of the semiconductor photoelectrodes but also to accept electrons from I^- ions effectively, delicate tuning of both the LUMO and HOMO

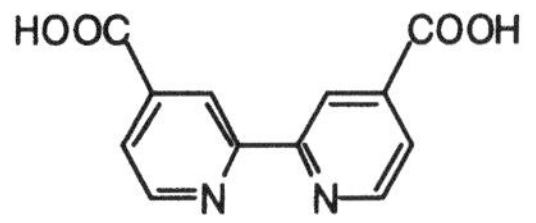

4,4'-Dicarboxy-2,2'-bipyridine

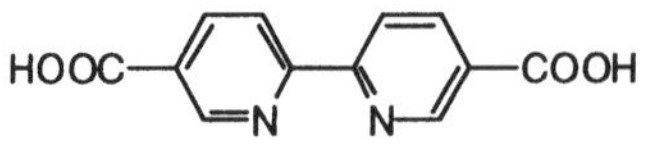

5,5'-Dicarboxy-2,2'-bipyridine

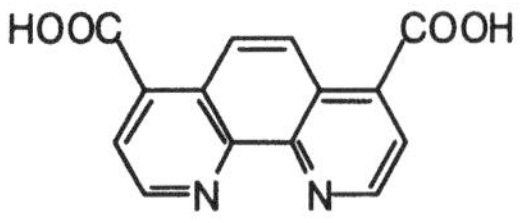

4,7-Dicarboxy-1,10-phenanthroline

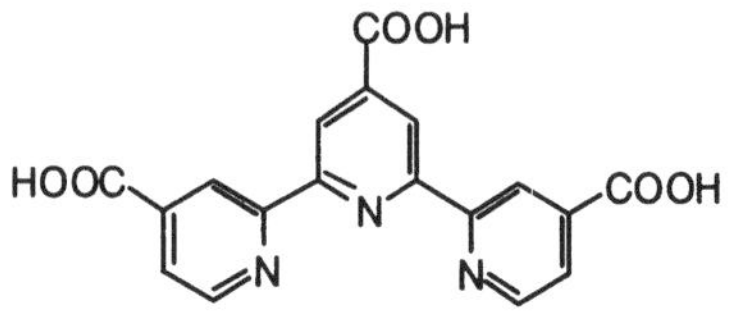

4,4',4"-Tricarboxy-2,2':6',2"-terpyridine

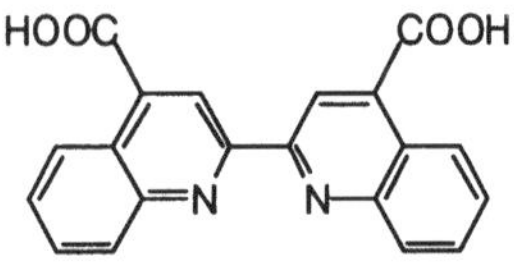

4,4'-Dicarboxy-2,2'-biquinoline

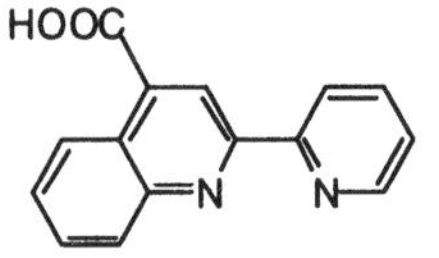

4-Carboxy-2-(2'-pyridyl)quinoline

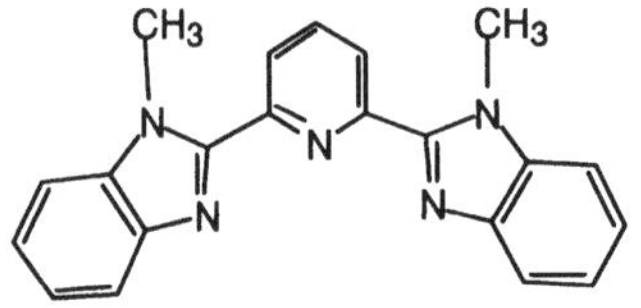

2,6-Bis(1-methylbenzimidazol-2-yl)pyridine

Figure 13   Molecular structures of various kinds of ligand for Ru complex photosensitizers.

levels of metal complexes is very important in developing new, efficient dye photosensitizers.

New metal complexes other than Ru have also been investigated. These include Fe complexes [117,118], Os complexes [119–122], Re complexes [123], and Pt complexes [124]. A nanocrystalline TiO_2 solar cell sensitized by a square-planar platinum(II) complex containing 4,4'-dicarboxy-2,2'-bipyridine and quinoxaline-2,3-dithiolate ligands showed an efficiency of 2.6% ($J_{sc} = 6.14$ mA/cm^2

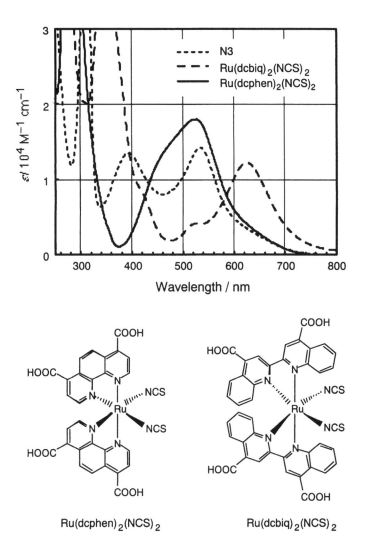

Figure 14 Molecular structures of newly synthesized Ru dye photosensitizers. Ru(dc-phen)$_2$(NCS)$_2$ and Ru(dcbiq)$_2$(NCS)$_2$, and their absorption properties in ethanolic solution. The y axis is represented by molar absorption coefficient, ε.

and V_{oc} = 0.60 V) under simulated AM 1.5 solar irradiation [124]. However, dyes that exhibit higher efficiency than the Ru dye photosensitizers have not been developed. From these results, it is considered that the HOMO level of Ru complexes derived from the d-orbital of the Ru metal center [i.e., the redox potential of Ru(II)/Ru(III)] is matched best to the I_3^-/I^- couple for accepting electrons effectively.

Porphyrin [125–128] and phthalocyanine [129] derivatives are also employed as photosensitizers in DSSCs. A nanocrystalline TiO_2 solar cell sensitized by Cu chlorophyllin produced 2.6% efficiency (J_{sc} = 9.4 mA/cm² and V_{oc} = 0.52 V) under 100 mW/cm² [127]. In order to develop new, efficient, metal dye photosensitizers, both an increase in the absorption coefficient of the metal complex and a greater red shift of the absorption spectrum is required.

2. Organic and Natural Dye Photosensitizers

Organic dyes having appropriate HOMO and LUMO levels to the redox potential of iodine and the conduction-band level of the semiconductor, respectively, could be utilized as well. As described in Section I, organic dyes such as 9-phenylxanthene dyes have been used as photosensitizers in the early research. Generally, organic dyes have several advantages as photosensitizers:

1. They have a variety of structures for molecular design.
2. They are relatively less expensive than metal complexes.
3. They have large absorption coefficients due to allowed $\pi-\pi^*$ transitions.

Many DSSCs using organic dye photosensitizers have been reported, and some structures are shown in Fig. 15 [90,91,132–141].

A nanocrystalline SnO_2 solar cell sensitized by a perylene derivative produced 0.9% efficiency under AM 1.5 (J_{sc} = 3.26 mA/cm² and V_{oc} = 0.45 V) [132]. A nanocrystalline TiO_2 solar cell (1 cm²) sensitized by eosin Y, oneof the 9-phenylxanthene dyes, produced 1.3% efficiency (J_{sc} = 2.9 mA/cm², V_{oc} = 0.66 V, and ff = 0.67) [131 A mercurochrome/ZnO solar cell (0.09 cm²) attained 2.5% efficiency (J_{sc} = 7.44 mA/cm², V_{oc} = 0.52 V, and ff = 0.64) under AM 1.5 (100 mW/cm²) [88,89].

Cyanine and merocyanine dyes have also been tested as dye photosensitizers [132–137]. A nanocrystalline TiO_2 solar cell sensitized by the special merocyanine dye as shown in Fig. 15 revealed a good efficiency of 4.2–4.5% (J_{sc} = 9.7 mA/cm², V_{oc} = 0.62 V, ff = 0.69, cell area = 0.25 cm²)) under AM 1.5 (100 mW/cm²) [135]. Aggregates of the merocyanine dye formed on the TiO_2 surface resulted in a wide expansion of the absorption area, especially in the longer-wavelength region. Therefore, the light-harvesting efficiency was improved [136]. We synthesized new coumarin derivatives that can absorb visible light from 400 to 700 nm and prepared a nanocrystalline TiO_2 solar cell using

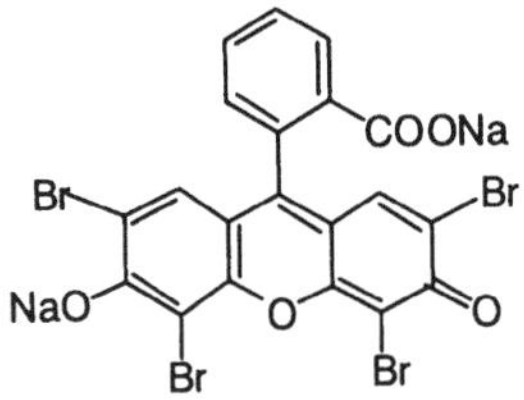

Eosin Yellow

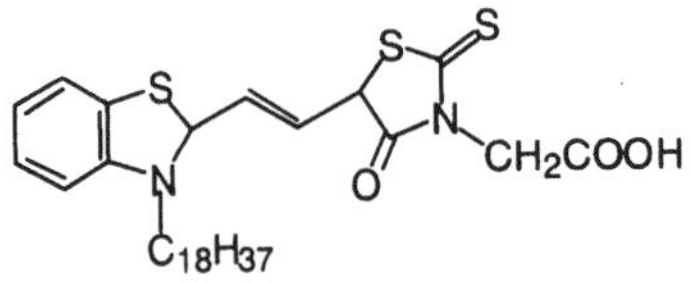

Merocyanine dye

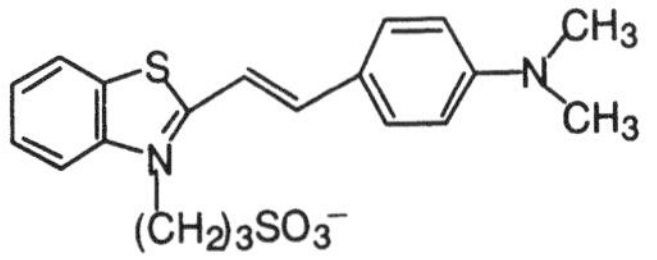

Cyanine dye

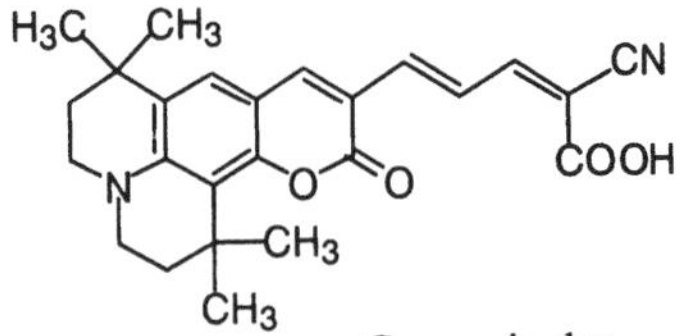

Coumarin dye

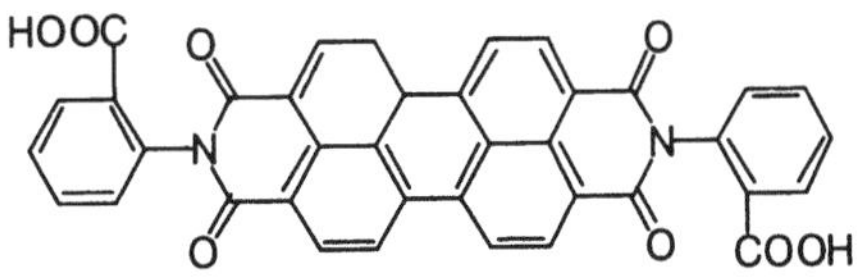

Perylene dye

Figure 15   Molecular structures of various organic dye photosensitizers.

this dye. The structure of new coumarin dye is shown in Fig. 15. Under AM 1.5 irradiation, an efficiency of 5.6% was obtained ($J_{sc} = 13.8$ mA/cm^2, $V_{oc} = 0.63$ V, ff $= 0.64$, cell area $= 0.25$ cm^2) [138]. Furthermore, we established the efficiency (η) of 6.0% by the optimization of electrolyte solution, as shown in Fig. 16. This is the best efficiency that has been achieved using organic dye-sensitized solar cells. A maximum IPCE of 76% was obtained at 470 nm. The photocurrent of this solar cell is almost equal to that of the N3 dye/TiO$_2$ solar cell, indicating a promising prospect for organic dye photosensitizers. Design and development of new organic dyes with absorption in the near-IR region and large absorption coefficients are needed to improve the performance of the DSSC based on organic dye photosensitizers.

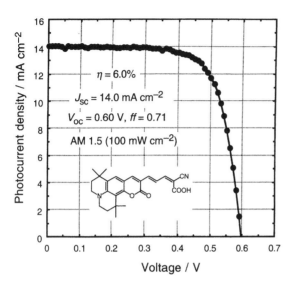

Figure 16 Photocurrent–voltage curve obtained for a nanocrystalline TiO_2 solar cell sensitized by coumarin dye.

In addition to organic dyes, natural dyes extracted from plants can be used as photosensitizers [36,140,141]. A nanocrystalline TiO_2 solar cell using a santalin dye extracted from red sandalwood can produce 1.8% efficiency under 80 mW/cm^2 irradiation [141]. Cherepy et al. reported that a nanocrystalline TiO_2 solar cell using flavonoid anthocyanin dyes extracted from blackberries could convert sunlight to electrical power with an efficiency of 0.6% (J_{sc} = 1.5–2.2 mA/cm^2 and V_{oc} = 0.4–0.5 V) under AM 1.5 [36].
 The maximum IPCE was 19% at the peak of the visible absorption band of the dye. They also observed a fast-electron injection of below 100 fsec from cyanine dye into the conduction band of TiO_2 by time-resolved transient absorption spectroscopy [36].

C. New Electrolyte Systems

Recently, room-temperature ionic liquids (molten salts) have been extensively studied in order to replace volatile organic solvents by them in electrochemical devices such as batteries. Interest in these materials is stimulated by their properties (e.g., high ionic conductivity, good electrochemical stability, and low volatility). Among these properties, the low volatility is the most critical for ensuring the long-term stability of electrochemical devices. Room-temperature ionic liquids can be used as alternative materials for the liquid electrolytes in DSSCS [82,142]. Ionic liquids used in the DSSC include imidazolium derivatives, such as 1-hexyl-3-methylimidazolium iodide (HMImI) [82] and 1-ethyl-3-methylimidazolium bis(trifluoromethylsulfonyl)imide (EMIm-TFSI) [142]. Matsumoto et al. reported that an N3-dye-sensitized TiO$_2$ solar cell using an EMIm salt with hydrofluoride anions, H$_3$F$^-$ or H$_4^-$, as the electrolyte exhibit an efficiency (η) of 1.2% under AM 1.5 (J_{sc} = 5.8 mA/cm^2, V_{oc} = 0.65 V, and ff = 0.56) [142]. If the viscosity of these ionic liquids could be decreased down to those of organic solvents, the performance of solar cell with ionic liquids will be improved by an increase of ionic mobility of the electrolyte.

D. Solid-State and Quasi-Solid-State DSSC

It is very attractive to develop solid-state or quasi-solid-state DSSCs. This is because they have some merits in long-term stability and economical production of DSSCs. In the case of the DSSC using a liquid electrolyte solution, the cell must be sealed perfectly in order to avoid leakage of electrolyte solution. By the present sealing technology using a polymer sealant, it is very difficult to avoid leakage of vaporized solvent of electrolyte solution at a relatively higher temperature such as 80°C (e.g., outdoor roof in the summer). In addition, the solid-state DSSC would allow an easier assembling system for integrated DSSC fabrication.

Grätzel and co-workers reported an N3-dye-sensitized nanocrystalline TiO$_2$ solar cell using a hole-transport material such as 2,2′,7,7′-tetrakis (*N*, *N*-di-*p*-methoxyphenyl-amine) 9,9′-spirobifluorene (OMeTAD) as shown in Fig. 17, as a solid electrolyte [143]. OMeTAD was spin-coated on the surface of the N3 dye/TiO$_2$ electrode and then Au was deposited by vacuum evaporation as the counterelectrode. The cell efficiency was 0.7% under 9.4 mW/cm^2 irradiation, and 3.18 mA/cm^2 of J_{sc} was obtained under AM 1.5 (100 mW/cm^2) [143]. The maximum IPCE was 33% at 520 nm. The rate for electron injection from OMeTAD into cations of N3 dyes has been estimated as 3 psec, which is faster than that of the I$^-$ ion case [144].

On the other hand, Tennakone and co-workers utilized a p-type semiconductor material, such as CuI (band gap,-3.1 eV), as a hole conductor and produced a solid-state DSSC [141,145,146]. Acetonitrile solution of CuI was dropped onto the surface of a dye-coated TiO$_2$ film, which was heated up to approximately 60°C and then the solution penetrated into the film. After evaporation of the acetonitrile, CuI was deposited into a nanoporous TiO$_2$ film. The Au-coated TCO substrate as the counterelectrode was pressed onto the surface of the TiO$_2$/dye/CuI film. In the system using the santalin dye photosensitizer, an efficiency of 1.8% was obtained under irradiation of 80 mW/cm^2 [141] and the efficiency reached 4.5% for the TiO$_2$/N3 dye/CuI/Au system. These results suggested that a highly efficient solid-state DSSC could be produced [145]. In these systems,

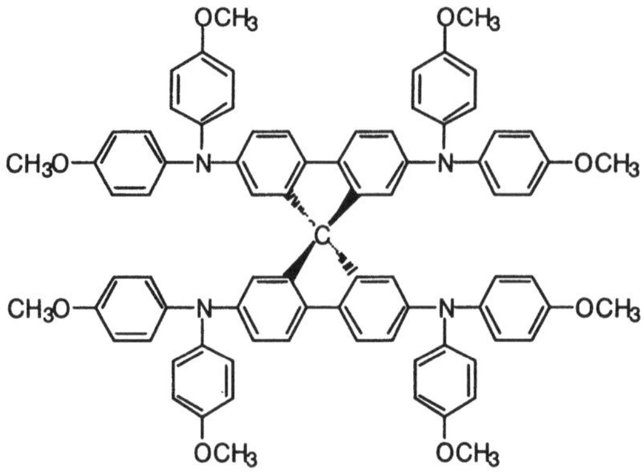

OMeTAD

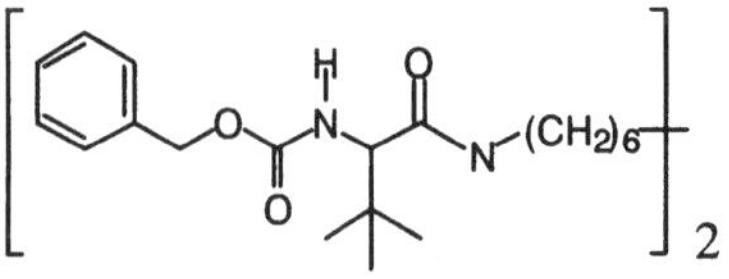

L-valine derivative galator

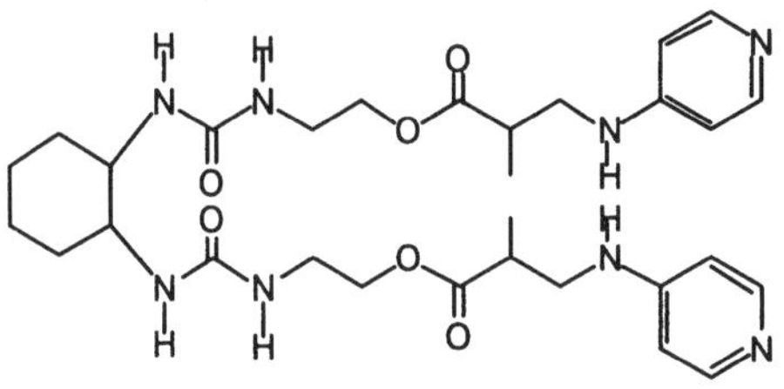

Gelator reported by Toshiba

Figure 17   Molecular structures of a solid-state electrolyte OMeTAD and gelators.

CuI could be partly in contact with TiO_2 directly; therefore, the efficiency decreased by the recombination of injected electrons with CuI. In order to increase cell performance, direct contact between the TiO_2 film and CuI must be minimized. Solid-state DSSCs have been studied using other organic and inorganic hole conductor materials, such as p-type CuSCN [147,148], polypyrrole [149], and polyacrylonitrile [95].

Quasisolidification of the electrolyte using gelator materials is another method for replacing liquid electrolytes in DSSCs. Gelation can be accomplished by adding a gelator material into the electrolyte solution without any other changes in the components of the electrolyte. Yanagida and co-workers investigated the gelation of the electrolyte solution using L-valine derivatives, as shown in Fig. 17, as a gelator and measured the solar cell performance of DSSCs with various gel electrolytes [150]. The gelator was added at a concentration of 0.1 M and then dissolved at 90–140°C. The gel solution was dropped onto dye-coated TiO_2 film followed by cooling. Interestingly, the performance of the DSSC using a gel electrolyte is almost the same as that using a liquid electrolyte. Almost the same long-term stability of the sealed cell using the gel electrolyte as that of the sealed cell using a liquid electrolyte was obtained.

Recently, Hayase and co-workers (Toshiba Co.) reported the construction of a highly efficient DSSC using a gel electrolyte [151,152]. The chemical structure of one of their gelator materials is shown in Fig. 17. The gelator was dissolved in the electrolyte at a high temperature and then, the gel solution was dropped onto the dye-coated TiO_2 photoelectrode surface and then cooled. The composition of electrolyte they reported is a mixture of a gelator (0.1 g), 1-methyl-3-propylimidazolium iodide (10 g), I_2 (0.1 g), and 1,2,4,5-tetrakisbromomethylbenzene (0.1 g). Gelation took place by condensed polymerization between the gelator material and 1,2,4,5-tetrakisbromomethylbenzene compounds. A high efficiency, 7.3% (J_{sc} = 17.6 mA/cm^2, V_{oc} = 0.60 V, and ff = 0.68), was obtained for an N3-dye-sensitized TiO_2 solar cell with the gel electrolyte under AM 1.5 irradiation. This value is almost the same as 7.8% for DSSCs, with liquid electrolyte solution. They concluded that the resistance of the gel electrolyte did not increase because no change of the fill factor was observed. The photocurrent increased linearly with increasing incident light intensity of up to 100 mW/cm^2, as well as the DSSC with a liquid electrolyte solution. This suggests that gelation of the electrolyte does not suppress diffusion of I^- and I_3^- ions in the electrolyte solution.

V. RESEARCH TOWARD COMMERCIALIZATION

A. Stability of the DSSC

In order to commercialize the DSSC successfully, the single cell and the integrated modules should have a long-term stability. In this section, we introduce studies of the photochemical, chemical, and physical stability of the DSSC, recent investigations concerning the long-term stability of DSSC are introduced also.

1. Photochemical and Physical Stability of the Cell Materials

The photochemical and thermal stabilities of Ru complexes have been investigated in detail [8,153–156]. For example, it has been reported that the NCS ligand of the N3 dye, cis-Ru(II)(dcbpy)$_2$(NCS)$_2$ (dcbpy = 2,2′-bipyridyl-4,4′-dicarboxylic acid), is oxidized to produce a cyano group (—CN) under irradiation in methanol solution. It was measured by both ultraviolet–visible (UV-vis) absorption spectroscopy and nuclear magnetic resonance (NMR) [8,153]. In addition, the intensity of the infrared (IR) absorption peak attributed to the NCS ligand starts to decrease at 135°C, and decarboxylation of N3 dyes occurs at temperatures above 180°C [155]. Desorption of the dye from the TiO$_2$ surface has been observed at temperatures above 200°C.

It has been considered that the high stability of the dye in a DSSC system could be obtained by the presence of I$^-$ ions as the electron donor to dye cations. Degradation of the NCS ligand to the CN ligand by a intramolecular electron-transfer reaction, which reduces consequently the Ru(III) state to the Ru(II) state, occurs within 0.1–1 sec [153], whereas the rate for the reduction of Ru(III) to Ru(II) by the direct electron transfer from I$^-$ ions into the dye cations is on the order of nanoseconds [30]. This indicates that one molecule of N3 dye can contribute to the photon-to-current conversion process with a turnover number of at least 10^7–10^8 without any degradation [153]. Taking this into consideration, N3 dye is considered to be sufficiently stable in the redox electrolyte under irradiation.

We should also consider the photoelectrochemical and chemical stability of the solvent. Organic solvents employed in DSSCs are, for example, propylene carbonate, acetonitrile, propionitrile, methoxyacetonitrile, methoxypropionitrile, and their mixtures. It is known that carbonate solvents, such as propylene carbonate, decompose under illumination, resulting in the formation of a carbon dioxide bubble in the cell. Methoxyacetonitrile (CH$_3$O–CH$_2$CN) reacts with trace water in the electrolyte to produce the corresponding amide (CH$_3$O–CH$_2$CONH$_2$), which decreases the conductivity of the electrolyte [157]. Acetonitrile and propionitrile are considered to be relatively stable, giving 2000 hr of stability under the dark condition at 60°C [157].

The stability of sputtered Pt on TCO substrate which is used as a counterelectrode has also been investigated. It was reported that the electrocatalytically active Pt layer did not seem to be chemically stable in an electrolyte solution of LiI, I$_2$, and methoxypropionitrile [158].

2. Long-Term Cell Stability

Grätzel and co-workers observed that a DSSC composed of a N3-dye-sensitized nanocrystalline TiO$_2$ photoelectrode has a good long-term stability, which is due to an effective photoinduced electron transfer from the Ru dye photosensitizer into the conduction band of the TiO_2 photoelectrode as well as from the iodine redox mediator to the oxidized dye photosensitizer. The turnover number of one dye photosensitizer molecule reaches 500 million, corresponding to continuous stability for 10 years under irradiation.

The long-term stability of the DSSC for commercial applications is currently being investigated at EPFL, Solaronix S.A. in Switzerland, the Netherlands Energy Research Foundation (ECN), INAP in Germany, and NIMC (the former AIST) [8,16,153, 157,159–161]. Such data are shown in Table 4. For example, 7000 hr of cell stability, which corresponds to 6 years of outdoor use, has been obtained under 1000 $\bar{W}/m^2$ with a UV cutoff filter, as shown in Fig. 18 [153]. Späth and co-workers conducted DSSC stability tests at Solaronix with polymer-sealed devices containing viscous electrolytes with a high boiling point, such as glutaronitrile [159]. They discovered that no significant degradation of stability occurred over a period of 9600 hr of continuous illumination at 35°C, indicating the chemical stability of components as well as the physical stability of polymer

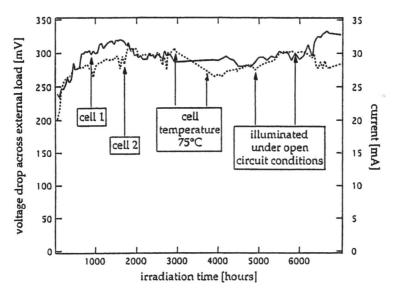

Figure 18 Stability test carried out with two sealed DSSCs over 7000 hr of continuous illumination with visible light (polycarbonate 395-nm cutoff filter) at 1000 W/m^2 light intensity. The photocurrent and voltage drop measured across an external load resistor of 10 Ω are plotted as a function of irradiation time. Cell 1 (solid line) was continuously illuminated at 35°C; the same for cell 2 (broken line) except that it was operated for a 700-hr period at 75°C and for 1000 hr at an open circuit. (From Ref. 153.)

Arakawa and Hara

Table 4 Test of Long-Term Stability of Dye-Sensitized TiO_2 Solar Cells Conducted at Various Institutions

Institute and Ref.	Dye	Test conditions	Components	Term	Results
EPFL [8]	N3	150 W W-halogen lamp UV cut off, 50°C	LiI/LiI$_3$ in PC or NMO	10 months	J_{sc}: 20–30% decreased initially. Passed charge: 10^5 C cm^{-2}, 10^7 turnovers
EPFL [15]	N3	800 W m^{-2} Xe lamp UV cut off	TBAI, I$_2$ in AN Surlyn + waterglass	100 days	J_{sc} increased and V_{oc} decreased for the first 20 days. The efficiency is constant for 100 days
EPFL [153]	N3	AM 1.5 (1000 W m^{-2}), UV cut off, 35°C	KI, I$_2$ in GN	7000 hrs	J_{sc} increased by 20–30% during the first 1000 hrs, thereafter reaching a plateau value
ECN, Solaronix [159]	N3	Fluorescent lamp (1000 W m^{-2}), UV cut off, 35°C	KI, I$_2$ in GN Surlyn 1702	9600 hrs	J_{sc} increased and V_{oc} decreased for the first 2000 hrs Passed charge: 103680 C cm^{-2}.
ECN, INAP, Solaronix [161]	N3	Sulphur lamp (2–3 sun), UV cut off, 20°C, $\eta = 2\%$	HMImI, LiI, I$_2$, TBP in MPN Surlyn	8300 hrs	V_{oc} decreased (50 mV) and J_{sc} increased
ECN, INAP, Solaronix [161]	N3	UV (10 mW cm^{-2}), 20°C	HMImI, MgI$_2$, I$_2$, TBP in AN Surlyn	1500 hrs	J_{sc} and V_{oc} were constant
INAP [157]	N3	Sulphur lamp (2.5 sun) UV cut off, 17°C	LiI, I$_2$, TBP in MPN	10000 hrs	J_{sc} was constant after initial decrease
NIMC-SOC	N3	AM 1.5 (1000 W m^{-2}) UV cut off, 20°C, $\eta = 5\%$	DMPImI, LiI, I$_2$, TBP in AN, PN, MPN	4500 hrs	J_{sc} decreased 5% and V_{oc} was constant 1.3×10^7 turnovers
NIMC	Merocyanine	AM 1.5 (1000 W m^{-2}) UV cut off, 20°C, $\eta = 3\%$	DMPImI, LiI, I$_2$ in MAN	1500 hrs	J_{sc} and V_{oc} were constant 1×10^6 turnovers

PC: propylene carbonate; NMO: 3-methyl-2-oxazolidinone; TBAI: tetrabutylammonium iodide; AN: acetonitrile; PN: propionitrile; GN: glutaronitrile; MAN: methoxyacetonitrile; MPN: 3-methoxypropionitrile; TBP: tert-butylpyridine; HMImI: 1-hexyl-3-methylimidazolium iodide; DMPImI: 1,2-dimethyl-3-propylimidazolium iodide; SOC: Sumitomo Osaka Cement Co. Ltd.

sealant. In addition, long-term stability of a small cell for more than 10,000 hr has also been accomplished under no UV light conditions at 17°C at 2.5 sun using an electrolyte of 0.5 M LiI, 0.05 M I$_2$, and 0.3 M TBP in methoxypropionitrile [157]. Stability tests under UV irradiation have also been conducted [161]. The addition of MgI$_2$ to the electrolyte can significantly improve stability to UV light, resulting in a stable photovoltaic performance for more than 1500 hr under UV irradiation [161]. The detailed mechanism of the UV-stabilizing effect due to MgI$_2$ has not been elucidated.

A sealed DSSC with a merocyanine dye photosensitizer (one of the organic dyes) also exhibited a good long-term stability in a preliminary test under continuous AM 1.5 irradiation with a 420-nm cutoff filter. We obtained a stability of more than 1500 hr, corresponding to a turnover number of more than 10 million, as shown in Table 4 [136].

These results strongly indicate that DSSCs show sufficient physical and chemical stability during extended periods of illumination. Nevertheless, stability tests at high temperatures and high humidity must be carried out for outdoor applications.

B. Module Fabrication and Other Issues Pertinent to Commercialization

The sheet resistance of transparent conducting oxide (TCO) substrates (i.e., FTO) is relatively high, which makes the limitation on enlarging a single cell size of DSSC only up to 1 cm^2. Increased sheet resistance of the TCO substrate on scale-up of the DSSC leads to loss of efficiency, especially the fill factor. Therefore, scale-up of the DSSC using a modular system has been investigated [162,163]. A module consists of many unit cells with two TCO glass substrates coated with TiO$_2$ or platinum, the electrolyte solutions, and materials for interconnections. The electrolyte solution is composed of iodine, iodide, both of which could dissolve metal materials, and organic solvent. Therefore, a standard conductor such as silver wire and plate cannot be utilized or it has to be coated by protecting materials. In addition, the use of a liquid electrolyte requires a perfect seal in between the two electrodes as well as the outside of the module. The use of a glass frit for sealing the electrolyte solution has been investigated [162]. An efficiency of 7% was achieved using a module consisting of 12 interconnected cells with a total area of 112 cm^2 (7.6% for a 3-cm^2 cell and 8% for a 1-cm^2 cell) [162]. A continuous process for the fabrication of a monolithic series connecting DSSC modules using laser scribing has been proposed by Kay and Grätzel, as shown in Fig. 19 [16].

Recently, polymer substrates instead of glass have been utilized in constructing DSSC. This expands possible commerical applications [164–167]. Polymer substrates allow roll-to-roll production, which can achieve high throughput.

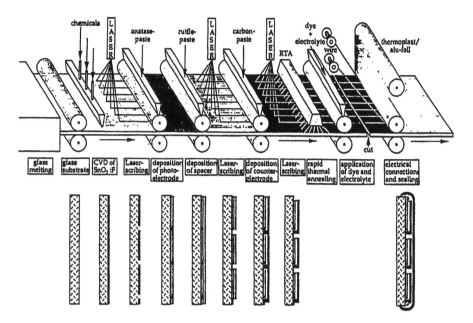

Figure 19 Continuous process for the fabrication of monolithic series connecting DSSC modules. (From Ref. 16.)

When a polymer film is used as a substrate, aqueous TiO₂ paste without organic surfactants is sintered at relatively low temperatures, with approximately 150°C being sufficient to produce mechanically stable TiO₂ films. Sommeling et al. at ECN used an ITO-coated poly(ethylene terephthalate) (PET) film as a substrate and prepared a plastic DSSC [164–167]. A cell performance with a J_{sc} of 15 μA/cm², V_{oc} of 0.48 V, and ff of 0.67 was obtained at an illumination intensity of 250 lux. This performance is sufficient for a power supply for indoor applications such as watches and calculators. Under AM 1.5 irradiation, a V_{oc} of 0.7 V and J_{sc} of 2 mA/cm² were obtained.

Recently, the DSSC has been used for an educational demonstration of solar energy-to-electricity conversion system because of its simple fabrication [12,168]. Students could purchase DSSC kits including all components such as TCO-coated glass, TiO₂ electrodes, blackberries (i.e., dye), and electrolyte solution [169,170] and easily demonstrate an artificial photosynthetic process. For detailed studies, it is possible to purchase raw materials, including Ru dye photosensitizers, TiO₂ paste, and sealing materials, from Solaronix S. A. [171].

VI. SUBJECTS FOR THE FUTURE

Since Grätzel and co-workers reported the development of highly efficient and novel DSSCs in 1991, researchers and engineers worldwide have intensively investigated the working mechanism, new materials, and commercialization for DSSC. A maximum efficiency of 10.4% has been obtained under AM1.5 in the laboratory in 1997. In addition, satisfactory long-term stability of sealed cells has been achieved under relatively mild test conditions such as low temperatures (65°C) and no UV exposure. It might be possible to achieve commercial DSSC production in the near future for indoor applications, such as calculators and watches. For expanded commercial applications, however, there are several problems which confront us. Overcoming these problems greatly brings DSSC close to expanded commercialization.

A. Improvement of Solar Energy Conversion Efficiency

In Japan, the DSSC is recognized as one of candidates for second-generation solar cells coming after conventional solar cells such as single-crystal Si cells in the market. The second-generation solar cells are expected to be able to provide electricity at a cost of 10–15 yen/kWh, which is the present cost of electricity coming from fossil fuel in Japan. We estimated the efficiency which is possible to provide such an inexpensive electricity by a DSSC. An efficiency of 15% is required to satisfy such conditions in Japan. In order to attain 15% efficiency, both J_{sc} and V_{oc} performances must be improved. Expansion of absorption area of dye photosensitizers into the near-IR region is necessary for improving J_{sc}. The absorption property of black dye, whose absorption threshold is close to 920 nm, is thought to be the optimal threshold for single-junction solar cells like GaAs solar cells. The development of new dye photosensitizers which are able to absorb in the near-IR region with higher absorption coefficients is also expected. Improvement of the IPCE performance of solar cells sensitized by black dye (or new dye photosensitizers) in the 700–900-nm range, which is relatively lower than that in the 500–600-nm range, would increase the J_{sc} from 20 to about 28 mA/cm^2. This would result in an overall efficiency of 15%. As described in Section III. B, the increase of the absorption coefficient of dye photosensitizers and light-scattering effect of the semiconductor film would improve the IPCE performance in the long-wavelength region. In addition, the energy gaps, ΔE_1 and ΔE_2 as driving forces for efficient electron-transfer processes as shown in Fig. 7 are considered to be some part of loss of energy absorbed by dye photosensitizers. If we could construct a good system performing with smaller energy gaps, ΔE_1 and ΔE_2, owing to the best molecular design of the dye photosensitizers, the efficiency would be improved. Improvement of V_{oc} is also very important for achievement of higher efficiency. As shown earlier, the decrease in V_{oc} is mainly attributed to recombination between injected electrons and the redox mediator, I_3^- (dark current). The molecular design for efficient dye photosensitizers and screening of adsorbates to block dark current, such as TBP and cholic acid, would improve the V_{oc}. The development of new semiconductor materials having a conduction-band level that is more negative than that of TiO_2 and new redox mediators having redox potentials that are more positive than that of the I^-/I_3^- redox would also increase the V_{oc} of the DSSC. At present, however, good semiconductor materials and redox mediator exceeding the combination between the TiO_2 electrode and the iodine redox have not been found.

B. Long-Term Stability for Outdoor Applications

As described in Section V, A. 2, satisfactory long-term stability of sealed cells has been already achieved under relatively mild test conditions (low temperatures and no UV exposure). For outdoor applications, additional stability tests under more rigorous conditions will be required (e.g., high temperature such as 80°C, high humidity in the atmosphere, and UV exposure).

C. Solid Electrolyte

As described in Section IV.D, the development of solid-state electrolytes is important for developing DSSCs with long-term stability and it is critical for commercialization. Several organic and inorganic materials, such as OMeTAD, polypyrrole, CuSCN, and CuI, have been investigated as solid electrolytes. However, at the present time, the performance of solid-state DSSCs using these materials are inferior to that for DSSC with liquid electrolytes at present. It is very difficult to form a good solid–solid interface, which can provide a high solar cell performance, because a nanoporous TiO_2 electrode gives a high surface area which should be contacted with a solid-state electrolyte. On the other hand, liquid electrolytes can easily penetrate into the nanoporous TiO_2 electrode. One of factors determining the high performance of DSSCs is an easy formation of a solid–liquid interface between the nanoporous TiO_2 electrode and a liquid electrolyte. Therefore, the electron-transfer process occurs mostly at the interface. Low electron (or hole) conductivity of the solid electrolyte materials would also decreases low performance of solid-state DSSCs. CuI is one of the attractive materials for a solid electrolyte in a DSSC and 4% efficiency was attained using the TiO_2/dye/CuI system. For commercial applications, improvement of the cell performance and long-term stability tests of the cell are required because stability of the CuI system has not been clarified. The development of new solid electrolyte materials is expected.

REFERENCES

1. Gerischer, H.; Tributsch, H. *Ber. Bunsen. Phys. Chem.* **1968**, *72*, 437–445.
2. Tributsch, H.; Gerischer, H. *Ber. Bunsen. Phys. Chem.* **1969**, *73*, 251–260.

3. Tsubomura, H.; Matsumura, M.; Nomura, Y.; Amamiya, T. *Nature* **1976**, *261*, 402.

4. Matsumura, M. Ph,D. Thesis, Osaka University, 1979.

5. Anderson, S.; Constable, E. C.; Dare-Edwards, M. P.; Goodenough, J. B.; Hammett, A.; Seddon, K.; Wright, R. D. *Nature* **1979**, *280*, 571–573.

6. Dare-Edwards, M. P.; Goodenough, J. B.; Hamnett, A.; Seddon, K. R.; Wright, R. D. *Faraday Discuss. Chem. Soc.* **1980**, *70*, 285–298.

7. O'Regan, B.; Grätzel, M. *Nature* **1991**, *353*, 737–740.

8. Nazeeruddin, M. K.; Kay, A.; Rodicio, I.; Humphry-Baker, R.; Muller, E.; Liska, P.; Vlachopoulos, N.; Grätzel, M. *J. Am. Chem. Soc.* **1993**, *115*, 6382–6390.

9. Nazeeruddin, M. K.; Péchy, P.; Grätzel, M. *Chem. Commun.* **1997**, 1705–1706.

10. Nazeeruddin, M. K.; Péchy, P.; Renouard, T.; Zakeeruddin, S. M.; Humphry-Baker, R.; Comte, P.; Liska, P.; Cevey, L.; Costa, E.; Shoklover, V.; Spiccia, L.; Deacon, G. B.; Bignozzi, C. A.; Grätzel, M. *J. Am. Chem. Soc.* **2001**, *123*, 1613–1624.

11. Smestad, G.; Bignozzi, C. A.; Aragazzi, R.; *Solar Energy Mater. Solar Cells* **1994**, *32*, 259–272.

12. Smestad, G. *Solar Energy Mater. Solar Cells* **1998**, *55*, 157–178.

13. Kalyanasundaram, K.; Grätzel, M. In *Photosensitization and Photocatalysis Using Inorganic and Organometallic Compounds*; Kalyanasundaram, K.; Grätzel, M., Eds. Kluwer Academic: Dordrecht, **1993**, pp. 247–271.

14. Smestad, G. *Solar Energy Mater. Solar Cells* **1994**, *32*, 273–288.

15. Hagfeldt, A.; Grätzel, M. *Chem. Rev.* **1995**, *95*, 49–68.

16. Kay, A.; Grätzel, M. *Solar Energy Mater. Solar Cells* **1996**, *44*, 99–117.

17. Bignozzi, C. A.; Schoonover, J. R.; Scandola, F. In *Molecular Level Artificial Photosynthetic Materials*; Meyer, G. J.6, Ed. Wiley: New York, **1997**, pp. 1–97.

18. Kalyanasundaram, K.; Grätzel, M. *Coord. Chem. Rev.* **1998**, *77*, 347–414.

19. Grätzel, M. *Curr. Opin. Colloid Interf. Sci.* **1999**, *4*, 314–321.

20. Hagfeldt, A.; Grätzel, M. *Acc. Chem. Res.* **2000**, *33*, 269–277.

21. Argazzi, R.; Bignozzi, C. A.; Heimer, T. A.; Castellano, F. N.; Meyer, G. J. *Inorg. Chem.* **1994**, *33*, 5741–5749.

22. Finnie, K. S.; Bartlett, J. R.; Woolfrey, J. L. *Langmuir* **1998**, *14*, 2744–2749.

23. Nazeeruddin, M. K.; Amirnasr, M.; Comte, P.; Mackay, J. R.; McQuillane, A. J.; Houriet, R.; Grätzel M. *Langmuir* **2000**, *16*, 8525–8528.

24. Murakoshi, K.; Kano, G.; Wada, Y.; Yanagida, S.; Miyazaki, H.; Matsumoto, M.; Murasawa, S. *J. Electroanal. Chem.* **1995**, *396*, 27–34.

25. Sayama, K.; Sugihara, H.; Arakawa, H. *Chem. Mater.* **1998**, *10*, 3825–3832.

26. Liu, Y.; Hagfeldt, A.; Xiao, X.-R.; Lindquist, S.-E. *Solar Energy Mater. Solar Cells* **1998**, *55*, 267–281.

27. Hara, K.; Horiguchi, T.; Kinoshitra, T.; Sayama, K.; Arakawa, H. *Solar Energy Mater. Solar Cells* **2001**, *70*, 151–161.

28. Vlachopoulos, N.; Liska, P.; Augustynski, J.; Grätzel, M. *J. Am. Chem. Soc.* **1998**, *110*, 1216–1220.

29. Zaban, A.; Ferrere, S.; Gregg, B. A. *J. Phys. Chem. B* **1998**, *102*, 452–460.

30. Tachibana, Y.; Moser, J. E.; Grätzel, M.; Klug, D. R.; Durrant, J. R. *J. Phys. Chem.* **1996**, *100*, 20,056–20,062.

31. Rehm, J. M.; McLendon, G. L.; Nagasawa, Y.; Yoshihara, K.; Moser, J.; Grätzel, M. *J. Phys. Chem.* **1996**, *100*, 9577–9578.

32. Kamat, P. V.; Bedja, I.; Hotchandani, S.; Patterson, L. K. *J. Phys. Chem.* **1996**, *100*, 4900–4908.

33. Nasr, C.; Liu, D.; Hotchandani, S.; Kamat, P. V.; *J. Phys. Chem.* **1996**, *100*, 11,054–11,061.

34. Hannappel, T.; Burfeindt, B.; Storck, W.; Willig, F. *J. Phys. Chem. B* **1997**, *101*, 6799–6802.

35. Heimer, T. A.; Heilweil, E. J. *J. Phys. Chem. B* **1997**, *101*, 10,990–10,993.

36. Cherepy, N. J.; Smestad, G.; Grätzel, M.; Zhang, J. Z. *J. Phys. Chem. B* **1997**, *101*, 9342–9351.

37. Murakoshi, K.; Yanagida, S.; Capel, M.; Castner, E. W., Jr. In *Interfacial Electron Transfer Dynamics of Photosensitized Zinc Oxide Nanoclusters*; Moskovits, M.; Ed. American Chemical Society: Washington, DC, **1997**, pp. 221–238.

38. Ellingson, R. J.; Asbury, J. B.; Ferrere, S.; Ghosh, H. N.; Sprangue, J. R.; Lian, T.; Nozik, A. J. *J. Phys. Chem. B* **1998**, *102*, 6455–6458.

39. Asbury, J. B.; Wang, Y. Q.; Lian, T. *J. Phys. Chem. B* **1999**, *103*, 6643–6647.

40. Heimer, T. A.; Heilweil, E. J.; Bignozzi, C. A.; Meyer, G. J. *J. Phys. Chem. A* **2000**, *104*, 4256–4262.

41. Tachibana, Y.; Haque, S. A.; Mercer, I. P.; Durrant, J. R.; King, D. R. *J. Phys. Chem. B* **2000**, *104*, 1198–1205.

42. Huber, R.; Spörlen, S.; Moser, J. E.; Grätzel, M.; Wachtveitl, J. *J. Phys. Chem. B* **2000**, *104*, 8995–9003.

43. Iwai, S.; Hara, K.; Murata, S.; Katoh, R.; Sugihara, H.; Arakawa, H. *J. Chem. Phys.* **2000**, *113*, 3366–3373.

44. Willig, F. In *Surface Electron Transfer Processes*; Miller, J. R. Eds. VCH: New York, **1995**, pp. 167–309.

45. Moser, J. E.; Grätzel, M. *Chimica* **1998**, *52*, 160–162.

46. Asbury, J. B.; Hao, E.; Wang, Y.; Lian, T. *J. Phys. Chem. B* **2000**, *104*, 11,957–11,964.

47. Asbury, J. B.; Hao, E.; Wang, Y.; Ghosh, H. N.; Lian, T. *J. Phys. Chem. B* **2001**, *105*, 4545–4557.

48. Haque, S. A.; Tachibana, Y.; Klug, D. R.; Durrant, J. R. *J. Phys. Chem. B* **1998**, *102*, 1745–1749.

49. Haque, S. A.; Tachibana, Y.; Willis, R. L.; Moser, J. E.; Grätzel, M.; Klug, D. R.; Durrant, J. R. *J. Phys. Chem. B* **2000**, *104*, 538–547.

50. Kuciauskas, D.; Freund, M. F.; Gray, H. B.; Winkler, J. R.; Lewis, N. S. *J. Phys. Chem. B* **2001**, *105*, 392–403.

51. Nasr, S.; Hotchandani, Kamat, S; P. V. *J. Phys. Chem. B* **1998**, *102*, 4944–4951.

52. Huang, S. Y.; Schlichthörl, G.; Nozik, A. J.; Grätzel, M.; Frank, A. J. *J. Phys. Chem. B* **1997**, *101*, 2576–2582.

53. Kumar, A.; Santangelo, P. G.; Lewis, N. S. *J. Phys. Chem.* **1992**, *96*, 834–842.

54. Stanley, A.; Matthews, D. *Austr. J. Chem.* **1995**, *48*, 1293–1300.

55. Matthews, D.; Infelta, P. P.; Grätzel, M. *Solar Energy Mater. Solar Cells* **1996**, *44*, 119–155.

56. Salafsky, J. S.; Lubberhuizen, W. H.; van Faassen, E.; Schropp, R. E. I. *J. Phys. Chem. B* **1998**, *102*, 766–769.

57. Duffy, N. W.; Peter, L. M.; Rajapakse, R. M. G.; Wijayantha, K. G. U. *J. Phys. Chem. B* **2000**, *104*, 8916–8919.
58. Cahen, D.; Hodes, G.; Grätzel, M.; Guillemoles, J. F.; Riess, I. *J. Phys. Chem. B* **2000**, *104*, 2053–2059.
59. Gregg, B. A.; Pichot, F.; Ferrere, S.; Fields, C. L. *J. Phys. Chem. B* **2001**, *105*, 1422–1429.
60. Cao, F.; Oskam, G.; Meyer, G. J.; Searson, P. C. *J. Phys. Chem. B* **1996**, *100*, 17,021–17,027.
61. Solbrand, A.; Lindström, H.; Rensmo, H.; Hagfeldt, A.; Lindquist, S.-E. *J. Phys. Chem. B* **1997**, *101*, 2514–2518.
62. Solbrand, A.; Henningsson, A.; Södergren, S.; Lindström, H.; Hagfeldt, A.; Lindquist, S.-E. *J. Phys. Chem. B* **1999**, *103*, 1078–1083.
63. Sommeling, P. M.; Rieffe, H. C.; van Roosmalen, J. A. M.; Schönecker, A.; Kroon, J. M.; Wienke, J. A.; Hinsch, A. *Solar Energy Mater. Solar Cells* **2000**, *62*, 399–410.
64. Hoyer, P.; Weller, H. *J. Phys. Chem. B* **1995**, *99*, 14,096–14,100.
65. Schwarzburg, K.; Willig, F. *Appl. Phys. Lett.* **1991**, *58*, 2520–2522.
66. de Jongh, P. E.; Vanmaekelbergh, D. *Phys. Rev. Lett.* **1996**, *77*, 3427–3430.
67. de Jongh, P. E.; Vanmaekelbergh, D. *J. Phys. Chem. B* **1997**, *101*, 2716–2722.
68. Könenkamp, R.; Henniger, R.; Hoyer, P. *J. Phys. Chem.* **1993**, *97*, 7328–7330.
69. Wahl, A.; Augustynski, J. *J. Phys. Chem. B* **1998**, *102*, 7820–7828.
70. Nelson, J. *Phys. Rev. B* **1998**, *59*, 15,374–15,380.
71. van de Lagemaat, J.; Park, N.-G.; Frank, A. J. *J. Phys. Chem. B* **2000**, *104*, 2044–2052.
72. Kopidakis, N.; Schiff, E. A.; Park, N.-G.; van de Lagemaat, J.; Frank, A. J. *J. Phys. Chem. B* **2000**, *104*, 3930–3936.
73. van de Lagemaat, J.; Frank, A. J. *J. Phys. Chem. B* **2000**, *104*, 4292–4294.
74. Barbé, C. J.; Arendse, F.; Comte, P.; Jirousek, M.; Lenzmann, F.; Shklover, V.; Grätzel, M. *J. Am. Ceram. Soc.* **1997**, *80*, 3157–3171.
75. Nippon Aerosil, Minato-ku, Tokyo 107, Japan. [Catalogue].
76. Park, N.-G.; van de Lagemaat, J.; Frank, A. J. *J. Phys. Chem. B* **2000**, *104*, 8989–8994.
77. Sonics and Materials Inc., Newton, CT, USA. [Catalogue].
78. Usami, A. *Chem. Phys. Lett.* **1997**, *277*, 105–108.
79. Ferber, J.; Luther, J. *Solar Energy Mater. Solar Cells* **1998**, *54*, 265–275.
80. Rothenberger, G.; Comte, P.; Grätzel, M. *Solar Energy Mater. Solar Cells* **1999**, *58*, 321–336.
81. Bonhôte, P.; Dias, A.-P.; Papageorgiou, N.; Kalyanasundaram, K.; Grätzel, M. *Inorg. Chem.* **1996**, *35*, 1168–1178.
82. Papageorgiou, N.; Athanassov, Y.; Armand, M.; Bonhote, P.; Pettersson, H.; Azam, A.; Grätzel, M. *J. Electrochem. Soc.* **1996**, *143*, 3099–3108.
83. Papageorgiou, N.; Maier, W. F.; Grätzel, M. *J. Electrochem. Soc.* **1997**, *144*, 876–884.
84. Redmond, G.; Fitzmaurice, D.; Grätzel, M. *Chem. Mater.* **1994**, *6*, 686–691.
85. Rensmo, H.; Keis, K.; Lindström, H.; Södergren, S.; Solbrand, A.; Hagfeldt, A.; Lindquist, S.-E. *J. Phys. Chem. B* **1997**, *101*, 2598–2601.

86. Rao, T. N.; Bahadur, L. *J. Electrochem. Soc.* **1997**, *144*, 179–185.
87. Keis, K.; Lindgren, J.; Lindquist, S.-E.; Hagfeldt, A. *Langmuir* **2000**, *16*, 4688–4694.
88. Hara, K.; Horiguchi, T.; Kinoshita, T.; Sayama, K.; Arakawa, H. *Chem. Lett.* **2000**, 316–317.
89. Hara, K.; Horiguchi, T.; Kinoshita, T.; Sayama, K.; Sugihara, H.; Arakawa, H. *Solar Energy Mater. Solar Cells* **2000**, *64*, 115–134.
90. Equchi, K.; Koga, H.; Sekizawa, K.; Sasaki, K. *J. Ceram. Soc. Jpn.* **2000**, *108*, 1067–1071.
91. Nasr, S.; Kamat, P. V.; Hotchandani, S. *J. Phys. Chem. B* **1998**, *102*, 10,047–10,056.
92. Burnside, S.; Moser, J. E.; Brooks, K.; Grätzel, M. *J. Phys. Chem. B* **1999**, *103*, 9328–9332.
93. He, J.; Lindström, H.; Hagfeldt, A.; Lindquis, S.-E. *J. Phys. Chem. B* **1999**, *103*, 8940–8943.
94. Tennakone, K.; Kumara, G. R. R. A.; Kottegoda, I. R. M.; Perera, V. P. S. *Chem. Commun.*, **1999**, 15–16.
95. Tennakone, K.; Senadeera, G. K. R.; Perera, V. P. S.; Kottegoda, I. R. M.; De Silva, L. A. A. *Chem. Mater.* **1999**, *11*, 2474–2477.
96. Zaban, A.; Chen, S. G.; Chappel, S.; Gregg, B. A. *Chem. Commun.* **2000**, 2231–2232.
97. Wang, Z.-S.; Huang, C.-H.; Huang, Y.-Y.; Hou, Y.-J.; Xie, P.-H.; Zhang, B.-W.; Cheng, H.-M. *Chem. Mater.* **2001**, *13*, 678–682.
98. Heimer, T. A.; Bignozzi, C. A.; Meyer, G. J. *J. Phys. Chem.* **1993**, *97*, 11,987–11,994.
99. Argazzi, R.; Bignozzi, C. A.; Heimer, T. A.; Castellano, F. N.; Meyer, G. J. *Inorg. Chem.* **1994**, *33*, 5741–5749.
100. Nazeeruddin, M. K.; Muller, E.; Humphry-Baker, R.; Vlachopoulos, N.; Grätzel, M. *J. Chem. Soc., Dalton Trans.* **1997**, 4571–4578.
101. Schklover, V.; Nazeeruddin, M.-K.; Zakeeruddin, S. M.; Barbe, C.; Kay, A.; Haibach, T.; Steurer, W.; Hermann, R.; Nissen, H.-U.; Grätzel, M. *Chem. Mater.* **1997**, *9*, 430–439.
102. Argazzi, R.; Bignozzi, C. A.; Hasselmann, G. M.; Meyer, G. J. *Inorg. Chem.* **1998**, *37*, 4533–4537.
103. Liska, P.; Vlachopoulos, N.; Nazeeruddin, M. K.; Comte, P.; Grätzel, M. *J. Am. Chem. Soc.* **1998**, *110*, 3686–3687.
104. Shklover, V.; Ovchinnikov, Y. E.; Braginsky, L. S.; Zakeeruddin, S. M.; Grätzel, M. *Chem. Mater.* **1998**, *10*, 2533–2541.
105. Ruile, S.; Kohle, O.; Pettersson, H.; Grätzel, M. *New J. Chem.* **1998**, 25–31.
106. Zakeeruddin, S.; Nazeeruddiin, M. K.; Humphry-Baker, R.; Grätzel, M. *Inorg. Chem.* **1998**, *37*, 5251–5259.
107. Jing, B.; Zhang, H.; Zhang, M.; Lu, Z.; Shen, T. *J. Mater. Chem.* **1998**, *8*, 2055–2060.
108. Sugihara, H.; Singh, L. P.; Sayama, K.; Arakawa, H.; Nazeeruddin, M. K.; Grätzel, M. *Chem. Lett.* **1998**, 1005–1006.
109. Lees, A. C.; Evrard, B.; Keyes, T. E.; Vos, J. G.; Kleverlaan, C. J.; Alebbi, M.; Bignozzi, C. A. *Eur. J. Inorg. Chem.* **1999**, 2309–2317.

110. Thompson, D. W.; Kelly, C. A.; Farzad, F.; Meyer, G. J. *Langmuir* **1999**, *15*, 650–653.

111. Islam, A.; Hara, K.; Singh, L.-P.; Kato, R.; Yanagaida, M.; Murata, S.; Takahashi, Y.; Sugihara, H.; Arakawa, H. *Chem. Lett.* **2000**, 490–491.

112. Yanagida, M.; Singh, L.-P.; Sayama, K.; Hara, K.; Katoh, R.; Islam, A.; Sugihara, H.; Arakawa, H. *J. Chem. Soc., Dalton Trans.* **2000**, 2817–2822.

113. Schwarz, O.; van Loyen, D.; Jockusch, S.; Turro, N. J.; Durr, H. *J. Photochem. Photobiol. A: Chem.* **2000**, *132*, 91–98.

114. Hara, K.; Sugihara, H.; Yanagida, M.; Sayama, K.; Arakawa, H. *Langmuir* **2001**, *17*, 5992–5999.

115. Takahashi, Y.; Arakawa, H.; Sugihara, H.; Hara, K.; Islam, A.; Katoh, R.; Tachibana, Y.; Yanagida, M. *Inorg. Chim. Acta* **2000**, *310*, 169–174.

116. Aranyos, V.; Grennberg, H.; Tingry, S.; Lindquist, S.-E.; Hagfeldt, A. *Solar Energy Mater. Solar Cells* **2000**, *64*, 97–114.

117. Ferrere, S.; Gregg, B. A. *J. Am. Chem. Soc.* **1998**, *120*, 843–844.

118. Yang, M.; Thompson, D. W.; Meyer, G. J. *Inorg. Chem.* **2000**, *39*, 3738–3739.

119. Alebbi, M.; Bignozzi, C. A.; Heimer, T. A.; Hasselmann, G. M.; Meyer, G. J. *J. Phys. Chem. B* **1998**, *102*, 7577–7581.

120. Trammell, S. A.; Meyer, T. J. *J. Phys. Chem. B* **1999**, *103*, 104–107.

121. Sauvé, G.; Stephen, M. E.; Doig, J.; Lauermann, I.; Pomykal, K.; Lewis, N. S. *J. Phys. Chem. B* **2000**, *104*, 3488–3491.

122. Sauvé, G.; Cass, M. E.; Coia, G.; Doig, S. J.; Lauermann, I.; Pomykal, K. E.; Lewis, N. S. *J. Phys. Chem. B* **2000**, *104*, 6821–6836.

123. Hasselmann, G. M.; Meyer, G. J. *Z. Phys. Chem.* **1999**, *212*, 39–44.

124. Islam, A.; Sugihara, H.; Arakawa, H. *New J. Chem.* **2000**, *24*, 343–345.

125. Kay, A.; Grätzel, M. *J. Phys. Chem.* **1993**, *97*, 6272–6277.

126. Kay, A.; Humphry-Baker, R.; Grätzel, M. *J. Phys. Chem.* **1994**, *98*, 952–959.

127. Boschloo, G. K.; Goossens, A. *J. Phys. Chem.* **1996**, *100*, 19,489–19,494.

128. Tennakone, K.; Kumara, G. R. R. A.; Wijayantha, K. G. U.; Kottegoda, I. R. M.; Perera, V. P. S.; Aponsu, G. M. L. P. *J. Photochem. Photobiol. A: Chem.* **1997**, *108*, 175–177.

129. Nazeeruddin, M. K.; Humphry-Baker, R.; Grätzel, M.; Murrer, B. A. *Chem. Commun.* **1998**, 719–720.

130. Ferrere, S.; Zaban, A.; Gregg, B. A. *J. Phys. Chem.* **1997**, *101*, 4490–4493.

131. Sayama, K.; Sugino, M.; Sugihara, H.; Abe, Y.; Arakawa, H. *Chem. Lett.* **1998**, 753–754.

132. Khazraji, A. C.; Hotchandani, S.; Das, S.; Kamat, P. V. *J. Phys. Chem. B* **1999**, *103*, 4693–4700.

133. Wang, Z.-S.; Li, F.-Y.; Huang, C.-H. *Chem. Commun.* **2000**, 2063–2064.

134. Wang, Z.-S.; Li, F.-Y.; Huang, C.-H.; Wang, L.; Wei, M.; Jin, L.-P.; Li, N.-Q. *J. Phys. Chem. B* **2000**, *104*, 9676–9682.

135. Sayama, K.; Hara, K.; Mori, N.; Satsuki, M.; Suga, S.; Tsukagoshi, S.; Sugihara, H.; Arakawa, H. *Chem. Commun.* **2000**, 1173–1174.

136. Sayama, K.; Tsukagoshi, S.; Hara, K.; Ohga, Y.; Sinpou, A.; Abe, Y.; Suga, S.; Arakawa, H. *J. Phys. Chem. B* **2002**, *106*, 1363–1371.

137. Sayama, K.; Hara, K.; Ohga, Y.; Sinpou, A.; Suga, S.; Arakawa, H. *New J. Chem.* **2001**, *25*, 200–202.

138. Hara, K.; Sayama, K.; Ohga, Y.; Sinpou, A.; Suga, S.; Arakawa, H. *Chem. Commun.* **2001**, 569–570.

139. Gao, F. G.; Bard, A. J.; Kispert, L. D. *J. Photochem. Photobiol. A: Chem.* **2000**, *130*, 49–56.

140. Tennakone, K.; Kumarasinghe, A. R.; Kumara, G. R. R. A.; Wijayantha, K. G. U.; Sirimanne, P. M. *J. Photochem. Photobiol. A: Chem.* **1997**. *108*, 193–195.

141. Tennakone, K.; Kumara, G. R. R. A.; Kottegoda, I. R. M.; Perera, V. P. S.; Weerasundara, P. S. R. S. *J. Photochem. Photobiol. A: Chem.* **1998**, *117*, 137–142.

142. Matsumoto, H.; Matsuda, T.; Tsuda, T.; Hagiwara, R.; Ito, Y.; Miyazaki, Y. *Chem. Lett.* **2001**, 26–27.

143. Bach, U.; Lupo, D.; Comte, P.; Moser, J. E.; Weissortel, F.; Salbeck, J.; Spreitzer, H.; Grätzel, M. *Nature* **1998**, *395*, 583–585.

144. Bach, U.; Tachibana, Y.; Moser, J. E.; Haque, S. A.; Durrant, J. R.; Grätzel, M.; Klug, D. R. *J. Am. Chem. Soc.* **1999**, *121*, 7445–7446.

145. Tennakone, K.; Kumara, G. R. R. A.; Kottegoda, I. R. M.; Wijayantha, K. G. U.; Perera, V. P. S. *J. Phys. D: Appl. Phys.* **1998**, *31*, 1492–1496.

146. Tennakone, K.; Perera, U. P. S. Kottegoda, I. R. M.; Kumara, G. R. R. A. *J. Phys. D: Appl. Phys.* **1999**, *32*, 374–379.

147. O'Regan, B.; Schwartz, D. T. *J. Appl. Phys.* **1996**, *80*, 4749–4754.

148. Kumara, G. R. R. A.; Konno, A.; Senadeera, G. K. R.; Jayaweera, P. V. V.; De Silva, D. B. R. A.; Tennakone, K. *Solar Energy Mater. Solar Cells* **2001**, *69*, 195–199.

149. Murakoshi, K.; Kogure, R.; Wada, Y.; Yanagida, S. *Solar Energy Mater. Solar Cells* **1998**, *55*, 113–125.

150. Kubo, W.; Murakoshi, K.; Wada, Y.; Hanabusa, K.; Shirai, H.; Yanagida, S. *Chem. Lett.* **1998**, 1241–1241.

151. Mikoshiba, S.; Sumino, H.; Yonetsu, M.; Hayase, S.; Proceedings of the 16th European Photovoltaic Solar Energy Conference, **2000**.

152. Mikoshiba, S.; Sumino, H.; Yonetsu, M.; Hayase, S. Japanese Patent Application 160427, **2001**.

153. Kohle, O.; Grätzel, M.; Meyer, A. F.; Meyer, T. B. *Adv. Mater.* **1997**, *9*, 904–906.

154. Durham, B.; Casper, J. V.; Nagle, J. K.; Meyer, T. J. *J. Am. Chem. Soc.* **1982**, *104*, 4803–4810.

155. Amirnasr, M.; Nazeeruddin, M. K.; Grätzel, M. *Thermochim. Acta* **2000**, *348*, 105–114.

156. Grünwald, R.; Tributsch, H. *J. Phys. Chem. B* **1997**, *101*, 2564–2575.

157. Kern, R. *Opto-Electron. Rev.* **2000**, *8*, 284–288.

158. Olsen, E.; Hagen, G.; Lindquist, S.-E. *Solar Energy Mater. Solar Cells* **2000**, *63*, 267–273.

159. Späth, M.; Kroon, J. M.; Sommeling, P. M.; Wienke, J. A.; van Roosmalen, J. A. M.; Meyer, T. B.; Meyer, A. F.; Kohle, O. Proceedings of the 14th European Photovoltaic Solar Energy Conference 1997.

160. Rijnberg, E.; Kroon, J. M.; Wienke, J.; Hinsch, A.; van Roosmalen, J.; Scholtens, B.; de Vries, J. G.; de Koster, C. G.; Duchateus, A.; Maes, I.; Hinderickx, H.

Proceedings of the 2nd World Conference and Exhibition on Photovoltaic Solar Energy Conversion, 1998.

161. Hinsch, A.; Kinderman, R.; Späth, M.; Rijnberg, E.; van Roosmalen, J. A. M. Proceedings of the 16th European Photovoltaic Solar Energy Conference, 2000.
162. Hanke, K. P. *Z. Phys. Chem.* **1999**, *212*, 1–9.
163. Pettersson, H.; Gruszecki, T.; *Solar Energy Mater. Solar Cells* **2001**, *70*, 203–212.
164. Sommeling, P. M.; Späth, M.; van Roosmalen, J. A. M. Proceedings of the 2nd World Conference and Exhibition on Photovoltaic Solar Energy Conversion 1998.
165. Sommeling, P. M.; Späth, M.; van Roosmalen, J. A. M.; Meyer, T. B.; Meyer, A. F. Proceedings of the 16th European Photovoltaic Solar Energy Conference, 2000.
166. Sommeling, P. M.; Späth, M.; Kroon, J.; van Roosmalen, J. A. M. the 13th International Conference on Photochemical Conversion and Storage of Solar Energy, 2000.
167. Hagfeldt, A. Abstract of the 4th NIMC International Symposium on Photoreaction Control and Photofunctional Materials, 2001.
168. Smestad, G. P.; Grätzel, M. *J. Chem. Educ.* **1998**, *75*, 752–756.
169. http://www.solideas.com/solrcell/cellkit.html.
170. http://www.nisinoda-electronics.co.jp/.
171. http://www.solaronix.ch/.

5

Spectroscopy and Dynamics of Layered Semiconductor Nanoparticles

David F. Kelley

Kansas State University, Manhattan, Kansas, U.S.A.

I. INTRODUCTION

There are many different types of layered semiconductors and semimetals. Their bulk crystallographic, electrical, and optical properties have been extensively studied and several texts and review articles are available [1–4]. Most of these materials are binary compounds: metal chalcogenides and halides. All of these materials are characterized as having highly anisotropic physical and electrical properties. Among the most heavily studied materials are the dichalcogenides of the group IVB, VB, VIB, and VIII transition metals and the group IIIA and IVA metals. Much of this interest comes from their use as catalysts. In particular, MoS_2 is widely used a hydrodesulfurization catalyst. There has recently been great interest in nanoparticles (quantum dots) and "inorganic fullerenes" of these materials, particularly the group VIB dichalcogenides. The layered structures of these materials are analogous to graphitic carbon. Extending this analogy, inorganic fullerenes are the structural analogs of carbon nanotubes. As such, inorganic fullerenes have extremely interesting structural, physical, and electronic properties [5–13]. The most extensively studied inorganic fullerene materials are those in the MoS_2 family, specifically MoS_2, $MoSe_2$, WS_2, and WSe_2. Very recently, inorganic fullerenes of GaSe has also received theoretical attention [14]. Inorganic fullerenes have either single-layer or multilayer tube-type morphologies. The electronic structures of single-wall and multiwall inorganic fullerenes approximate two-dimensional thin films of these materials. However, the curvature necessary to produce the tube morphology also results in a modest perturbation of the electronic structure.

Quantum dots are essentially "zero dimensional"; their small size in all three dimensions results in significant perturbation of the electronic structure. As a result, the electronic, spectroscopic, and photophysical properties of quantum dots can be very different than those of the bulk material or of thin films. The quantum confinement effects that define quantum dots occur in semiconductor nanoparticles when the particle size becomes less than or comparable to that of a photogenerated exciton in the bulk material. Only recently have the electronic, spectroscopic, and photophysical properties of layered semiconductor quantum dots been carefully studied. Despite the characterization of semiconductor nanoparticles as being "zero-dimensional" quantum dots, the crystal structure may result in not all dimensions being equivalent (i.e., the anisotropic nature of bulk material is reflected in the optical properties of the quantum dot). This is particularly true for nanoparticles of layered semiconductors. Detailed optical and dynamical studies of layered semiconductor nanoparticles have been largely confined to MoS_2-type materials and are only now being expanded to other types of layered semiconductors, specifically GaSe and InSe. This review will focus on the spectroscopic and photophysical properties of MoS_2-type nanoparticles.

II. ELECTRONIC AND CRYSTAL STRUCTURE OF MoS_2-TYPE SEMICONDUCTORS

Materials in the MoS_2 family (MoS_2, $MoSe_2$, WS_2 and WSe_2), are very similar to each other, exhibiting the same crystal structures and similar physical, electronic, and spectroscopic properties. These materials have layered-crystal structures consisting of two-dimensional hexagonally close-packed sheets of sulfur or selenium atoms on each side of a sheet of molybdenum or tungsten atoms; see Fig. 1. The chalcogenide atoms in each pair of sheets are directly on top of each other with the molybdenum atoms occupying the trigonal prismatic interstitial sites [15]. These compounds form two polytypes, 2H and 3R, differing only in how the chalcogenide–metal–chalcogenide trilayers stack on each other. The 2H polytype is most common, with the crystal structure of $P6_3/mmc$ D_{6h}^4. In the 2H case, there are two layers in a unit cell, with the metal atoms of a given layer sitting directly above the chalcogenide atoms of the adjacent layer. In the MoS_2 case, the distance between Mo atoms within an xy plane is 3.16 Å and the length of the unit cell along the z axis is 12.3 Å. $MoSe_2$, WS_2, and WSe_2 also have similar lattice constants. This crystal structure is depicted in Fig. 1.

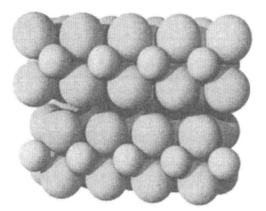

Figure 1 Crystal structure of MoS_2. The smaller and larger spheres indicate molybdenum and sulfur atoms, respectively.

MoS_2-type materials are indirect band-gap semiconductors. The energies of the indirect (momentum forbidden) and direct (momentum allowed) band-gap transitions are given in Table 1. The electronic structure of these materials may be qualitatively understood in terms of the crystal structure.

A highly simplified picture of bonding and electronic transitions in MoS_2-type materials results from the consideration of the ligand field splitting of the metal d-orbitals by the trigonal prismatic environment of the chalcogenides [22,23]. In the trigonal prismatic environment, the lowest energy orbital is the d_{z^2}, followed by the degenerate d_{xy} and $d_{x^2-y^2}$ with the degenerate and d_{xz}- and d_{yz}-orbitals having the highest energy. In the ligand field model, the metal d-orbitals do not mix with the chalcogenide s- and p-orbitals and the metal d-

Table 1 Transition Energies of MoS_2-Type Materials

	Indirect gap (eV) [17]	A Exciton	B Exciton
MoS_2	1.23	1.88 eV, 660 nm [18,19]	2.06 eV, 602 nm [18,19]
$MoSe_2$	1.09	1.57 eV, 790 nm [19]	1.82 eV, 681 nm [19]
WS_2	1.35	1.95 eV, 636 nm [20]	2.36 eV, 525 nm [20]
WSe_2	1.20	1.71 eV, 725 nm [21]	2.30 eV, 539 nm [21]

Note: Similar values are given by Evans [16].

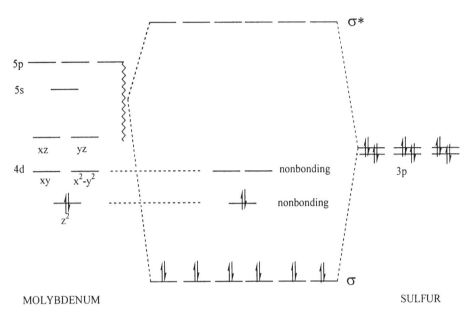

Figure 2   Zeroth-order molecular orbital picture of the bonding in MoS_2.

orbitals are therefore nonbonding. In this model, a Mo^{4+} ion has a d^2 electronic configuration and only the lowest energy d_{z^2}- orbital is filled. The same result is obtained in the zeroth-order molecular orbital picture of the MoS_2 bonding and is depicted in Fig. 2. Indeed, one may think of the stability of this electronic configuration as being why the material has a layered crystal structure with the molybdenum atoms in a trigonal prismatic environment [24]. A similar argument explains why other layered transition metal dichalcogenides have the metal in an octahedral geometry (e.g., PtS_2, in which the Pt^{4+} has a d^6 electron configuration).

In the above molecular orbital model, the band-gap transition is nominally $d_{z^2} \rightarrow d_{xz}, d_{x^2-y^2}$. This is a nonbonding-to-nonbonding transition and this argument has been used to explain the high photostability of MoS_2-type semiconductors as photoelectrodes [23]. This simple picture is more or less borne out by augmented spherical wave (ASW) band structure calculations [25,26]. The orbital nature of the states involved in the indirect and direct transitions was calculated. These and other calculations indicate that the top of the valence band is at Γ, and the bottom of the conduction band is midway between Γ and K [7,25–27], defining the indirect transition. The orbitals involved in this lowest excited state are mostly the metal d_{xz}, $d_{x^2-y^2}$, and d_{z^2}-orbitals, but also have significant Se character (for $MoSe_2$). Specifically, Γ_4^- (top of valence band) consists of 57%

d_{z^2}, 35% Se p_z, 3% Mo s and 5% Se s and 0.55ΓK (bottom of the conduction band) consists of 38% d_{xy}, $d_{x^2-y^2}$, 9% Mo d_{z^2}, 4% Se s, 26% Se $p_x p_y$, 9% Se d_{xy}, d_{yz}, and 5% Se d_{xy}, $d_{x^2-y^2}$. Thus, these calculations corroborate that the lowest excited state has a large Mo $d_{z^2} \rightarrow d_{xy}$, $d_{x^2-y^2}$ component. They also indicate that the top of the valence band at Γ_4^- has antibonding character (Mo d_{z^2} and Se p_z). Excitation to the band edge promotes an electron from this orbital to the conduction band, further stabilizing the excited state. The reactivity of MoS_2-type materials is greatest at coordinately unsaturated sites, specifically, at step edges and corners, where metal atoms are exposed. It is widely believed that these sites are involved in the catalytic activity of MoS_2. Edge and corner metal atoms do not have as well-defined symmetry and the orbitals are more mixed. *Ab initio* calculations show that edge and, especially, corner (coordinately unsaturated) Mo atoms have larger positive charges [28] and have states near the Fermi level [29], Surface and edge states within the band gap can act as electron and/or hole trap states. As such, the presence of these states can dramatically alter the electron and hole recombination and interfacial charge-transfer dynamics, as will be discussed below.

The lowest-energy-resolved features in the absorption spectrum of MoS_2-type semiconductors are the A and B excitons, shown in Fig. 3 for the case of MoS_2 [30]. The dissociation limit of peaks built off of the A exciton correspond to the direct band edge. Calculations indicate that these transitions are at K and that the A and B excitons correspond to K4 → K5 and K1 → K5, respectively [25,26]. The lowest-energy direct-transition excited states have considerably different orbital character than the band-edge state. K1 and K4 correspond to 84% and 82% d'_{xy}, $d_{x^2-y^2}$ respectively corresponds to 77% d_{z^2}.

The energy separation between the A and B exciton peaks is about 1500 cm^{-1} in the case of MoS_2. The A and B excitons arise from the same transition, which is split by interlayer interactions. This is seen from the spectra of thin films and inorganic fullerenes and band structure calculations. Films of WSe_2 show only a single exciton peak for thinnest films, having a thickness of 13 Å (one unit cell) [31]. Inorganic fullerenes are the MoS_2 analog of carbon nanotubes and consist of single or multiple layers of S–Mo–S sheets, curved into a nanotube morphology. MoS_2 inorganic fullerenes show convergence of A and B excitons to a single peak for structures having only one or two layers [7]. Band structure calculations (full potential linear muffin tin orbitals) reproduce this behavior [7]. Specifically, the spectra were calculated for MoS_2 with increasing interlayer distances. The energies of the K4 and K1 points converge when the interlayer distance is increased by 7%. This agrees with the linear combination of atomic orbitals (LCAO) calculations of the MoS_2 band structure considering only a single layer linear combination of atomic orbitals (LCAO) [27].

Thin-film spectra show that the A exciton energy depends on the thickness of the film. These spectra show that the A exciton in MoS_2 shifts about 0.15 eV to the blue as the film is thinned from bulk to 13 Å, one unit cell, two layers

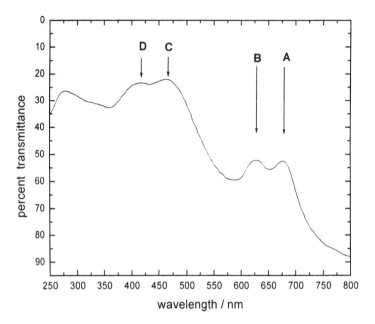

Figure 3 Room-temperature absorption spectrum of MoS_2. The spectrum is taken with E perpendicular to the crystallographic c axis and is therefore sensitive only to xy polarized transitions. (Adapted from Ref. 30.)

[31]. This thickness dependence may be described by a simple, effective mass quantum confinement theory down to a thickness of about five layers [32,33]. This theory describes the blue shift of the exciton in terms of "particle-in-a-box" considerations of the electron and hole. The electron and hole masses enter into this theory and are taken to be the effective masses of these carriers. The effective masses are given by $m_e = \hbar/(\partial^2 E/\partial k^2)$, where E and k are the energy and wavevector, respectively, of the band into which the carrier is promoted. The energy bands and, hence, the effective masses are well defined only for an infinite lattice, and this treatment is expected to break down for sufficiently small sizes. This breakdown at the smallest sizes is observed in many types of quantum dots and in the thinnest WSe_2 films. The theory typically predicts larger quantum confinement effects than are observed. Similar results are obtained from inorganic fullerenes, even though the inorganic fullerene spectra are complicated by curvature effects. The curvature of MoS_2-type nanotubes also alters the electronic structure, resulting in "curved excitons." This curvature also results in incommensurate lattices, which also affects the electronic structure [7].

Obtaining the polarizations of different optical transitions is of great value in assigning the transitions and comparing experimental results with theory. In layered semiconductors, optical transitions are polarized either in the plane of the layers (*xy* polarized) or perpendicular to that plane (*z* polarized). Polarized spectra are typically obtained from oriented single crystals. Because layered materials cleave easily between the layers, high-quality single crystals with an exposed basal (*xy*) plane are readily obtained. Absorption spectra taken at normal incidence to the cleaved face of these crystals are sensitive only to transitions polarized in the *xy* plane. Because it is more difficult to cleave these crystals perpendicular to the basal planes, limited spectroscopic information about *z*-polarized optical transitions is available. The A and B excitons are *xy* (in plane) polarized, at least in the cases of WS_2 and WSe_2 [34]. The same polarization would presumably be found for the MoS_2 case, but good single-crystal spectroscopic data are lacking [35]. In addition to the *xy*-polarized transitions, there are several *z*-axis-polarized transitions at higher energies. WSe_2 shows a strong transition $Q_2^+ \rightarrow Q_2^-$, at 429 nm and weaker transitions at 500, 486, and 470 nm [34]. Comparisons with these results are valuable in assigning the transitions observed in nanoparticles of these materials, as discussed below.

III. SYNTHESIS OF MoS₂-TYPE NANOPARTICLES

MoS_2-type nanoparticles have been produced by a variety of different methods, resulting in either supported or free-standing solution-phase particles. Efforts to produce nanoparticles on solid supporting surfaces have primarily been directed at understanding the catalytic properties of MoS_2 in its ground electronic state. Several different methods of producing well-defined supported particles and highly dispersed polycrystalline samples have been developed. Single trilayer MoS_2 nanoparticles have been grown on a Au (1,1,1) surface, using the surface as a template [36]. These particles are supported on an atomically flat silver surface and are very well characterized by scanning tunneling microscopy (STM). The particles have a triangular shape with the different types of edges clearly visible in the STM images. These STM images also permit characterization of individual atomic edge defects. Because of the template synthesis on a silver surface, no optical spectra were obtained. Recent methods of producing highly dispersed MoS_2 include the thermal decomposition of $(NH_4)_2MoS_4$ [37–39], the reduction of $(NH_4)_2MoS_4$ [40], sonochemical syntheses [38], and the decomposition of $Mo(CO)_6$ in the presence of H_2S [41]. In all cases, small, polydisperse, supported particles are produced. The optical properties of these particles have not been reported.

Early efforts to produce unsupported, solution-phase nano-sized MoS_2 and WS_2 particles involved dissolution in of bulk material in hot acetonitrile, followed by filtration [42]. This procedure produced polydisperse particles in the 1.0–3.5-nm range. Absorption spectra are unresolved, but indicate the presence of quantum confined particles. Alternatively, small particles may be obtained by sonication of MoS_2 or WSe_2 in water under a hydrogen/argon atmosphere [43]. The spectra show that these particles are also polydisperse.

The first synthesis of relatively monodisperse, unsupported MoS_2 nanoparticles was reported by Wilcoxon et al. [44,45]. This synthesis is based on the reaction of $MoCl_4$ with H_2S in an inverse micelle solution. The synthetic method is easily generalized to $MoSe_2$, WS_2, and WSe_2. Inverse micelle syntheses of colloids are very common. What makes this synthesis somewhat unusual is that the inverse micelles are anhydrous. Typically, the inverse micelles consist of a hydrocarbon solvent, a surfactant (typically AOT) and a substantial amount of water. AOT micelles are usually characterized by their AOT-to-water ratio. $MoCl_4$ and WCl_4 are very air and water sensitive, making the usual synthetic methods in an aqueous inverse micelle impossible. The syntheses of MoS_2-type nanoparticles therefore make use of anhydrous and anaerobic ternary inverse micelles to solublize the metal chloride. These solutions are typically ternary surfactant–consurfactant–solvent systems. The solvent is an alkane (typically octane), the surfactant is a tetra-alkyl quaternary ammonium salt, and the cosurfactant is an alcohol (typically hexanol). The most commonly used surfactants are tridodecyl-methyl-ammonium chloride or iodide, although quaternary amines having one or two long alkyl groups can also be used. The purpose of the inverse micelle is to dissolve the metal chloride and thereby provide for homogeneous (rather that heterogeneous) reaction. Following dissolution, H_2S is injected into the rapidly stirring solution. Reaction proceeds rapidly and particle diameters from 2.5 to 4.5 nm are obtained. Control of the particle size distribution is discussed below. Elemental analysis on MoS_2 particles indicates an Mo:S ratio of 1:2.5. For a stoichiometric particle, the ratio would, of course, be 1:2.0. The ratio being in excess of this value is probably due to excess sulfur on the particle edges. The initial syntheses and detailed characterization have focused on MoS_2 nanoparticles. The absorption spectrum of WS_2 nanoparticles are also reported and it is stated that $MoSe_2$ and WSe_2 particles have been grown. These MoS_2 particles were initially characterized by absorption and emission spectroscopy, transmission electron microscopy (TEM and high-resolution TEM), dynamic light scattering, and x-ray diffraction (XRD). TEM and light-scattering results were used to determine particle sizes. XRD results on aggregated particles were used to confirm that the crystal structure was the same as bulk $2H-MoS_2$. Electron-diffraction measurements were also performed on aggregates of the largest particles and also confirmed crystal structure. These studies showed that it is possible to chromatographically purify the particles. Subsequently, the syntheses and characterization of WS_2, WSe_2, and $MoSe_2$ nanoparticles have been reported [46,47]. These studies and further spectroscopic studies of MoS_2 reported experimental electron-diffraction results for MoS_2 nanoparticles and bulk MoS_2 and compared these results with electron-diffraction simulations. This comparison shows that these nanoparticles consist of single trilayer (S–Mo–S) disks [48]. This is an important point because the particle morphology plays a large role in its spectroscopy and dynamical properties. As stated earlier, MoS_2 commonly exists as one of two different polytypes, 2H and 3R, which differ in stacking of the S–Mo–S sheets. The more common is the 2H polytype, in which the unit cell consists of two S–Mo–S trilayers, with the molybdenum atoms of one layer situated directly above the sulfur atoms of the adjacent layer [16]. Electron-diffraction simulations show that the diffraction patterns for the 2H and 3R crystal structures converge as the size of the structures decreases along the crystallographic c axis. These electron-diffraction simulations show that four small-angle reflections, having indices of (1,1,0), (3,0,0), (2,2,0), and (4,1,0) remain as the crystal thickness is decreased to a single unit cell. Other very intense reflections calculated for bulk $2H–MoS_2$ [specifically, (0,0,2), (1,0,0), and (1,0,2)] broaden out and are lost. This matches the experimental nanoparticle result. Diffraction rings at radii corresponding to the (1,1,0), (3,0,0), (2,2,0), and (4,1,0) reflections are observed, and nothing is observed at radii corresponding to the (0,0,2), (1,0,0), and (1,0,2) reflections. The above comparison of the experimental and calculated diffraction patterns is the basis for the conclusion that the MoS_2 nanoparticles consist of a single trilayer type of structure [48]. This same conclusion was reached for WS_2, $MoSe_2$, and WSe_2 nanoparticles of comparable size [46,47]. This conclusion makes good sense in terms of the bonding; the only thing holding the trilayer sheets to each other is the comparatively weak van der Waals force. When the particles are sufficiently small, this binding is weak and the sheets will spontaneously exfoliate. This result is also consistent with studies of other layered materials, which exfoliate into single covalently bound sheets [49,50].

IV. SIZE-DEPENDENT NANOPARTICLE THERMODYNAMICS

A. Metal Nanoparticles and Theory

Nanoparticles are often synthesized as free-standing particles in solution using colloid chemistry methods, as described earlier. There are several different methods that can be used to control the size distribution of semiconductor and metal nanoparticles and obtain monodisperse samples. These methods fall into three broad classes, although there is considerable overlap among them. Monodisperse nanoparticles may be obtained by (1) controlling the kinetics of nucleation and subsequent crystal growth [51,52], (2) synthesizing polydisperse samples and selecting a portion of that distribution [53–55], or (3) adjusting the synthetic conditions so that a monodisperse size distribution corresponds to a thermodynamic minimum [56]. In most semiconductor nanoparticle syntheses, narrow size distributions are obtained by separating the time scales of nucleation and subsequent crystal growth. Conditions are adjusted such that a burst of nucleation occurs upon mixing of the reactants, followed by much slower growth. Because all nucleation occurs simultaneously and growth occurs at the same rate for all of the nascent nanoparticles, a narrow size distribution is obtained in the kinetically controlled limit. By controlling the solution monomer concentration during the particle growth phase, it is possible to vary the growth rates of different sized particles, thereby focusing or defocusing the size distribution [51]. Following synthesis, the particles are often precipitated and resuspended in a different solvent. At this point, the size distribution may be narrowed by size-selective precipitation. This makes use of the fact that the nanoparticle solubility is dependent on its size. A miscible solvent in which the particles are not soluble is added to the nanoparticle solution until precipitation occurs. This selectively precipitates the largest particles, thereby narrowing the size distribution in both the precipitated and solution phases. Typically, these methods result in nanoparticle size distributions that do not correspond to a thermodynamic minimum. However, the crystal growth kinetics as well as the particle solubility are strongly influenced by the particle thermodynamics and thermodynamics may therefore play a key role in all three methods [51]. Central in determining the role of thermodynamics in the particle size distribution is the extent to which the particles are labile. Nonlabile materials may have size distributions that are far from thermodynamic equilibrium, but are stable for long periods of time.

In some cases, the particle size distribution is determined solely by thermodynamics; reaction kinetics play no role. Thermodynamic control of the static, equilibrium nanoparticle size distribution requires that there be a free energy minimum corresponding to a specific nanoparticle size and that the particles be sufficiently labile that this distribution is realized. This occurs in surface-supported germanium nanoparticles [57,58] and in solution-phase metal nanoparticles [56,59–61]. In the case of solution-phase gold particles, all particle sizes are stable at room temperature, and the thermodynamic minima size distributions are realized only at elevated temperatures. The thermodynamic minima in the size distribution result from the combined effects of surface and bulk energies of the particle and binding energies of ligands adsorbed onto the particle and in solution. The particle surface energy is usually positive. This means that making larger particles lowers the surface-to-volume ratio and is, therefore, thermodynamically more favorable. In addition, the surface free energy increases with the surface curvature, which is the Gibbs–Thomson effect [52]. This variation of total surface energy with particle size leads to the well-known phenomenon of Oswald ripening, in which larger particles grow at the expense of smaller ones. Surface-energy considerations result in a thermodynamic minimum corresponding to the bulk material. A thermodynamic minimum occurrs at a finite size only if the presence of surface-binding ligands in solution is also considered. Unlike the surface free energy, the free energy associated with surface-binding ligands is lowered as the particle size decreases. These two contributions to the free energy (surface and ligand energies) have different functional dependencies on the particle size. The net result is a free energy minimum corresponding to a finite particle size, and that size depends on the ligand concentration. The most extensively studied example of a thermodynamically controlled nanoparticle size distribution in solution is that of thiol-capped gold particles. Polydisperse gold nanoparticles are stable in the presence of alkylthiols at room temperature. However, when these nanoparticles are heated in the presence of thiols in solution ("digestive ripening"), monodisperse particles are formed, with the average size determined by the thiol concentration. A high thiol concentration stabilizes a larger total surface area and thereby results in smaller particles. The qualitative theory of how thermodynamics controls the size distribution of gold particles in the presence of alkyl thiols is based on simple scaling arguments [56]. The basic idea is as follows. Particles establish a size distribution which minimizes the free energy of the entire system. In the gold–thiol system, thiols bind quite strongly to the surface of the gold, and the total free energy has contributions from both the surface energy of the gold particles and the adsorbed thiols. The surface free energy of the gold particles varies with the number of atoms per particle, for a constant total number of gold atoms. Specifically,

$$\mu_{surf} = n_s(\mu_s - \mu_b) = (constant)n^{-1/3}(\mu_s - \mu_b) = \frac{A}{n^{1/3}} + \frac{B}{n^{2/3}} \tag{1}$$

where n is the number of atoms per particle, n_s is the total number of surface atoms, and μ_s and μ_b are the surface and bulk free energies per atom, respectively. n_s is equal to the (number of particles) $\times$ (number of surface atoms/particle). We define N as the total number of atoms, which is constant, so that the number of particles is N/n. The number of surface atoms on a given particle scales as the square of the particle radius, or as $n^{2/3}$. Thus, n_s is proportional to the total surface area and scales as $n^{-1/3}$ [second part of Eq. (1)]. The surface energy per area (or the energy per surface atom) depends on the surface radius of curvature. The surface energy is taken to be a constant plus a term that is inversely proportional to the radius of curvature. Thus, the energy per surface atom ($\mu_s - \mu_b$) scales as $C + D/n^{1/3}$, where C and D are constants. Thus, we finally get $\mu_{surf} = A'/n^{1/3} + B/n^{2/3}$, where A and B are constants, as indicated in Eq. (1). Both producing surface and decreasing the surface radius of curvature cost energy, and A and B are positive constants. The thiol binding energy is given by

$$\mu_{bind} = (No. \text{ of bound thiols}) \times \left(\frac{Binding\ energy}{Thiol}\right) \tag{2}$$

The number of bound thiols is proportional to the total number of surface atoms, n_s, and hence scales as $n^{-1/3}$ at a constant solution concentration. If the thiol-binding energy is taken to be surface-curvature independent, then $\mu_{bind} = A'/n^{1/3}$. Binding of the thiols is energetically favorable and A' is a negative constant. Thus, combining Eqs. (1) and (2), the total free energy depends on the particle size and we get

$$\mu_{tot} = \mu_{surf} + \mu_{bind} = \frac{A + A'}{n^{1/3}} + \frac{B}{n^{2/3}} \tag{3}$$

If the thiol binding is sufficiently strong, then the quantity $A + A'$ is negative and Eq. (3) has a minimum at $n = [-2B/(A + A')]^3$. This corresponds to the thermodynamically favored nanoparticle size. It is important to note the free energy minimum results from the combination of thiol binding favoring a high surface area and therefore small particles, along with the radius-dependent surface energy favoring larger particles. Analogous functional dependencies may be developed for two-dimensional nanoparticles and applied to systems such as MoS$_2$, as discussed below.

B. Thermodynamic Control of MoS$_2$-Type Nanoparticle Sizes

The physical and chemical factors controlling the size distribution of MoS$_2$ particles have been examined recently [48]. These studies found that the size distribution depends critically on the synthetic conditions, specifically, the halide counterion of the tetra-alkyl ammonium surfactant and the pH. The experimental observations may be summarized as follows. Nanoparticles synthesized in a didodecyl-dimethyl-ammonium bromide (DDAB)-based micelle (DDAB/hexanol/octane) have an unresolved spectrum and a polydisperse ($\cong 2$ to ≥ 15 nm) size distribution. In contrast, nanoparticles synthesized using tridodecyl-methyl-ammonium iodide (TDAI) as the surfactant exhibit a well-resolved peak at 362 nm and a comparatively narrow size distribution centered at 3.5 nm. Alternatively, monodisperse 3.5-nm particles may be obtained by the addition of TDAI to a polydisperse nanoparticle sample. In this case, after about 2 days, a resolved peak at about 362 nm forms and TEM images reveal 3.5-nm monodisperse particles. These samples are almost indistinguishable from those synthesized using TDAI as the surfactant. 8-nm particles may also be obtained by using the appropriate synthetic conditions. Specifically, 8-nm nanoparticles are synthesized by dissolving MoCl$_4$ in either a TDAI or dodecyltrimethyl ammonium chloride (DTAC) ternary inverse micelle solution, followed by the injection of about 4 molar equivalents of (NH$_4$)$_2$S in the form of a 20% aqueous solution. TEM images reveal these samples to be relatively monodisperse and the absorption spectrum exhibits a well-resolved peak at 473 nm. Alternatively, an anhydrous base, such as NaCN or Na$_2$S, may be added to a polydisperse sample. This results in no immediate

Figure 4 TEM images of 3.5-nm (left) and 8-nm (right) MoS$_2$ nanoparticles. (From Ref. 48.)

change in the sample, but in a few days, the spectrum develops a well-resolved 470-nm peak and the TEM images show monodisperse 8-nm particles. TEM images of 3.5-, and 8-nm particles are shown in Fig. 4. The above observations indicate that the nanoparticle size distributions are controlled by thermodynamics, rather than the kinetics of nucleation and crystal growth.

The scaling considerations applied to three-dimensional the gold–thiol system described above can be adapted to two-dimensional MoS$_2$ particles and applied here. It is reasonable to assume that in the MoS$_2$ case, solution-phase ligands bind to the exposed molybdenum atoms at the nanoparticle edges; these are the most reactive sites. Thus, in the case of two-dimensional particles, the edge of the (assumed circular) nanoparticles has the role of the surface in the three-dimensional (assumed spherical) gold nanoparticles. The ligands which bind to the edges are I$^-$ and SH$^-$ in the syntheses using TDAI and (NH$_4$)$_2$S or other bases (Na$_2$S or NaCN), respectively. Because MoS$_2$ particles are two dimensional, the edge area per particle scales as $n^{1/2}$, and the total edge area at constant N scales as $n^{-1/2}$. In analogy with the three-dimensional case, there is an energy term which scales inversely with the particle radius. Analogous to Eq. (3), the total free energy may be evaluated for the two-dimensional case, and we get

$$\mu_{\text{tot}} = \mu_{\text{surf}} + \mu_{\text{bind}} = \frac{A+A'}{n^{1/2}} + \frac{B}{n} \tag{4}$$

The presence of a strong binding ligand results in the quantity $A + A'$ being negative and Eq. (4) has a minimum at $n = [-2B/(A + A')]^2$. A stronger binding ligand will result in a more negative value for A' and thus smaller particles. The experimental results indicate the extent to which the chemical potential is lowered by HS^- adsorption is somewhat less than in the case of I^-. Thus, the quantity $A + A'$ is not as negative in the SH^- case compared to I^- and smaller particles are formed in the I^- case. One caveat needs to be mentioned. Although the final particle size distribution is controlled by thermodynamics, this is not to say that kinetics play no role in these syntheses. The synthesis of 8-nm particles[via $(NH_4)_2S$] initially produces 4.5-nm particles, which subsequently disappear, giving rise to 8-nm particles. Why the synthesis with $(NH_4)_2S$ initially produces metastable 4.5-nm particles and only later gives the more stable 8-nm particles is not well understood. The initial formation of 4.5-nm particles must be due to both kinetic and thermodynamic factors.

V. SIZE-DEPENDENT POLARIZATION SPECTROSCOPY

Polarization spectroscopy can be a powerful tool for the assignment of electronic states [62,63]. The extent to which it is useful depends on the symmetry of the system involved. Nanoparticles or molecules in a liquid sample are randomly oriented and the overall absorbance is isotropic. In many optical polarization studies, anisotropic optical properties are recovered by crystallizing the sample so that all of the molecules have the same orientation. However, nanoparticle samples typically cannot be crystallized without aggregation or large interparticle interactions occurring. Anisotropy in a subset of the solution-phase sample can be produced by photoexcitation with polarized light. Polarized light photoselects those molecules or nanoparticles having their absorption oscillator most closely aligned with the electric vector of the excitation light. The probability of photoexcitation is proportional to $\cos^2\phi$, where ϕ is the angle between the electric vector of the light and the absorption oscillator. (The above statement is strictly correct only in the limit where a negligible fraction of the molecules are excited. Excitation with intense laser pulses can result in saturation effects, making the photoexcited population less anisotropic.) This anisotropy is then probed either by emission or absorption (i.e., the polarization of the emitted light or of a transient absorption is then measured). The emission measurements may be either static or time resolved. These measurements permit the determination of the relative orientations of the photoexcited and probed oscillators. The parallel and perpendicularly polarized components of the emission or transient absorption define the anisotropy, from which the relative orientations of the oscillators may be determined. Photoselection theory gives the anisotropy for an incoherent pump/probe emission experiment as [62,63]

$$r \equiv \frac{I_{par} - I_{perp}}{I_{par} + 2I_{perp}} = \frac{2}{5} P_2 (\cos \theta) \exp\left(\frac{-t}{\tau_{rot}}\right)$$

$$= \frac{1}{5} (3 \cos^2\theta - 1) \exp\left(\frac{-t}{\tau_{rot}}\right)$$

(5)

where I_{par} and I_{perp} are the time-dependent probe (emission or absorption) intensities having polarizations parallel and perpendicular to the polarization of the excitation light, respectively, θ is the angle between the pump and probe oscillators, and τ_{rot} is the rotational diffusion time. Nanoparticles are sufficiently large that they rotationally diffuse slowly compared to most other processes of interest. Thus, the rotational diffusion factor may be ignored in many cases. Disklike nanoparticles have two equivalent (degenerate) axes defined by the basal planes and the unique axis perpendicular to that plane. Several specific cases of excitation and probe oscillator alignment are of particular interest for these particles. An anisotropy of 0.40 is obtained only when both pump and probe oscillators are linear and aligned (i.e., $\theta = 0°$). An example of this is when the absorption oscillator is linearly polarized and emission occurs from the same oscillator. Because these oscillators are linear and nondegenerate, this can only apply to the unique axis, the crystallographic c axis. If both oscillators are two dimensional and in the xy plane, then one must integrate over all orientations in the plane and an anisotropy of 0.10 is obtained. The other situation relevant to disklike particles is if one oscillator is in plane and the other is perpendicular to the plane. In this case, $\theta = 90°$ and the anisotropy is -0.20. Of particular interest is the case where the sample is probed by a transient absorption measurement and the excited state has no absorbance at the probe wavelength. In this case, the absorbance change is negative; the transition is bleached. This may be thought of as "burning a polarization hole" in the absorption spectrum. When the excitation and probe wavelengths are the same, the anisotropy of the bleach will have a value of 0.40 or 0.10, depending on if the absorption oscillator is linear or planar, respectively. One caveat about transient bleach measurements need to be mentioned. The results can easily be confused by the presence of an excited-state absorption. The presence of both the bleached ground-state absorption and the excited-state absorption may result in complicated polarization dynamics that cannot be interpreted in terms of a single oscillator. This possibility is most easily addressed by probing the transient absorption at a slightly different wavelength, where the ground state has no absorption. If the excited-state absorption is broad (which is usually the case), then this type of control experiment is needed to reveal the presence of such an absorption.

The above ideas have been applied to several different sizes of MoS_2 nanoparticles. Specifically, relatively monodisperse samples of 3.5-, 4.5-, and 8-nm particles are easily accessible. The absorption spectra of these sizes of MoS_2 nanoparticles are collected in Fig. 5.

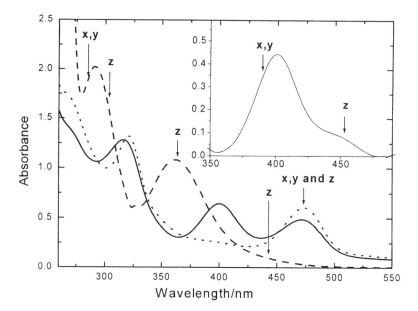

Figure 5 Absorption spectra of 3.5-, 4.5-, and 8-nm-diameter MoS₂ nanoparticles. The spectra of 3.5- and 8-nm particles are shown as dotted and dashed curves, respectively. The spectrum of a mixture of 8- and 4.5-nm particles is shown as a solid curve. Comparison of the spectra of the 8-nm particles and the 4.5- and 8-nm mixture yields the spectrum of 4.5-nm particles (inset). The polarizations of several of the lowest energy transitions are also indicated. (From Ref. 48.)

The 3.5- and 8-nm nanoparticles show well-resolved peaks at 362 and 473 nm, respectively, as well as other features at higher energies. The 4.5-nm particles show a well-resolved peak at 400 nm and a shoulder at 450 nm. It is tempting to assume that in each case, the lowest energy absorption corresponds to the lowest allowed transition (the A exciton) in bulk MoS₂. Polarization spectroscopy can be used to determine if this is the case. The lowest allowed transitions in bulk material, the A and B excitons, are polarized perpendicular to the crystallographic c axis. If the lowest allowed transition correlates to the A exciton, then it would be expected to also be a planar (xy polarized) oscillator. However, the results of polarization studies reveal that the actual situation is more complicated. A combination of time-resolved polarized emission and one-color time-resolved polarized absorption (transient bleach) studies facillitate assignment of the polarizations of the observed nanoparticle transitions. The 3.5-nm particles are emissive and the polarization of the several of the lowest transitions may be determined from polarized emission and/or polarized absorption measurements. The emission anisotropy decays for the 3.5-nm particles are shown in Fig. 6. These data were obtained following 310-nm excitation and with 450-nm detection. The anisotropy starts out at close to 0.4 and subsequently decays, indicating that both the emitting and absorbing oscillators must be linear and therefore z-axis polarized.

Excitation-wavelength-dependent emission polarization studies indicate the presence of an overlapping xy polarized transition in the bluer part of the 290–315-nm range, as indicated in Fig. 5. The combination of static absorption, time-resolved emission, and emission quantum yield measurements suggests that the emitting state has the same polarization (z axis, linear), but is not the same state as that giving rise to the 362-nm absorption peak. These assignments for the 3.5-nm particles are summarized in Fig. 5.

The 8- and 4.5-nm particles are at most weakly emissive and the polarizations of spectral features assigned to these particles are determined by polarized bleach measurements. The bleach anisotropy was determined using femtosecond pulses, with the probe delayed a few picoseconds from the pump. This delay ensures electronic and vibrational relaxation as well as relaxation of optical Kerr effects induced in the solvent. As a control, transient absorption experiments were performed with excitation at 475 nm and detection at 550 nm. This detection wavelength is to the red of the wavelengths at which a bleach would be observed and provides a measure of the transient absorption intensity in this general spectral

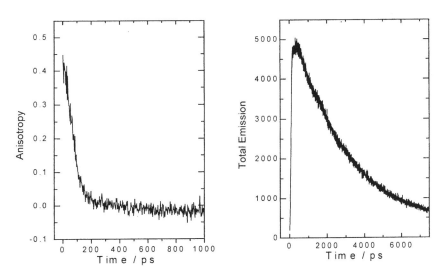

Figure 6 Emission anisotropy (left) and total emission (right) of MoS$_2$ nanoparticles following 312-nm excitation. The detection wavelength was 425 nm.

range. Only an extremely small transient absorption is observed (at least a factor of 10 smaller than the magnitude of the observed bleach signals), indicating that the polarizations of the bleach transients give an accurate indication of the polarization of the transition being excited. The transient bleach results indicate that excitation of the 400-nm peak of the 4.5-nm nanoparticles is xy polarized. Similar studies on the 450-nm shoulder indicate that excitation of this feature induces a transient anisotropy that is very close to 0.40. Thus, the 450-nm shoulder is z-axis polarized. Bleach anisotropy measurements following excitation of 8-nm particles at 475 or 490 nm show little or no anisotropy. The most likely explanation of the lack of polarization is that there is mixing of xy- and z-polarized transitions at comparable energies. These assignments are also summarized in Fig. 5.

The polarization results described above make it possible to establish the correlations of the transitions observed in these nanoparticles and in bulk MoS_2. In order to establish this correlation, we must consider the polarizations of several of the lowest transitions in bulk MoS_2. The lowest allowed transitions in bulk MoS_2 are the A and B excitons, at about 660 and 602 nm, respectively [64]. Both of these absorption features are polarized perpendicular to the z axis, in the xy plane. Band structure calculations indicate that these two features correspond to the same transition, with the approximately 1500-cm^{-1} splitting caused by the interaction of adjacent trilayer sheets [7,26]. Thus, only a single transition is expected for the present case of single trilayer nanoparticles. The spectroscopy of MoS_2 thin films and inorganic fullerenes shows that z-axis quantum confinement has only a small effect on this transition energy. The A and B excitons shift only about 0.1–0.15 eV from the bulk energy in the thinnest (a few trilayers thick) inorganic fullerenes or thin films [7,31]. Thus, in the absence of quantum confinement effects, interlayer interactions split a single transition at about 15900 cm^{-1} (630 nm) into the observed A and B exciton transitions.

Although these lowest xy-polarized transitions are comparatively well understood, assignment of the lowest z-polarized transition is somewhat problematic. The z-axis-polarized transitions in bulk MoS_2 are at an energy higher than the A and B excitons and have not been studied. Polarized absorptions in MoS_2 and related materials are typically studied using reflection spectroscopy of single crystals. Related literature on this subject indicates producing MoS_2 single crystals, cut perpendicular to the basal plane, and of sufficient optical quality is difficult [19,34]. However, the approximate energy of the lowest z-polarized transition may be inferred from the unpolarized amorphous thin film or colloidal particle spectra [30,43], knowing the energies of the xy-polarized transitions from single-crystal spectra. Specifically, this comparison suggests that there is a z-polarized transition in the (approximately) 540-nm region.

Quantum confinement affects the energies of the xy- and z-axis-polarized transitions very differently in these highly anisotropic nanoparticles. The bulk

MoS_2 and nanoparticle state assignments shown in Fig. 5 may be understood in terms of elementary considerations of the nodal planes in electronic excited states and different degrees of quantum confinement along the z axis and in the xy plane. Photoexcitation produces a nodal plane which is perpendicular to polarization of the light. The presence of a nodal plane increases the spatial extent of the wavefunction perpendicular to that nodal plane. Thus, an excited state corresponding to an xy-polarized transition will generally have a larger xy spatial extent than one corresponding to a z-polarized transition; that is, xy quantum confinement will have the largest effects on xy-polarized transitions. In general, reducing the xy dimensions of the particle will cause a larger blue shift of the transitions that are xy polarized than those that are z polarized. This effect can be seen by comparing the absorption spectrum of bulk MoS_2 with that of the 8-, 4.5-, and 3.5-nm particles and these comparisons are summarized in Table 2.

The quantum confinement shifts of the lowest xy- and z-polarized transitions are given as $\Delta(xy)$ and Δz, respectively. The lowest z-polarized transition in bulk MoS_2 is taken to be at 540 nm, as discussed above. The lowest observed xy-polarized transition correlates to the A and B excitons, and the lowest observed z-polarized transition correlates to the shoulder at 540 nm in bulk MoS_2. Table 2 shows that as the particle size is reduced, quantum confinement causes both the xy- and z-polarized transitions to move to higher energies. As expected from the above considerations, reducing the particle dimension in the xy plane has a larger effect on the energies of the xy-polarized transition than the z-polarized transition. The values in Table 2 indicate that the effect of quantum confinement in the xy plane is about twice as large for transitions that are xy compared to z-polarized. Figure 5 and Table 2 indicate that a reversal of the ordering of the excited states occurs as the particles get smaller. In the 3.5- and 4.5-nm particles, the lowest allowed transition is z polarized, whereas in bulk MoS_2, it is xy polarized. The crossover appears to occur in particles that are about 8 nm in diameter.

Table 2 Observed Transition and Quantum Confinement Energies in MoS_2 Nanoparticles

	xy (observed)		z (observed)		$\Delta(xy)$	Δz
	nm	$1000\ cm^{-1}$	nm	$1000\ cm^{-1}$	$(1000\ cm^{-1})$	$(1000\ cm^{-1})$
Bulk	660,602	15.88[a]	$\cong 540$[b]	18.5	0	0
8 nm	473	21.1	473	21.1	5.3	2.6
4.5 nm	400	25.0	440	22.7	9.1	4.2
3.5 nm	290	34.5	362	27.6	18.6	9.1

[a] Taken from the average of the A and B excitons at 660 and 602 nm, respectively.
[b] Estimated from the broad onset in the amorphous bulk MoS_2 spectra.

As an aside, it is of interest to note that the absorption spectra of these particles also bear on the particles' morphology, confirming the assignment made on the basis of electron-diffraction results. This is seen in the comparison of the spectrum of bulk MoS_2 with that of the 8-nm particles. The splitting between the A (660 nm) and B (602 nm) excitons in bulk MoS_2 is about 1500 cm^{-1}. These transitions are of comparable intensity, as one would expect from the fact that they result from interlayer interactions causing the splitting of a single transition [7,26]. In contrast, the spectrum of the 8-nm particles shows a single, well-resolved peak at 473 nm, with the next transition at about 340 nm. The separation between these transitions is about 8000 cm^{-1}, which is much too large for these peaks to correspond to the A and B excitons. A peak 1500 cm^{-1} to the blue of 475 nm would be at about 445 nm. Figure 5 shows that there is no evidence in the 8-nm particle absorption spectrum of the sort of splitting that gives rise to the A and B exciton in bulk MoS_2. Because band structure calculations indicate that the A–B exciton splitting is due to interlayer interactions [7,26], this result is consistent with the assignment that these particles are single trilayers.

VI. ELECTRON- AND HOLE-TRAPPING DYNAMICS

Photoexcitation of a semiconductor produces conduction-band electrons and valence-band holes, which may undergo several types of relaxation and recombination processes. If the photon energy exceeds the band-gap energy, the nascent electrons (holes) may relax to the bottom (top) of the conduction (valence) band. Carrier localization at a defect site (trapping) may compete with these relaxation processes. The relaxed electrons and holes may undergo radiative and/or nonradiative recombination. Trapping of either or both of the carriers often competes with radiative relaxation, quenching the band-edge emission. Trap states are localized electron or hole states having energies in the band gap. They are most often associated with defects in the crystal structure. Nanoparticles have extremely high surface-to-volume ratios. As a result, in semiconductor nanoparticles, these defects are usually associated with the surface. Because the trap states are at lower energy, radiative recombination of the trapped carriers results in red-shifted emission. Alternatively, nonradiative recombination of trapped carriers is often rapid, with the result being that the trapping process effectively quenches the nanoparticle luminescence. The above considerations make clear that the dynamics of electron and hole trapping are of central importance in determining the luminescence properties of a semiconductor nanoparticle. These dynamics may be elucidated using a variety of time-resolved optical spectroscopic methods. MoS_2-type nanoparticles have highly anisotropic structures, leading to anisotropic optical properties, as discussed above. The anisotropy of the optical properties can be exploited to determine the carrier-trapping dynamics. The basic idea is to use the emission intensity and anisotropy kinetics to determine the electron- and hole-trapping times. Such determinations are based on the premise that the localized electron and hole traps are located primarily at the nanoparticle edges and therefore lack the well-defined symmetry of band-edge electron and holes. It is therefore expected that although band-edge emission is strongly polarized, trapping of either carrier will result in at least some depolarization of the emission. The problem is then to separate the effects of electron and hole trapping. This may be accomplished through carrier-quenching and/or carrier-injection studies. In carrier-quenching studies, the roles of electron versus hole trapping can be assessed by adsorbing electron acceptors or electron donors (hole acceptors) onto the nanoparticles and observing the changes in the total emission and anisotropy kinetics following nanoparticle photoexcitation. This carrier-quenching approach has also been used to sort out electron and hole dynamics in CdSe [65]. 2,2′-Bipyridine binds onto the edges of MoS_2-type nanoparticles and can act as an electron acceptor. This molecule has been used in the electron-quenching studies on MoS_2 and WS_2 [66–68]. In the carrier-injection studies, electrons or holes are rapidly injected into the nanoparticle following photoexcitation of an adsorbed electron or hole donor [69]. The transients observed in this case are characteristic of only the injected carrier. This type of study also permits distinction between electron- and hole-trapping dynamics. Both types of study have been performed on MoS_2 and WS_2 nanoparticles, and these two types of particles yield qualitatively similar results.

Electron quenching along with emission polarization studies have been applied to the trapping dynamics of MoS_2 and WS_2 in the following way. Following excitation with polarized 312-nm light, the emission from both MoS_2 and WS_2 nanoparticles is polarized. Figure 6 shows that the emission anisotropy starts out with a value close to 0.40 and subsequently decays. The anisotropy has an initial fast, almost instrument-response-limited decay, followed by a slower, hundreds of picoseconds decay, and, finally, a very small anisotropy which decays on the nanosecond time scale. The initial $r \cong 0.40$ value indicates that photoexcitation produces a state which is a linear oscillator, as discussed above. The simplest interpretation of the anisotropy decay kinetics is in terms of a two-state model in which the initial and final states have anisotropies of 0.40 and 0.0, respectively. In this case, it is possible to relate the observed parallel and perpendicular emission intensities to the total emission intensities from the polarized ($r=0.4$) and unpolarized ($r=0.0$) oscillators. This is done by expressing the parallel and perpendicular emission intensities in terms of their components from each of these two different oscillators. Specifically,

$$I_{par} = I_{par,r=0.0} + I_{par,r=0.40} \tag{6a}$$

and

$$I_{per} = I_{per,r=0.0} + I_{per,r=0.40}$$

Using Eq. (5) and an anisotropy value of 0.40, we get that emission from the polarized and unpolarized oscillators have

$$I_{par,r=0.40} = 3I_{per,r=0.40} \qquad (6b)$$

and

$$I_{par,r=0.0} = I_{per,r=0.0}$$

The total population in either state is proportional to the total ($I_{par} + 2I_{per}$) time-dependent emission intensity from that state. The total time-dependent emission intensities in these states are given by

$$I_{tot,r=0.40} = (I_{par,r=0.40} + 2I_{per,r=0.40}) = \frac{5}{2}(I_{par} - I_{per})$$

$$I_{tot,r=0.0} = (I_{par,r=0.0} + 2I_{per,r=0.0}) = \frac{3}{2}(3I_{per} - I_{par}) \qquad (6c)$$

The above expressions assume that all of the polarized emission is from an oscillator having an anisotropy of 0.40. States having emission which is partially polarized will contribute to the apparent intensities of both the polarized ($r=0.40$) and unpolarized ($r = 0.0$) components. The extent of the contribution into each component will depend on the extent to which the emission is polarized. The polarized emission kinetics described by Eq. (6c), $(5/2)(I_{par} - I_{per})$, are shown in Fig. 7 and exhibit an instrument-limited rise followed by a triexponential decay. These kinetics are fit to components having decay times and relative amplitudes of 42 ps (60%), 275 ps (35%), and 3.0 ns (5%). Figure 7 shows that the unpolarized emission, $(3/2)(3I_{per} - I_{par})$, rises as the polarized emission decays. Similar results are obtained for WS_2. These studies do not permit assignment of the trapping dynamics. What dynamics correspond to each of the kinetic components is ambiguous, based on these results alone. This ambiguity has been resolved by electron quenching in the case of WS_2 and electron-injection studies in the case of MoS_2.

The electron-injection studies on MoS_2 nanoparticles involve the use of transient absorption spectroscopy on bare 3.5-nm nanoparticles and on these nanoparticles with an adsorbed cyanine dye, DTDCI. Photoexcitation of the nanoparticle produces a conduction-band electron and a valence-band hole. This band-edge state exhibits a strong absorption at 700 nm and the absorption kinetics of MoS_2 nanoparticles following 387.5-nm excitation are shown in Fig. 8. Several kinetic components can be seen. There is a fast transient observed in the first 10 ps which is associated with electron and/or hole relaxation to the band edge. This is followed by a decay having components of 42 ps (61%), 275 ps (17%), and 3 ns (22%). A calculated curve corresponding to these values is also shown in Fig. 8. These decay times correspond to the times observed in the polarized emission kinetics discussed above. Assignment of these kinetic components is

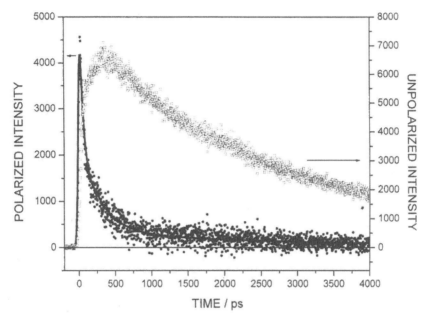

Figure 7   Polarized, $(5/2)(I_{par} - I_{per})$, and unpolarized, $(3/2)(3I_{per} - I_{par})$, components of the emission from MoS_2 nanoparticles following 312-nm excitation are indicated as closed and open circles, respectively. The detection wavelength was 425 nm. Also show is a calculated curve corresponding to a 42-ps (60%), 275-ps (35%), and 3.0-ns (5%) decay. (From Ref. 69.)

facilitated by comparison to the transient absorption kinetics observed following excitation of the DTDCI absorbed on the nanoparticle surface. This dye has an intense ground-state absorption in the 600–700-nm region and is known to inject electrons into semiconductors [70]. Following dye photoexcitation, an electron is injected from the excited-state dye into the nanoparticle conduction band. The absorption kinetics are observed at 660 nm (close to the absorbance maximum of the ground-state dye) and at 700 nm (on the red edge of the ground-state dye absorbance). The 660-nm probe kinetics are shown in Fig. 9. Several species can absorb in this wavelength range and therefore contribute to the observed time-dependent absorbance. Because the ground state of DTDCI has a large absorption at 660 nm, the net absorbance change following photoexcitation is negative; the ground-state dye is bleached. In addition to the ground-state bleach, there can be absorption from the excited-state dye, the dye cation, and the conduction band or trapped electron. The 660-nm kinetics show a small amplitude, a 14-ps increase in the absorption (a bleach decrease), followed by a slower bleach decay. The bare

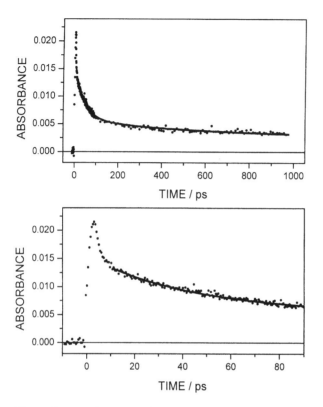

Figure 8   Transient absorption kinetics of MoS_2 nanoparticles following 387.5-nm excitation. The probe wavelength was 700 nm. A calculated curve decay having components of 42 ps (61%), 275 ps (17%), and 3 ns (22%) is also shown. Both panels show the same data, but on different timescales. (From Ref. 69.)

MoS_2 kinetics exhibit no 14-ps transient, so this kinetic component is assigned to an interaction of the nanoparticle with the dye. (We note that cyanine dyes undergo an excited-state isomerization [71–73] with a viscosity-dependent rate. There is no solvent viscosity dependence of the 12-ps transient seen in the MoS_2/DTDCI system, indicating that it is not associated with DTDCI isomerization.) Following the 14-ps transient, there is a slower, 225-ps decay as well as a long-lived nanosecond absorbance change. The important thing to note is that no 42-ps component is observed in the MoS_2/dye transient absorption kinetics. This establishes that the 42-ps transients seen in the polarized emission kinetics and in the bare nanoparticle absorption kinetics are not related to the electron dynamics and may, therefore, be assigned to hole trapping.

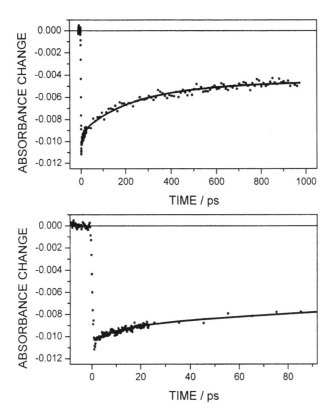

Figure 9 Transient absorption kinetics of MoS_2 nanoparticles with adsorbed DTDCI. Excitation was at 560 nm, which excites only the dye. The probe wavelength was 660 nm. A calculated curve having decay components of 12 ps, 225 ps, and 3.0 ns is also shown. Both panels show the same data, but on different time scales. (From Ref. 69.)

It is of interest to note that the 225-ps bleach recovery transient observed in the transient absorption kinetics is significantly shorter than the 275-ps decays observed in the polarized emission and the bare nanoparticle absorption kinetics. This difference, along with magnitude of this component at 660 and 700 nm, indicates that partial repopulation of the ground-state dye from the conduction-band electron occurs with this time constant. These dynamics are summarized in Scheme 1. From the difference of the bare nanoparticle and nanoparticle/dye conduction-band decays, it is possible to estimate the rates of conduction-band reverse electron transfer. These values (275 ps and 225 ps, respectively) give an estimate for the reverse electron-transfer time of about 1.2 ns. The ratio of the

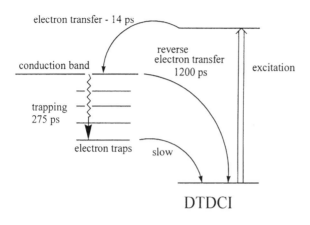

electron transfer - 14 ps

reverse
electron transfer
1200 ps

excitation

conduction band

trapping
275 ps

electron traps   slow

DTDCI

valence band

MoS$_2$

Scheme 1 Relaxation and trapping dynamics of MoS$_2$/DTDCI following photoexcitation of the dye.

trapping and reverse electron-transfer rates indicate that about 20% of the conduction-band electrons undergo reverse electron transfer, and about 80% get trapped. Following electron trapping, reverse electron-transfer from trap states occurs on a longer time scale.

The electron- and hole-trapping dynamics in the case of WS$_2$ are elucidated by electron-quenching studies, specifically by the comparison of polarized emission kinetics in the presence and absence of an adsorbed electron acceptor, 2,2′-bipyridine [68]. In the absence of an electron acceptor, WS$_2$ exhibits emission decay kinetics similar to those observed in the MoS$_2$ case. The polarized emission decays with 28-ps, 330-ps, and about 3-ns components. For carrier-quenching studies to resolve the dynamics of electron trapping, it is necessary that the electron acceptor quenches only conduction-band (not trapped) electrons. It is therefore first necessary to determine that electron transfer occurs only from the conduction band. The decay of the unpolarized emission (when both the electron and the hole are trapped) is unaffected by the presence of the 2,2′-bipyridine, indicating that electron transfer does not take place from trap states in the WS$_2$ case. Comparison of the polarized emission kinetics in the presence and absence of the electron acceptor indicates that electron transfer does occur from the conduction band. Specifically, this comparison reveals that the presence of 2,2′-bipyridine significantly shortens the slower decay component of the polarized emission from 330 ps to 200 ps. This establishes that there is still an electron in the conduction band during this transient (i.e., that the 330-ps transient is due to electron trapping). Shortening the decay time from 330 ps to 200 ps corresponds to an electron-transfer time of 500 ps. [The effect of adsorbed 4,4′,5,5′-tetramethyl-2,2′-bipyridine (tmb) on the WS_2 nanoparticle emission kinetics was also examined and found to have essentially no effect on either the polarized and unpolarized emission kinetics. tmb is expected to bind to the nanoparticles in the same way as 2,2′-bipyridine; however, tmb is 0.23 V more difficult to reduce than 2,2′-bipyridine [74]. This control experiment establishes that it is electron transfer to the adsorbed bipyridine that alters the emission kinetics, rather than the presence of the bipyridine altering the particle surface states.] The above results indicate that the slower component of the polarized emission decay and the corresponding rise in the unpolarized emission may be assigned to electron transfer from the conduction band in competition with electron trapping. Hole trapping partially depolarizes the emission, and subsequent electron trapping results in close to isotropic emission. This is the same conclusion as was obtained from electron-injection studies in the MoS_2 case and is summarized in Scheme 2.

     It is possible to assess the approximate depths of the electron and hole traps from spectral evolution data. This has been done in the case of WS_2 nanoparticles from the analysis of the unpolarized component of the emission [68]. The emission following hole trapping (but prior to electron trapping) is only partially polarized and the emission following electron trapping is unpolarized. Thus, the unpolarized

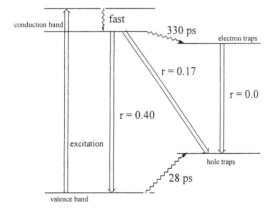

Scheme 2   Trapping dynamics of WS_2 nanoparticles following photoexcitation. Radiative and nonradiative processes are indicated with double-line arrows and wavy arrows, respectively.

emission has contributions from radiative electron–hole recombination, both preceding and following electron trapping. As electron trapping proceeds, the spectrum of the unpolarized emission will shift to the red. A reasonable estimate of the trap depth is simply the amount that the spectrum shifts. Time-dependent emission spectra may be reconstructed from wavelength-dependent kinetics, permitting the determination of the shift in the emission maximum as electron trapping proceeds. The reconstructed time-dependent depolarized emission spectra for WS_2 nanoparticles in octane are shown in Fig. 10.

The emission spectrum following hole trapping, but preceding significant electron trapping has a maximum at 430 nm. As electron trapping proceeds, the emission maximum shifts a small amount, to 434 nm. This shift is corresponds to an electron-trap depth of about 200 cm^{-1}. With the limited temporal resolution of time-correlated photon counting, estimating the depths of the hole traps is more problematic. However, the magnitude of the 28-ps decay component of the polarized emission is much larger on the blue edge of the spectrum (e.g., comparing 380- versus 420-nm kinetics), indicating considerable spectral evolution as hole trapping occurs and suggesting that the hole traps are comparatively deep.

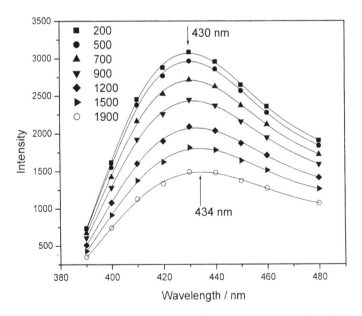

Figure 10 Reconstructed time-dependent emission spectra of WS_2 nanoparticles in octane. The times (in picoseconds) corresponding to each spectrum are also given. (From Ref. 68.)

Alternatively, very little of the time-integrated emission occurs prior to hole trapping and the depth of the hole traps may be roughly estimated from the Stokes shift of the static emission compared to the absorption spectrum. If the band edge is taken to be at about 380 nm and the static emission onset at about 410 nm, then the hole-trap depth is estimated to be about 2000 cm^{-1}. The central conclusion from these spectra is that hole traps are relatively deep, whereas electron traps are shallow, about 200 cm^{-1}. This analysis indicates that the electron traps have depths that are only slightly greater than thermal energy, at room temperature. With such shallow electron traps, one might expect that there would be a small but significant thermal population of the conduction band. Careful examination of long-time behavior of the polarized emission kinetics reveals that this is probably the case. The long-lived (3-ns) polarized emission decay approximately matches the decay of the unpolarized emission at long times, suggesting that states giving rise to each emission are in equilibrium.

It may be possible to understand the difference in electron- and hole-trap depths in terms of a simple exciton model of the traps. Traps at the nanoparticle edge are considered to result from the truncation of a partially ionic lattice. Thus, the model considers electron (hole) traps to be partially uncompensated positive (negative) charges at the nanoparticle edges. In this model, the trap depth is related to the magnitude of the uncompensated charge and the effective mass of the carrier. A large uncompensated charge trapping a carrier having a large effective mass results in a deep, highly localized trap state. Mobility and photocurrent measurements indicate that the mobility of the electron is much greater than that of the hole in bulk MX_2-type compounds (M = Mo, W, X = S, Se) [75]. Because the mobility is inversely proportional to the effective mass [76], this indicates a much smaller electron, compared to hole, effective mass. This conclusion is in agreement with calculations on the bulk materials [25]. It is therefore tempting to explain the difference in trap depths in terms of the difference in electron and hole effective masses. However, the situation may be more complicated than this simple consideration would imply. As discussed earlier, the lowest excited state of the nanoparticles is different than in the bulk material, and the electron and hole effective masses of the nanoparticle are unknown. The difference in nanoparticle-trap depths is probably due to both the carrier effective masses and the chemical nature of the surface defect causing the trap.

The results discussed above show that in both MoS_2 and WS_2 nanoparticles, electron trapping occurs much more slowly than hole trapping. It is possible to rationalize this result in terms of the electron- and hole-trapping reorganization energies and the extent of electron–phonon coupling. The small magnitude of the electron–phonon coupling results from the nature of the bonding in MX_2-type compounds. Band-gap excitation in these materials is largely a metal-centered, d–d, nonbonding-to-nonbonding transition. As a result, photoexcitation produces very little lattice distortion and there is very little electron–phonon coupling associated with band-gap excitation. If electron or hole trapping significantly distorts the local lattice, then there is a considerable reorganization energy associated with these processes. In addition, a significant contribution to the reorganization energy may result from interaction of the electron or hole with the surrounding solvent. These considerations suggest that in the case of electron trapping, the shallow traps along with the lattice distortion result in small Franck–Condon factors for electron trapping. Larger Franck–Condon factors occur only following thermal excitation of the phonon modes. In other words, the significant reorganization energy and small trapping energy result in an energetic barrier to electron trapping. Hole traps are much deeper and the barrier may be smaller in this case. This is basically a Marcus theory consideration [77–79]. The barrier for electron or hole trapping is given by

$$\Delta G^{\neq} = \frac{\lambda}{4} \left(\frac{1 + \Delta G}{\lambda} \right)^2 \tag{7}$$

where λ is the reorganization energy and ΔG is the trapping energy. Equation (7) shows that if λ is comparatively large and ΔG is small, then there is a significant barrier to trapping. As trapping becomes more exothermic (i.e., the magnitude of the trapping energy gets larger and approaches the same magnitude as the reorganization energy), the barrier gets smaller and the trapping rate gets larger. In this model, electron trapping is in the Marcus normal region, and hole trapping may be in either the normal or inverted region. The above model assumes the electron- and hole-trapping reorganization energies are comparable, and this may not be the case. The electron traps are very shallow and trapped electrons may be quite delocalized compared to trapped holes. If this is the case, then bond distortions associated with electron trapping would be expected to be comparatively small, resulting in a smaller reorganization energy. This does not change the qualitative considerations. Even if the reorganization energy is considerably larger in the case of hole trapping, the dramatically larger hole-trapping energy results in this process being faster than electron trapping.

Trapping in these nanoparticles is very slow compared to other types of semiconductor nanoparticles. In most other types of semiconductor nanoparticle (such as SnO_2, TiO_2, Fe_2O_3, and, most notably, CdS and CdSe), trapping occurs on the time scale of a few picoseconds or less [80–92]. It is reasonable to speculate that the reason that slow trapping is observed in MoS_2 and WS_2 nanoparticles is related to the weak electron–phonon coupling in these materials. Thus, the slow trapping in the present case may be due to the same considerations that explain the differences between the electron- and hole-trapping rates. However, electronic factors may also be important in determining the trapping rates, and the slow trapping observed in MoS_2 and WS_2 nanoparticles may also be due to electronic overlap considerations. The roles of electronic and vibrational factors in controlling trapping rates in MoS_2 and WS_2 nanoparticles are not clear at this point.

VII. INTERFACIAL ELECTRON-TRANSFER DYNAMICS

Charge separation is the primary photochemical event in many processes having great practical importance (e.g., photocatalysis and solar-energy conversion). Semiconductor nanoparticles are ideally suited as the active species in these processes because of their high degree of photostability. This is especially true of the MoS_2-type semiconductors, which are particularly photostable. Charge separation may occur as a result of interfacial transfer of either the photogenerated electron or hole. The electron is initially in the conduction band and later in surface-trap states and interfacial electron transfer may occur from either of these states. Because in the case of WS_2 nanoparticles (and presumably other MoS_2-type nanoparticles), the electron traps are shallow [68], the conduction-band and trap states are not very different energetically. In most cases, this small energetic difference is expected to have little effect on the electron-transfer dynamics. However, the nanoparticle conduction-band and trap states have different spatial extents and therefore may have very different couplings to adsorbed electron acceptors. Thus, the interfacial electron-transfer dynamics may be very different before and after trapping has occurred. The interfacial transfer rates of trapped electrons on MoS_2 nanoparticles to adsorbed electron acceptors has been studied [93]. In this case, the electron-transfer rates were slow compared to trapping and could be accurately determined only following electron trapping.

An important concern in interfacial electron-transfer dynamics is the variation of the electron-transfer rate with energetic driving force. Determination of the electron-transfer rate as a function of driving force allows the results to be interpreted in the framework of Marcus theory. There are two obvious ways in which the driving force may be varied: by varying reduction potential of the electron acceptor and by varying the oxidation potential of the conduction band or trapped electron. The former may be accomplished by putting electron-donating or electron-accepting substituents on the electron acceptor and the latter may be accomplished by varying the size of the nanoparticle. There are caveats associated with both of these ways of varying the driving force. When choosing electron acceptors having different reduction potentials, it is important that the nanoparticle–acceptor electronic coupling remain constant. Variations in the coupling would also cause changes in the electron-transfer rates, confusing the changes caused by variations in the energetics. Two different electron acceptors were used in the MoS_2–acceptor studies [93]:2,2'-bipyridine (bpy) and 4,4',5,5'-tetramethyl-2,2'-bipyridine (tmb). The methyl groups on tmb are electron donating, with the result being that tmb is 0.29 V harder to reduce than is bpy [94]. It would be expected, however, that the bonding and hence the electronic coupling would be very close to the same in these two cases. Two different sizes of nanoparticles were used, having lowest-energy absorption maxima at 473 and 362 nm. Variation of the nanoparticle size changes the energetics of both the valence and conduction bands. The extent to which the potentials of the conduction and valence bands shift due to quantum confinement is related the effective masses of the electron and hole [32,95]. In the case of bulk MoS_2-type materials, the electron effective mass is considerably less than that of the hole. However, as discussed earlier, the nature of the nanoparticle lowest excited state changes below a size of about 8 nm, and the electron and hole effective masses in particles below this size are unknown. In the absence of detailed information, the simplest approximation is that the difference in the energies of the absorption peaks of the two sizes of nanoparticles reflects the difference in the conduction-band energies. With this approximation, the potential of the conduction band is easily calculated. Particles having absorption maxima at 473 and 362 nm have the conduction band of the smaller particles at a potential about 0.80 V more negative than for the larger particles. To the extent that the electron-trap depths are the same in the two sizes of particles, 0.80 V is also the difference in the redox potential of the electron-trap states. Determination of the electron transfer rates is also not completely straightforward, because the decay of the trapped electrons and holes is nonexponential. This is due to the distribution of electron–hole separations [66]. However, the average electron-transfer time may be taken as the 1/e time of the trapped-carrier emission decay. With these approximations, electron-transfer times may be determined as a function of the electron-transfer driving force. The results are shown in Fig. 11. Marcus normal and inverted regions are evident. These results cannot be quantitatively interpreted in terms of simple classical Marcus theory[Eq. (7)]; no value of λ adequately fits the experimental result. The difficulty is that classical Marcus theory ignores excitation of quantized vibrational modes. However, the experimental results may be quantitatively interpreted in terms of Marcus theory if a quantized vibrational receiving mode is included. Specifically, nanoparticle-to-acceptor electron-transfer results in the formation of a bipyridine radical anion. The bipyridine radical anion has an electron in the lowest aromatic π^*-orbital, resulting in an expansion of the aromatic rings. This is a high-frequency vibration (about 1450 cm^{-1}) and must be treated quantum mechanically [79,96]. With the inclusion of this quantum mechanical mode, the expression for the electron transfer rate is

$$k_{ET} = \frac{V^2}{\hbar}\left(\frac{\pi}{\lambda kT}\right)^{1/2} \sum_n \exp(-s)\frac{s^n}{n!}\exp\left(\frac{-\Delta G_n^*}{kT}\right) \tag{8}$$

and

$$\Delta G_n^* = \frac{\lambda}{4}\left(1 + \frac{\Delta G + n\hbar\omega}{\lambda}\right)^2$$

where λ is the classical reorganization energy and ω is the angular vibration frequency of the quantized (in this case, ring breathing) mode. The parameter s

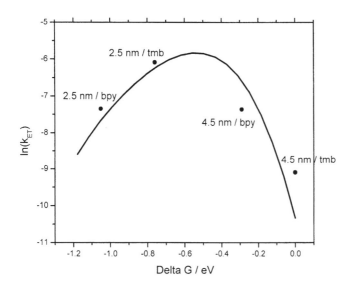

Figure 11 Plot of the log of the electron transfer rate as a function of the driving force. Also shown is a Marcus theory curve calculated from Eq. (8), taking $V = 2.12$ cm^{-1} and $\lambda = 3225$ cm^{-1}.

is related to the extent to which the addition of a π^* electron excites the ring breathing mode, specifically, $s = M\omega/2\hbar(\Delta Q)^2$, where ΔQ is the displacement in the vibrational coordinate. Fitting of the absorption and emission spectra of the MLCT transition of $Ru^{II}(bipyridine)_3$ indicates that for the bpy $\rightarrow$ bpy$^-$ process, $s = 1.1$ [96]. A Marcus theory curve calculated from the above equation with λ taken to be 3225 cm^{-1} is also shown in Fig. 11. The above results were obtained in acetonitrile solvent at room temperature. Electron-transfer reorganization energies in polar solvents vary with the nature of the electron-transfer reaction involved. However, for a reaction involving charge transfer to an electron-accepting ligand, solvent reorganization energies of a few thousand wavenumbers are typical. For example, electron transfer from the ruthenium to the bipyridine in $Ru^{II}(bipyridine)_3$ in acetonitrile has a λ value of about 1000 cm^{-1} [74,97,98]. The observed 3225-cm^{-1} value is roughly typical, or perhaps somewhat larger that what would be expected for polar solvent reorganization energies in electron-transfer reactions. This implies that at least some of the reorganization energy is due to the solvent, rather than relaxation of the nanoparticle. Consistent with this interpretation, the reaction in octane occurs slowly, if at all [99]. The above considerations are quite analogous to those that apply to electron-transfer reactions in molecular systems. In this sense, the nanoparticle–to-acceptor electron-transfer reaction is very similar to other solvent-mediated electron-transfer reactions.

VIII. CONCLUSIONS

There are two very broad, general conclusions resulting from the above review. The first is that MoS_2-type nanoparticles are very different than other types of semiconductor nanoparticles. Nanoparticles of several different types of semiconductors, such as CdSe, CdS, and InP, have been extensively studied. Experimental and theoretical studies have elucidated much of their spectroscopy, photophysics, and dynamics. The results reviewed above are, in many places, in sharp contrast with those obtained on other types of quantum dots. This does not come as a surprise. The properties of the bulk semiconductor are reflected in those of the nanoparticle, and properties of layered semiconductors are vastly different from those of semiconductors having three-dimensional crystal structures. Although the electronic and spectroscopic properties of nanoparticles are strongly influenced by quantum confinement effects, the differences in the semiconductors cause there to be few generalizations about semiconductor quantum dots that can be made.

The other conclusion that comes from the above review is that there is still much work on these systems to be done. There is still only a rudimentary understanding of the size-dependent electronic structure and dynamics. There is a need for further synthetic, spectroscopic, dynamical, and theoretical work. Perhaps the area in which progress most needs to be made is theory. Further electronic structure calculations would be a tremendously valuable guide for further experimental studies. It is hoped that the work done to this point will stimulate further studies in all areas.

ACKNOWLEDGMENT

This work was supported by a grant from the Department of Energy (grant No. DE-FG03-00ER15037).

REFERENCES

1. Grasso, V. *Electronic Structure and Electronic Transitions in Layered Materials*; D. Reidel: Dordrecht, 1986.
2. Levy, F. *Crystallography and Crystal Chemistry of Materials with Layered Structures*; D. Reidel: Dordrecht, 1976.
3. Wieting, T. J.; Schluter, M. *Electrons and Phonons in Layered Crystal Structures*; D. Reidel: Dordrecht, 1979.

4. Lee, P. A. *Physics and Chemistry of Materials with Layered Crystal Structures*; D. Reidel: Dordrecht, 1976.
5. Feldman, Y.; Frey, G. L.; Homyonfer, M.; Lyakhovitskaya, V.; Margulis, L.; Cohen, H.; Hodes, G.; Hutchison, J. L.; Tenne, R. *J. Am. Chem. Soc.* **1996**, *118*, 5362.
6. Feldman, Y.; Lyakhovitskaya, V.; Tenne, R. *J. Am. Chem. Soc.* **1998**, *120*, 4176.
7. Frey, G. L.; Elani, S.; Homyonfer, M.; Feldman, Y.; Tenne, R. *Phys. Rev. B.* **1998**, *57*, 6666.
8. Frey, G. L.; Tenne, R.; Matthews, M. J.; Dresselhaus, M. S.; Dresselhaus, G. *Phys. Rev. B.* **1999**, *60*, 2883.
9. Homyonfer, M.; Alperson, B.; Rosenberg, Y.; Sapir, L.; Cohen, S. R.; Hodes, G.; Tenne, R. *J. Am. Chem. Soc.* **1997**, *119*, 2693.
10. Rothschild, A.; Popovitz-Biro, R.; Lourie, O.; Tenne, R. *J. Phys. Chem. B.* **2000**, *104*, 8976.
11. Rothschild, A.; Sloan, J.; Tenne, R. *J. Am. Chem. Soc.* **2000**, *122*, 5169.
12. Tenne, R.; Homyonfer, M.; Feldman, Y. *Chem. Mater.* **1998**, *10*, 3225.
13. Zak, A.; Feldman, Y.; Alperovich, V.; Rosentsveig, R.; Tenne, R. *J. Am. Chem. Soc.* **2000**, *122*, 11,108.
14. Côté, M.; Cohen, M. L.; Chadi, D. J. *Phys. Rev. B.* **1998**, *58*, R4277.
15. Balchin, A. A. In *Crystallography and Crystal Chemistry of Materials with Layered Structures*; Levy, F., Ed.; D. Reidel: Dordrecht, 1976, Vol. 2, p. 1.
16. Evans, B. L. In *Physics and Chemistry of Materials with Layered Structures*, Lee, P. A., Ed.; D. Reidel: Dordrecht, 1976, Vol. 4, pp. 1.
17. Kam, K. K.; Parkinson, B. A. *J. Phys. Chem.* **1982**, *86*, 463.
18. Acrivos, J. V.; Liang, W. Y.; Wilson, J. A.; Yoffe, A. D. *J. Phys. C.* **1971**, *4*, L18.
19. Beal, A. R.; Hughes, H. P. *J. Phys. C.* **1974**, *12*, 881.
20. Ballif, C.; Regula, P. E.; Remskar, M.; Samsjines, R.; Levy, F. *Appl. Phys. A: Solids Surface* **1996**, *62*, 543.
21. Beal, A. R.; Liang, W. Y.; Hughes, H. P. *J. Phys. C.* **1976**, *9*, 2449.
22. Tributsch, H. *J. Electrochem. Soc.* **1978**, *125*, 1086.
23. Gerischer, H. In *Topics in Applied Physics*; Seraphin, B. O., Ed.; Springer-Verlag: Berlin, 1979, Vol. 31, p. 117.
24. Kertesz, M.; Hoffman, R. *J. Am. Chem. Soc.* **1984**, *106*, 3453.
25. Coehoorn, R.; Haas, C.; deGroot, R. A. *Phys. Rev. B.* **1987**, *35*, 6203.
26. Coehoorn, R.; Haas, C.; Dijkstra, J.; Flipse, C. J. E.; deGroot, R. A.; Wold, A. *Phys. Rev. B.* **1987**, *35*, 6195.
27. Kobayashi, K.; Yamauchi, J. *Phys. Rev. B.* **1995**, *51*, 17,085.
28. Li, Y.-W.; Pang, X.-Y.; Delmon, B. *J. Phys. Chem. A.* **2000**, *104*, 11,375.
29. Tan, A.; Harris, S. *Inorg. Chem.* **1998**, *37*, 2205.
30. Frindt, R. F., Yoffe, A. D. *Proc. R. Soc. Ser. A.* **1963**, *273*, 64.
31. Casadori, F.; Frindt, R. F. *Phys. Rev. B.* **1970**, *2*, 4893.
32. Brus, L. E. *J. Chem. Phys.* **1983**, *79*, 5566.
33. Brus, L. E. *J. Phys. Chem.* **1986**, *90*, 2555.
34. Liang, W. Y. *J. Phys. C.* **1973**, *6*, 551.
35. Liang, W. Y. *Phys. Lett. A.* **1967**, *24*, 573.
36. Helveg, S.; Lauritsen, J. V.; Laegsgaard, E.; Stensgaard, I.; Norskov, J. K.; Clausen, B. S.; Topsoe, H.; Besenbacher, F. *Phys. Rev. Lett.* **2000**, *84*, 951.

37. Bezverkhy, I.; Afanasiev, P.; Lacroix, M. *Inorg. Chem.* **2000**, *39*, 5416.

38. Mdleleni, M. M.; Hyeon, T.; Suslick, K. S. *J. Am. Chem. Soc.* **1998**, *120*, 6189.

39. Alonso, G.; Aguirre, G.; Rivero, I. A.; Fuentes, S. *Inorg. Chim. Acta.* **1998**, *274*, 108.

40. Afanasiev, P.; Xia, G.-F.; Berhault, G.; Jouguet, B.; Lacroix, *M. Chem. Mater.* **1999**, *11*, 3216.

41. Close, M. R.; Peterson, J. L.; Kugler, E. L. *Inorg. Chem.* **1999**, *38*, 1535.

42. Peterson, M. W.; Nenadovic, M. T.; Rajh, T.; Herak, R.; Micic, O. I.; Goral, J. P.; Nozik, A. J. *J. Phys. Chem.* **1988**, *92*, 1400.

43. Gutierres, M.; Henglein, A. *Ultrasonics* **1989**, *27*, 259.

44. Wilcoxon, J. P.; Samara, G. A. *Phys. Rev. B.* **1995**, *51*, 7299.

45. Wilcoxon, J. P.; Newcomer, P. P.; Samara, G. A. *J. Appl. Phys.* **1997**, *81*, 7934.

46. Huang, J. M.; Laitinen, R.; Kelley, D. F. *Phys. Rev. B.* **2000**, *62*, 10,995.

47. Huang, J. M.; Kelley, D. F. *Chem. Mater.* **2000**, *12*, 2825.

48. Chikan, V.; Kelley, D. F. *J. Phys. Chem. B,* **2002**, *106*, 3794.

49. Kaschak, D. M.; Johnson, S. A.; Hooks, D. E.; Kim, H.-N.; Ward, Mallouk, T. E. *J. Am. Chem. Soc.* **1998**, *120*, 10,887.

50. Kerimo, J.; Adams, D. M.; Barbara, P. F.; Kaschak, D. M.; Mallouk, T. E. *J. Phys. Chem. B.* **1998**, *102*, 9451.

51. Peng, X.; Wickham, J.; Alivisatos, A. P. *J. Am. Chem. Soc.* **1998**, *120*, 5343.

52. Sugimoto, T. *Adv. Colloid Interf. Sci.* **1987**, *28*, 65.

53. Chemsiddine, A.; Weller, H. *Ber Bunsen-Ges. Phys. Chem.* **1993**, *97*, 636.

54. Micic, O. I.; Sprague, J. R.; Curtis, C. J.; Jones, K. M.; Machol, J. L.; Nozik, A. J.; Giessen, H.; Fluegel, G.; Mohs, G.; Peyghambarian, N. *J. Phys. Chem.* **1995**, *99*, 7754.

55. Murray, C. B.; Norris, D. J.; Bawendi, M. G. *J. Am. Chem. Soc.* **1993**, *115*, 8706.

56. Leff, D. V.; Ohara, P. C.; Heath, J. R.; Gelbart, W. M. *J. Phys. Chem.* **1995**, *99*, 7036.

57. Williams, R. S.; Medeiros-Rebeiro, G.; Kamins, T. I.; Ohlberg, D. A. A. *J. Phys. Chem. B.* **1998**, *102*, 9605.

58. Williams, R. S.; Medeiros-Rebeiro, G.; Kamins, T. I.; Ohlberg, D. A. A. *Acc. Chem. Res.* **1999**, *32*, 425.

59. Lim, X. M.; Sorenson, C. M.; Klabunde, K. J. *Chem. Mater.* **1999**, *11*, 198.

60. Lim, X. M.; Sorenson, C. M.; Klabunde, K. J. *J. Nanoparticle Res.* **2000**, *2*, 157.

61. Lin, X. M.; Wang, G. M.; Sorenson, C. M.; Klabunde, K. J. *J. Phys. Chem. B.* **1999**, *103*, 5488.

62. Albrecht, A. *J. Mol. Spectrosc.* **1961**, *6*, 84.

63. Tao, T. *Biopolymers* **1969**, *8*, 609.

64. Evans, B. L.; Young, P. A. *Proc. R. Soc. A.* **1965**, *284*, 402.

65. Burda, C.; Link, S.; Mohamed, M.; El-Sayed, M. *J. Phys. Chem. B.* **2001**, *105*, 12,286.

66. Parsapour, F.; Kelley, D. F. in *Semiconductor Quantum Dots*; Materials Research Society Symposium Proceedings Vol. 571; Lee. H.; Moss, S.; Norris, D.; IIa, D. Eds. Warrendale, PA; 2000, p. 197.

67. Doolen, R.; Laitinen, R.; Parsapour, F.; Kelley, D. F. *J. Phys. Chem. B.* **1998**, *102*, 3906.

68. Huang, J. M.; Kelley, D. F. *J. Chem. Phys.* **2000**, *113*, 793.
69. Chikan, V.; Waterland, M. R.; Huang, J. M.; Kelley, D. F. *J. Chem. Phys.* **2000**, *113*, 5448.
70. Parkinson, B. A. *Langmuir* **1988**, *4*, 967.
71. Onganer, Y.; Yni, M.; Bessire, D. R.; Quitevis, E. L. *J. Phys. Chem.* **1993**, *97*, 2344.
72. DiPaolo, R. E.; Scaffardi, L. B.; Duchowicz, R.; Bilmes, G. M. *J. Phys. Chem.* **1995**, *99*, 13,796.
73. Chibsov, A. K.; Zakharova, G. V.; Gorner, H.; Sogulyaev, Y. A.; Muskalo, I. L.; Tolmachev, A. I. *J. Phys. Chem.* **1995**, *99*, 886.
74. Cooley, L. F.; Bergquist, P.; Kelley, D. F. *J. Am. Chem. Soc.* **1990**, *112*, 2612.
75. Kam, K. K.; Rath, D.; Parkinson, B. A. in *Photochemistry: Fundamental Processes and Measurement Techniques*; Electrochemical Society, Pennington, NJ, 1982, p. 537.
76. Kittel, C. *Introduction to Solid State Physics*, 7th Ed.; Wiley: New York, 1996.
77. Marcus, R. A. *Annu. Rev. Phys. Chem.* **1964**, *15*, 155.
78. Marcus, R. A. *J. Chem. Phys.* **1965**, *43*, 679.
79. Jortner, J.; Bixon, M. *J. Chem. Phys.* **1988**, *88*, 167.
80. Ernsting, N. P.; Kaschke, M.; Weller, H.; Katsikas, L. *J. Opt. Soc. Am. B.* **1990**, *7*, 1630.
81. Cavaleri, J. J.; Colombo, J.; Bowman, R. M.; *J. Phys. Chem.* **1998**, *102*, 1341.
82. Zhang, J. Z.; O'Neil, R. H.; Roberti, T. W.; McGowen, J. L.; Evans, J. E. *Chem. Phys. Lett.* **1994**, *218*, 479.
83. Zhang, J. Z.; O'Neil, R. H.; Roberti, T. W. *J. Phys. Chem.* **1994**, *98*, 3859.
84. Bawendi, M. G.; Wilson, W. I.; Rothberg, L.; Carroll, P. J.; Jedju, T. M.; Steigerwald, M. L.; Brus, L. E. *Phys. Rev. Lett.* **1990**, *65*, 1623.
85. Bawendi, M. G.; Carroll, P. J.; Wilson, W. L.; Brus, L. E. *J. Chem. Phys.* **1992**, *96*, 946.
86. Eychmüller, A.; Hässelbarth, A.; Katsikas, L.; Weller, H. *Ber. Bun. Ges. Phys. Chem.* **1991**, *95*, 79.
87. Kaschke, M.; Ernsting, N. P.; Weller, H.; Muller, U. *Chem. Phys. Lett.* **1990**, *168*, 543.
88. Klimov, V. I.; Schwarz, C. J.; McBranch, D. W.; Leatherdale, C. A.; Bawendi, M. G. *Phys. Rev. B.* **1999**, *60*, R2177.
89. Guyot-Sionnest, P.; Shim, M.; Matranga, C.; Hines, M. *Phys. Rev. B.* **1999**, *60*, R2181.
90. Klimov, V. I.; McBranch, D. W. *Phys. Rev. Lett.* **1998**, *80*, 4028.
91. Waggon, H.; Geissen, H.; Gindele, F.; Wind, O.; Fluegel, B.; Peyghambarian, N. *Phys. Rev. B.* **1996**, *54*, 17,681.
92. Underwood, D. F.; Kippeny, T.; Rosenthal, S. J. *J. Phys. Chem. B.* **2001**, *105*, 436.
93. Parsapour, F.; Kelley, D. F.; Craft, S.; Wilcoxon, J. P. *J. Chem. Phys.* **1996**, *104*, 4978.
94. Elliott, C. M.; Hershenhart, E. J. *J. Am. Chem. Soc.* **1982**, *104*, 7519.
95. Rossetti, R.; Nakahara S.; Brus, L. E. *J. Chem. Phys.* **1983**, *79*, 1086.
96. Chen, P.; Duesing, R.; Graff, D. K.; Meyer, T. J. *J. Am. Chem. Soc.* **1991**, *95*, 5850.
97. Malone, R. A.; Kelley, D. F. *J. Chem. Phys.* **1991**, *95*, 8970.
98. Waterland, M. R.; Kelley, D. F. *J. Phys. Chem. A.* **2001**, *105*, 4019.
99. Parsapour, F.; Kelley, D. F. unpublished results.

6

Mechanisms of the Photocatalytic Transformation of Organic Compounds

**Claudio Minero, Valter Maurino, and
Ezio Pelizzetti**
University of Torino, Torino, Italy

I. INTRODUCTION

The decontamination of polluted waters by photocatalytic treatment has been suggested as a viable, low-cost, and environmental friendly technique. The possibility of using solar energy as well as low-cost technology pushed to develop solar plants based on this process. The field results are very encouraging [1].

Understanding of the chemical pathways by which the degradation of organic pollutants is achieved is of considerable practical interest due to the possibility of increasing the degradation rate and of achieving the complete mineralization. The completion of mineralization will assure total degradation and environmental safe discharge, without the need of further treatment, and with reduced safety and/or health risks. The increase of the rate will decrease the solar-plant-collecting surface, reducing equipment costs, and both the power consumption and installation costs in the case of artificial light sources. In both cases, the need for improved photon utilization would require the calculation of the photon efficiency or the related quantum yield of the process. In this case, the notion of relative photon efficiency based on a reference compound (e.g., phenol [2]) is of little usage.

211

The knowledge of the chemical pathways of degradation is also of interest for forecasting the intermediate products, their time evolution, the treatment times, and the eventual toxicity of the effluent, because changing the process conditions could form different species at dissimilar concentrations. This problem would address the environmental compatibility of the process.

Finally, the chemical pathways of degradation are of fundamental chemical interest because the reaction mechanism proceeds predominantly through radicals in the presence of aqueous systems, a field largely open to investigation.

In early works, the reactivity associated with TiO_2 photocatalysis was assigned to free OH radicals, but more recently the oxidative species has been identified as bound HO radicals and photogenerated holes. Reaction of the substrate with these species would require either the direct adsorption of organic compounds to the semiconductor interface or Fickian diffusion of the substrate to the semiconductor surface under depletion conditions. However, these two pathways imply different mechanisms, and in some cases different reaction products or intermediates.

This chapter reviews the current literature on the reaction mechanism following the interaction of an organic substrate with the semiconductor particle. It has been organized in subject matters that directly follows from this introduction. Some previous reviews covering the literature before 1997 [3–5] may be consulted for a historical valuation of the progress in the field.

II. SURFACE LOCATION OF THE PROCESS

The basic mechanism was treated in a number of articles [6]. Following the light absorption, primary excited species are formed which can either recombine or migrate to the surface of the semiconductor, where several redox reactions may take place. The organic substrate reacts with formed active species (oxidant or reducing) depending on its initial oxidation state and the nature of substituents [7], forming radicals and other species that are further oxidized or reduced. Several complex networks of reaction have been reported on the basis of detailed chemical analyses of the time evolution of the substrate and formed intermediates or by-products [8–11].

Convincing evidence that photogenerated species do not migrate far from the catalyst surface has been derived from the study of the degradation of decafluorobiphenyl (DFBP), a substrate that is strongly absorbed (>99%) on alumina and TiO_2 [12]. When DFBP is absorbed on alumina particles and mixed with titania particles, the amount exchanged is very low (<5%). Irradiation of DFBP adsorbed on alumina in the presence of H_2O_2 (generating OH in solution) or titania colloids (supposed to generate OH free or bound to the surface on mobile particles) leads to degradation, whereas DFBP is not degraded when larger titania particles (P25, TiO_2 beads) are present. Pentafluorophenol, which is easily exchanged between the two oxides, is photodegraded in all cases. This makes it clear that active species formed upon irradiation do not migrate in solution, and the organic substrate may or not migrate to the catalyst surface. Only when both are present at the catalyst surface does the degradation take place.

Thus, the possibility of adsorption is of primary importance. Adsorption may originate either from chelating properties of the organic substrate toward surface metal species or, because of the low hydrophobicity of the metal oxide surface, from the expulsion of the organic molecules from the solution for entropy reasons. Because there is depletion of substrate at the catalyst surface when degradation takes place, migration from the solution is assisted by a concentration difference in the two environments.

Some additional complexity arises from the possibility of different adsorption sites and the presence of pores, which reflect in nonideal adsorption isotherms and mass-transfer problems. The mass transport can be relatively slow in pores and interparticle spaces [13], as it is the case of P25, for which, in suspension, there are particles ranging from 0.2 to 2 μm, formed by 30-nm-sized primary particles. In such spaces, the diffusion coefficient is comparable to liquid diffusion in zeolites.

III. KINETIC MODELS

Because of the complex network of reactions, even for a chosen substrate it is difficult to model the dependence of the degradation rate on the experimental parameters for the whole treatment time. All of the possible formed species must be identified; their kinetic constants for reaction either with photogenerated species or with other active species (radicals, other intermediates) have to be measured or estimated. Because this is an enormous task, kinetic modeling of the photocatalytic process is usually restricted to the analysis of the initial rate of degradation r_0. Even under the reduced complexity deriving from the initial rate simplification, the interpretation of the rates obtained in photocatalytic experiments is still under debate. This implies either the understanding of primary events or the modeling of the reactor. Although crucial for the interpretation of kinetic data and the resulting kinetic parameters of the model, this last point will not be further developed here. The reader could refer to some key articles [14–18].

The primary events are deeply described in the literature [5] and are summarized in Eqs. (1a)–(1f) for the oxidative pathways in which a substrate Red_1 is involved:

Charge separation

$$TiO_2 \xrightarrow{k_1\phi} e^- + h^+ \tag{1a}$$

Recombination

$$e^- + h^+ \xrightarrow{k_2} heat, light \tag{1b}$$

Interf. charge transfer

$$h^+ + Red_1 \xrightarrow{k_3} Ox_1^{\cdot} \tag{1c}$$

$$e^- + Ox_2 \xrightarrow{k_4} Red_2^{\cdot} \tag{1d}$$

Degradation

$$Ox_1^{\cdot} + h^+ \xrightarrow{k_5} Ox_1 \tag{1e}$$

$$Ox_1^{\cdot} + Ox_2 \xrightarrow{k_6} Ox_4^{\cdot} \tag{1f}$$

The primary photochemical act, following the near-UV (ultraviolet) light absorption by TiO_2 ($\lambda < 380$ nm), is the generation of electron – hole pairs in the bulk of the semiconductor [Eq. (1a)]. The charge carriers can either recombine in the lattice [Eq. (1b)] or migrate rapidly to the surface where they are ultimately trapped; the electron as a surface Ti(III), and the hole as a surface radical hydroxyl group, $\equiv TiO^{\cdot}$, following a release of H^+ depending on pH. In the presence of surface complexation by ligands other than OH^-, possible trapping by these surface species is also possible. When electron donors (reduced species Red_1, solvent) are present at the surface or absorbed, interfacial electron transfers may occur to holes and $\equiv TiO^{\cdot}$ forming $Ox_1^{\cdot}$ [Eq. (1c)] that is further oxidized either by active species [e.g., (1e)] or Ox_2 [Eq. (1f)]. Conduction-band electrons are scavenged by electron acceptors (usually oxygen in aerated systems or other oxidants, Ox_2) to form Red_2 [Eq. (1d)]. Characteristic times for the various steps range from 10 ps to 100 ns, except for electron scavenging by Ox_2 reported as slow as milliseconds [5]. Electron transfer to oxygen may be the rate-limiting step [19]. More efficient electron scavengers like peroxydisulfate may promote the overall degradation rate [20]. To complicate the system, species such as O_2, H_2O_2, and O_2^- may also be directly involved in the transformation of organic compounds (Red_1).

The common shape observed for r_0 as a function of the substrate or catalyst concentration depending on the primary events depicted in Eqs. (1a)–(1f) is normally reported of saturative type as if the rate would be determined by adsorption properties of the substrate on the catalyst surface according to a Langmuir isotherm:

$$r_0 = \frac{k_{\mathrm{LH}} \, KC_0}{1 + KC_0} \tag{2}$$

Equation (2), usually referred as the Langmuir–Hinshelwood (LH) equation, interprets the rate as the product of a specific rate constant k for reaction of photogenerated surface species with the absorbed substrate $r = k_{\mathrm{LH}} \{\mathrm{Red}_1\}$, the extent of absorption being determined by K. The role of other species was coherently interpreted as a competition for absorption, adding to the denominator the proper terms $(K_i C_i)$ for competing species i [17,21]. The role of the electron scavenger Ox_2 is often taken into account *assuming* a rate law of the form $r = k_{\mathrm{LH}} \{\mathrm{Red}_1\} \{\mathrm{Ox}_2\}$, which is not supported by any microscopic analysis of the system. An empirical rate law of the form $r = k\{\mathrm{Red}_1\}^m \phi^h$ was also proposed [22], where ϕ is the photon flux and $k_{\mathrm{LH}} = k\phi^n$. This last approach fit each individual kinetic experiment but with different power-law exponents m and n, suggesting an intriguing dependence of all experimental variables without explicitly expressing their intimate relationship.

In general, there is no correspondence between the value of K obtained from the fit of kinetic data through Eqs. (1a)–(1f) and dark adsorption measurements. The degradation rate of phenol (ph, poorly adsorbed) and nonylphenol (nph, strongly adsorbed) differs only by a factor of 3 [23]. Because it was demonstrated that the aromatic moiety is more susceptible of attack than the aliphatic chain, k_{LH} would be almost identical in the two cases. Owing to the large ratio of $K_{\mathrm{nph}}/K_{\mathrm{ph}}$ ($>>3$), it follows that the LH equation is inadequate.

Kormann et al. [24] have found that a two-site Langmuirian model is necessary to model the observed rate of reaction for chloroform concentration >1 mM and suggested two different binding sites.

The deficiency of the LH equation was demonstrated also for pentachlorophenol [25] and by Cunningham [26–28]. The quantum yields for photodegradation of salicylate and other strongly sorbing substituted benzenes were less than those measured for nonsorbing chlorophenols. For salicylic acid, which chemisorbs, the photo-oxidation rates in air-saturated solutions are independent of the salicylic acid concentration, although the surface excess increases with the concentration [29].

The approach based on the LH equation, or the related Eley–Rideal mechanism [30], is common in the literature today [5]. The equation is generally derived assuming equal steady-state concentrations of charge carriers or their concentration independent of substrate and the electron scavengers' concentration. The independence of the actual concentration of active species (either h^+ or e_{CB}) is also usually assumed by taking a mean time of survival of these species independent on other system concentrations. However, it is obvious from the oldest literature that the concentrations of the active species are mutually dependent. The first factor is the recombination in bulk. Bahnemann et al. derived the transient absorption spectra of photogenerated trapped holes and electrons using electron and hole scavengers, respectively [31]. Several kinetic models have been developed to generate the functional rate dependence as in Eq. (2). Turchi and Ollis, first of all, proposed four different mechanistic schemes involving an OH radical attack on the adsorbed substrate as the rate-determining step [32]. Although photodegradation studies were performed at a TiO_2 electrode operated at a constant current density, as a support to the relationship $r = k_{LH}\{Red_1\}$, there is the linear correlation of the photocatalytic reaction rate and the concentration of the strongly adsorbed 4-chlorocatecol [33].

Under the hypotheses of constant illumination intensity, fast reaction of the electron scavenger with photogenerated electrons, and steady-state conditions applied to e^-_{CB} and h^+_{VB}, a functional form like Eq. (2) was obtained without invoking adsorption [34]. The rate expression was given as in the LH model by reaction of surface-active species with the substrate, in which $k_{LH} = k\{h\}$, and where $\{h\}$ is the surface concentration of any oxidative active species. No assumptions were made on the steady-state concentrations of conduction-band electrons, valence-band holes, and other transient species. The rate is given by

$$r_0 = \frac{k_1 \phi k_4 k_3 \{Ox\}\{Red\}}{k_2 k_1 \phi + k_4 k_3 \{Ox\}\{Red\}} \tag{3}$$

where k_1 takes into account the photon absorption, the charge separation, and the migration of carriers to the catalyst surface; for the other constants refer to Eq. (1). Equation (3) shows a Langmuirian shape as a function of light flux, adsorbed substrate, and oxidant concentration, and independent of the partition properties of the species. These can be obviously taken into account by relating surface concentrations to bulk concentrations through proper adsorption isotherms. After a deep investigation of the implication for the time evolution of the substrate and the rate dependence on the experimental conditions, it was concluded that although experimental data can be fitted by an approximate kinetic solution which has the analytical form of a LH equation, the parameters of such working model equation do not have a physical significance [25]. This functional form was also empirically generalized to obtain a manageable two-parameter equation for application to reactor design [25].

However, under some circumstances, the rate is not only leveled but it is also inhibited as the substrate concentration increases. In a comprehensive study on 3-chloro-4-hydroxybenzoic acid and chlorophenols, Cunningham and colleagues [35] deduced that the rate can be strongly inhibited in consequence of (1) chemisorption-induced depletions of surface —OH groups, (2) adsorbate-enhanced hole–electron recombination on the TiO_2 surfaces, (3) mass transport limitations within the TiO_2 particle aggregates. They reported that, depending on

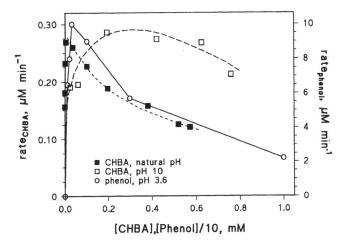

Figure 1 Degradation rate of phenol [36] and 3-chloro-4-hydroxybenzoic acid (CHBA) [35] as a function of its their concentration. (Adapted from Refs. 36 and 35, respectively; see the references for experimental conditions.)

experimental conditions for some substrates, the simple LH model fails. The rates may be non-Langmuirian, peaking at some optimal concentration.

To our experimental knowledge, a relationship more complex than a simple LH model is, indeed, the common behavior. Figure 1 reports the degradation rate of phenol [36], a poorly adsorbed compound, and that of CHBA [35], a strongly absorbed compound, as a function of their initial concentration. A peaked reaction rate is observed, in contrast to the saturative LH model, also for $CHCl_3$ and dodecane (see Fig. 2 in Ref. 37). For the photocurrents measured in photoelectrochemical oxidation experiments of methanol and salicylic acid on anatase film electrodes, a saturation curve for the poorly adsorbed methanol and a peak at an intermediate concentration for strongly adsorbed salicylic acid were also observed as a function of the substrate concentration [38].

None of the previous kinetic models is able to predict these peaked shapes, suggesting that some unreasonable simplification or some key reaction be neglected in the chemical model. The second criticism to the adsorption-based kinetic models relies, in fact, on the very nature of the primary processes. Competitive with the previous processes are the interfacial recombination ($e_{CB}^- + \equiv Ti—O^{\cdot} + H^+ \rightarrow \equiv TiOH$) and secondary surface trapping ($OH^{\cdot} + \equiv Ti—OH \rightarrow \equiv TiO^{\cdot} + H_2O$) [39]. In addition, when the degradation takes place, intermediate products like $Ox_1^{\cdot}$ and $Red_2^{\cdot}$ may be confined at the surface or in the catalyst pores, being able to back-react with active species, as recently outlined by Cunningham and colleagues [35]. These processes are summarized as follows:

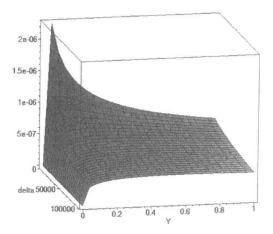

Figure 2   Model calculation according to Ref. 37 of the photocatalytic rate as a function of parameters $Y = \alpha_2$ and $Z = \alpha_3$ (see text), given $\phi = 0.001$ and $\alpha_1 = 0.01$. Note that the rate shows a maximum which position depends on the photon flux. At low Z, the rate shape is comparable to those observed in Fig. 1. At high Z, a Langmuirian shape is produced.

Back reactions

$$\text{Red}_2^{\bullet} + \text{h}^+ \ (\text{or} \equiv \text{Ti} - \text{O}^{\bullet}) \xrightarrow{\ k_7\ } \text{Ox}_2 \tag{1g}$$

$$\text{Ox}_1^{\bullet} + e_{\text{CB}}^- \xrightarrow{\ k_8\ } \text{Red}_1 \tag{1h}$$

These reactions would produce a net e^-/h^+ recombination and, consequently, net null cycles, decreasing the rate of degradation [37]. As for salicylic acid cited earlier, the formation of surface-adsorbed cation radicals was invoked to increase the charge-carrier recombination rate and thus lower the quantum yields for degradation [27]. Added redox reagents can also act as recombination sites, so that the overall photocatalytic rate could increase, decrease, or remain unchanged [40].

The hypothesis that electron scavenging by oxygen or other oxidants is faster than electron–hole recombination is perhaps too restrictive, as it is now recognized that this reaction, in particular with oxygen, can be rather slow (see Ref. 19 and the values of estimated rate constants in Ref. 41). When Ox_2 acts as a mere electron scavenger, reaction (1g) could be neglected. The rate

$$\text{Rate} = k_3\{\text{h}^+\}\{\text{Red}_1\} - k_8\{e^-\}\{\text{Ox}_1^{\bullet}\} \tag{4}$$

was evaluated on the basis of Eqs. (1a)–(1f) and (1h), assuming the steady state for $\{e^-\}$, $\{h^+\}$, and $\{Ox_1^{\cdot}\}$, giving an analytical solution for the initial rate [37] as a function of the following three parameters α_i:

$$\alpha_1 = k_4k_5\{Ox_2\} \tag{5a}$$

$$\alpha_2 = k_3k_8\{Red_1\} \tag{5b}$$

$$\alpha_3 = 16\phi k_2k_8k_5 \tag{5c}$$

which account for experimental conditions and catalytic system characteristics. The quantum yield $\eta = r_0/\phi$ (initial degradation rate of the primary compound/absorbed photon flux) will depend on α_i. As previously noted [34], it will depend on the system on which it is measured (substrate, catalyst, its type, texture, morphological form, and light intensity) and hardly permits relative estimation as in Ref. 2. Although quite involved, the equation proved successful for the engineering calculation of a lamp-based photocatalytic pilot plant able to treat 1000–2000 ppm TOC effluents at flux of 400–500 L/h^{-1} [42].

The main result of the reported kinetic analysis is that the rate *always* has a maximum as a function of α_2, given α_1 and α_3, suggesting that it is a general kinetic characteristic of the photocatalytic systems. The rate has a maximum only with respect to α_2. The maximum is sharpest for small values of α_3, and modest values of α_1 (i.e., low light flux or concentration of catalyst, or small recombination constant) and small values of α_2 (see Fig. 2). This agrees with the measurements of Mills [43] on the effect of the light intensity. Given α_2 as in Eq. (5), depending on the type of substrate, the observed rate maximum (at constant ϕ or α_3) *will appear at comparatively smaller concentrations of substrate, as the back reaction will be more important* (see data in Fig. 1). After the maximum, the rate decreases, as evidenced by recent observations of Cunningham and colleagues [35], who reported a negative-order dependence on $\{Red_1\}$. Some dependences of the rate versus α_2 are reported in Fig. 2. A more general analysis is reported in the original article [37].

The product $k_8\{Ox_1^{\cdot}\}$ in reaction (1h) will increase, resulting in a higher apparent k_8 if the outward diffusion ofOx$_1^{\cdot}$ is limited by confinement in restricted spaces (e.g., pores) or it is is tenaciously adsorbed. This explains the "strange" rate behavior observed for strongly adsorbed compounds cited earlier, and the reverse shape selectivity observed in the presence of titania supported zeolites [44]. According to the solution reported [37], the rate (or η) versus concentration has, in some cases (see Fig. 2), a Langmuirian shape showing that the predictions of the Langmuir–Hinshelwood equation are a coincidence. Following this kinetic model, the zero-order dependence of the observed rate (or quantum yield) on the concentration of catalyst and substrate is not basically due to the surface adsorption or availability of surface sites, which are not considered in the kinetic model, but to the back reaction of oxidation–reduction of intermediate products. Anyway, the decrease after the maximum along α_2 has no explanation in term of adsorption.

In the case that there is no back reaction (e.g., $k_8 = 0$), the rate $= k_3 \{h^+\}$ $\{Red_1\}$ is given by

$$\text{Rate} = \gamma\left\{\left(1 + \frac{\phi}{\gamma}\right)^{1/2} - 1\right\} \tag{6}$$

where $\gamma = k_3 k_4 \{Red_1\}\{Ox_2\} / (2k_2)$ refers to a given substrate (k_3), a given oxidant (k_4; e.g., O_2), and a given semiconductor (recombination rate k_2) under proper light flux/geometry of absorption (ϕ) [37]. This reduced kinetic solution shows that, in the absence of the back reaction [reaction (1h)], the rate increases with a square-root-like dependence on light intensity (ϕ), $\{Ox_2\}$, and $\{Red_1\}$. It is never possible that the rate has a maximum or a true plateau for some conditions, and possibly decreases at high $\{Red_1\}$. The rate increases as the oxidant becomes more powerful (k_4), the substrate is less recalcitrant to oxidation (k_3), or the recombination reaction is decreased (k_2). Thus, the Langmuirian or the bell shape of the rate is due to the back reaction [reaction (1h)], which dominates the kinetic performance and photon yield of photocatalytic systems.

IV. NATURE OF ACTIVE SPECIES

The previously outlined kinetic models are silent on the very nature of the active species at the surface of the catalyst, although the importance of the back reactions strongly depends on their intimate nature. Their sort and role for the degradation of organics on irradiated titania have been investigated in depth [3,21,45–47]. After the photoexcitation of the semiconductor, the primary oxidant species, h^+, reaches the surface and reacts with surface hydroxyl groups or molecular water. The trapped hole is usually described as an adsorbed hydroxyl radical($\equiv$Ti—O˙), whose nature is probably represented by the spin resonance between the hydroxyl and lattice —O—. However, the role of direct electron transfer to the hole versus ˙OH or $\equiv$Ti—O˙ radical oxidation of the organics was controversial [48–51]. Although most evidence favor the ˙OH radical mechanism, hole oxidation has been suggested for compounds lacking abstractable hydrogen and for some aromatic compounds [52].

In several instances, the same intermediates formed from OH radical generating systems (e.g., pulse radiolysis, Fenton's reagent, and UV/H_2O_2) are also involved in heterogeneous photocatalytic processes [21,53,54]. Indirect kinetic evidence point to the presence of a *surface-bound* hydroxyl radical on irradiated TiO_2 particles in aqueous suspensions [39]. Nonetheless, time-resolved diffuse reflectance spectroscopic measurements indicate the transient formation of inter- mediates derived from single electron transfer (SET) from a variety of organic and inorganic substrates [52]. Some *s*-triazine derivatives without alkyl groups (e.g., 2,4-diamino-6-chloro-1,3,5-triazine), which are reported as not reactive toward OH radicals [55], are slowly degraded through oxidative pathways to cyanuric acid under photocatalytic conditions [9]. Experiments with isotopically labeled 2,4-dichlorophenoxyacetic acid (2,4-D) [56] and the comparison of intermediates formed during either the photocatalytic degradation or the O_3/UV treatment of butanoic acid [57] indicate that the exact mechanism (OH versus SET to a hole) is influenced by both the ionization of the substrate and the surface charge of the photocatalyst, which, in turn, control the extent of surface complexation and hole oxidation potential.

Very recently, a major revision of the chemical pathways operating in the photocatalytic conversion of organic substrates was undertaken. These studies were carried out through very careful product analysis [58,59], by using selected substrates with similar chemical properties but different sorption behavior [60], or by surface modification through complexation by redox inactive species [61,62].

The exhaustive studies on 4-chlorophenol photocatalytic degradation [58] pointed out a heavy implication of SET steps in relation to the formation of ring-opening products. The well-established first step is mediated through OH-like chemistry, with the formation of both hydroquinone (ipso attack) and 4-chlorocatechol. The transformation proceeds further with the introduction of a third OH group. Tetrahydroxylated compounds are hardly detected. On the contrary, a series of muconic acid derivatives are present in relevant amounts, clearly deriving from the ring opening of trihydroxylated compounds (Fig. 3). Although possible for any aromatic compound, the ring cleavage is prominent with respect to hydroxylation only when ortho-hydroxy groups are present. The details of the ring-opening mechanism remain somewhat speculative. However taking into account that: (1) the ortho-hydroxy arrangement lead to strong adsorption [46,61] and (2) the more highly oxidized structures have inherently lower oxidation potentials, making electron transfer more favorable, the reported results are consistent with a degradation chemistry driven by a SET process.

After electron transfer to benzene derivatives, the adsorbed substrate is a radical cation or a semiquinone. This reactive specie can back react with a surface e_{CB}, reverting to the initial compound and acting as recombination center. Further transformations, and among them ring opening, should be promoted by reaction with O_2 or superoxide [63]. Although O_2 is surely more abundant, Pichat and co-workers have carried out a series of experiments using superoxide dismutase that seem to strongly indicate superoxide as a key reactant when a radical cation intermediate is envisaged [64,65].

These results are further confirmed by the comparison of products formed during the degradations of 1,2,4-trihydroxybenzene and its methoxylated products [60]. The capping of exposed benzene hydroxy groups causes a dramatic change in the products distribution in the early degradation steps. With the increase of benzene hydroxy groups capping, hydroxylation and demethylation become predominant over ring-opening chemistry. Taking into account that (1) there is

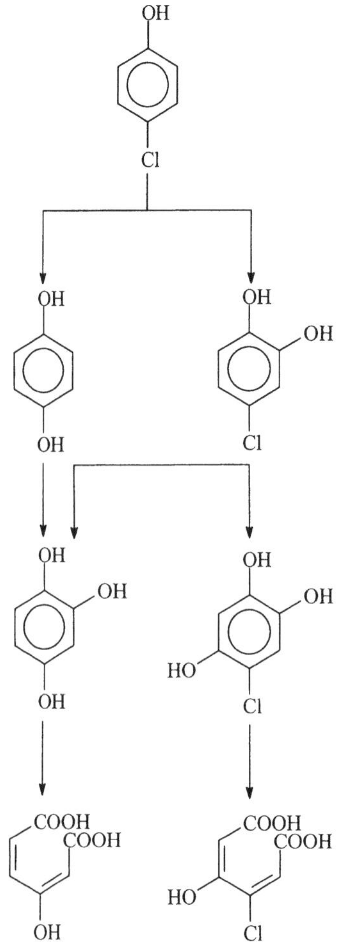

Figure 3 Major pathways of 4-chlorophenol degradation as derived from product analysis. (From Refs. 58 and 59.)

a limited influence of OH radical scavengers (alcohols) in the degradation kinetic of less capped substrates, which give predominantly ring opening intermediates, and (2) the less capped substrates are more tightly bound to the photocatalyst surface, one can assume a SET-driven chemistry for the less methoxylated substrates.

Another example of the role of SET is found in the transformation of carboxylic acids. Hole oxidation has been invoked to explain the TiO_2 photocatalyzed degradation of acidic compounds lacking abstractable hydrogen or C—C unsaturation (e.g., oxalate and trichloroacetate [50]). When SET and OH attacks are both possible, the prevailing pathway depends on the pH and on the surface complexation ability of the carboxylate anion. The photocatalytic degradation of 2,4-D over TiO_2 at pH 3 gives the expected products from SET from the carboxyl group. A near-stoichiometric yield of $^{14}CO_2$ from [carboxy-^{14}C]-2,4-D, high yields of 2,4-dichlorophenol and 2,4-dichlorophenol formate, and little inhibition by the OH radical scavenger are observed. At this pH, the extent of inner-sphere complexation between the TiO_2 surface and 2,4-D should be at maximum due to the low OH^- competition, the positively charged surface, and ionization degree in water greater than 50% for 2,4-D. At higher and lower pHs, the yields of $^{14}CO_2$ and the mentioned products are lower, indicating that the aromatic ring is involved. Under these conditions, the transformation rate is strongly inhibited by OH radical scavengers [56].

Interestingly, when carboxylate derivatives are considered, the SET process is irreversible (due to release of CO_2) and the degradation rate is maximum when this mechanism is supposed to be operating. On the contrary, when a radical cation or a semiquinone intermediate is invoked, as in the case of quinoline or hydroxy-methoxy-chloro benzenes [60,64], the rate is minimum due to the back reaction.

The key requirement for a SET step in the photocatalytic process seems to be the surface complexation of the substrate, according to an exponential dependence of the probability of electronic tunneling from the distance between the two redox centers [66]. However, as was pointed out in the preceding section on the key role of back reactions, the presence of a SET mechanism could be a disadvantage from an applicative point of view. If the formed SET intermediate (e.g., a radical cation) strongly adsorbs and/or does not transform irreversibly [e.g., by loss of CO_2 from a carboxylic acid or fast reaction with other species (e.g., superoxide or oxygen)], it can act as a recombination center, lowering the overall photon efficiency of the photocatalytic process.

From the above discussion, a major change in phototransformation mechanisms of organic substrates can be envisaged if the surface of TiO_2 is covered by a strongly adsorbed ligand with high inertness toward redox reactions. Fluoride ions fulfill both conditions [61]. The redox potential of the couple $F^·/F^-$ is 3.6 V, making the fluoride stable against oxidation by TiO_2 valence-band holes, even in acidic media. Taking into account the acid dissociation constants of HF and that of TiO_2, surface speciation calculations indicate that around pH = 3.6, the TiO_2 surface is covered by F^- with almost complete substitution of surficial OH groups. Competitive adsorption experiments with catechol indicate an efficient removal of surface cathecolate complexes by fluoride at pH 3.6 (see ref. 61 for details).

Light-induced transformations over fluorinated titania (TiO_2/F) *cannot* be initiated either by $\equiv Ti—O^\bullet$ (OH_{ads}), due to the lack of $\equiv Ti—OH$ groups, or by SET from a surface complexed substrate, due to the fluoride competition. In addition to these major effects, the adsorption of molecular oxygen can be affected also and the surface charge is dramatically decreased. The last effect may be important particularly for charged substrates and intermediates and for the possibility of interfacial electron transfer.

When the photocatalytic degradation of phenol is considered, the most striking features of the TiO_2/F with respect to naked TiO_2 are as follows:

1. The marked increase in the photodegradation rate, which shows a bell-shaped dependence on pH, reflecting the distribution of $\equiv Ti—F$. Because the degradation cannot be driven either by $\equiv TiO^\bullet$, or by SET, the processes should be mediated by OH_{free} formed through SET from the solvent (water).

2. The photodegradation rate dependence on phenol concentration in the TiO_2/F system shows a plateau in the $3 \times 10^{-4}\ M$ to $3 \times 10^{-3}\ M$ range, whereas for naked TiO_2, a maximum is reached around $2 \times 10^{-2}\ M$ of phenol followed by a decrease. This behavior is rationalized by the possibility of reductive back reactions of intermediates formed after the first oxidation step. The presence of fluoride could limit the occurrence of this detrimental effect, reducing the interaction of the formed intermediates with the surface. Moreover, also the change of the oxidation pathway changes the amount of the products.

Moreover, the oxidative transformation rates of compounds lacking abstractable (e.g., trichloroacetate and trifloroacetate) hydrogen are *inhibited* by the presence of fluoride, due to the suppression of the SET pathways [67].

The results reviewed show how the surface complexation by inorganic anions is a very useful tool for mechanistic studies in photocatalysis. The TiO_2/F system allowed the elucidation of the very nature of active species involved in photocatalysis and the quantification of their relative roles. A detailed kinetic analysis of competition experiment with different OH radical scavengers allowed the quantification of the relative role of direct (SET) and mediated (through OH radical, either free or adsorbed) oxidation pathways in the photocatalytic degradation of phenol. The results showed that the transformation of phenol over TiO_2/F proceed almost entirely through mediated oxidation by free OH, whereas on naked TiO_2, about 10% is due to a direct hole oxidation and 90% is attributable to $\equiv TiO^\bullet$ [62].

V. OXIDATIVE VERSUS REDUCTIVE PROCESSES

The previous discussion refers mainly to the key factors affecting the performance of a photocatalytic system in which oxidative pathways are predominant. Oxygen,

according to Eqs. (1e) and (1f), could acts as a mere scavenger or directly participate to the reaction scheme. Molecular oxygen is known to add very rapidly to
both aliphatic and aromatic radical species. After the direct electron transfer
between the organic molecule and the excited semiconductor, or processes mediated by adsorbed $^\bullet$OH, the formed organic radical may add O_2, forming more
oxygenated compounds on the route to CO_2 formation. The pathways involving
O_2 or $O_2^{\bullet-}$ are the sole processes operating in most of photocatalytic degradations
where routes for direct reduction/oxidation are not allowed.

It was pointed out that during the development of the photocatalytic transformations in the presence of oxygen, the average oxidation number of carbon
n_C in the system increases [7] at a rate depending on the initial n_C of the substrate.
Although for compounds in which n_C is low, direct reaction with hole or bound/
free hydroxyl radicals leads to oxidation (increase of n_C), for compounds in which
the carbon is at the highest oxidation number, the very redox nature of the photocatalytic process implies that reductive pathways should be operating. As the
final product in the presence of oxygen is CO_2, n_C should reach the $+4$ value.
This is evident for dodecane and CH_2Cl_2. For CCl_4, n_C should not change from
the reactant to the final product ($+4$ in CCl_4 and $+4$ in CO_2). However, in
this case, an initial decrease of n_C was observed. Evidently, but necessarily, the
reduction pathway is largely predominant in the early part of the process. In the
case of $CHCl_3$, a smooth variation of n_C was observed during the initial steps of
reaction. This is indicative of concomitant oxidative and reductive pathways,
even in the presence of oxygen. For benzoquinone, for which $n_C = 0$, hydroquinone was formed with an 80% yield with respect to the initial p-benzoquinone
concentration after 1 min of irradiation [8]. Quinonoid structures appear to be
efficient e_{CB}^- scavengers, even in aerated solutions.

The dependence of the degradation rate, or the rate of CO_2 formation, on
the partial pressure of oxygen [68,69] was reported to follow a Langmuirian
behavior. This dependence is very well described by the kinetic model developed
by Minero (see above) [37]. However, this is usually valid when oxygen has
the role of an electron scavenger (i.e., in the presence of compounds having
medium–low values of n_C, such as CH_2Cl_2 [11], C_6H_{14}, phenol, 2- [69], and 4-
chlorophenol [18]. In the other cases, the presence of oxygen can be detrimental,
as it is the case for CCl_4. In this case, oxygen is competitive for reduction with
the organic and, as a consequence, has a detrimental effect [11].

When nitrogen-containing organic compounds are considered, the formation of ammonium ions from nitroderivatives (e.g., nitrobenzene [8] and tetranitromethane [70]) cannot be justified by photocatalytic transformation of nitrogen-
containing inorganic species and should be assigned, even in the presence of
oxygen, to reduction processes occurring at the nitrogen substituent before it
is released. This has been demonstrated by investigating the interconversion of
nitrogen-containing benzene derivatives in the set nitrobenzene, nitrosobenzene,

phenylhydroxylamine, and aniline [8]. The analysis of product evolution at short irradiation times revealed that a large part of the photocatalytic reactions took place at the nitrogen-containing substituents and involved e^- and h^+ in comparable extent.

A simple conceptual scheme was developed [7] using stoichiometric concepts for transformations in which the average oxidation number of carbon remains unchanged. These can be achieved by hydrolysis and disproportionation reactions involving the carbon atom or some linear combination of them. Although the stoichiometric reactions do not imply any particular degradation pathway, because photocatalysis operates through electron-transfer reactions (concurrently oxidative and reductive), the consecutive redox reactions with e^- and $^\bullet OH$ or with $^\bullet OH$ and e^-, formally equivalent to reaction with OH^-, are equivalent to hydrolysis. This *photocatalytic-induced hydrolysis* was demonstrated to be 10^6–10^8 times faster than the corresponding thermal process for chloromethanes [11,71].

The photocatalytic-induced hydrolysis implies that, in stoichiometric reactions, every water molecule corresponds to concurrent reactions with $^\bullet OH/e^-/$ H^+ (in any order) and that the ratio of reaction stoichiometric coefficients has to be equal to that between the concentrations of products found in the experimental system. This conceptual frame was demonstrated to be useful for understanding complex degradation pathways of organic molecules.

REFERENCES

1. Blanco, J.; Malato, S.; Fernandez, P.; Vidal, A.; Morales, A.; Trincado, P.; Oliveira, J. C.; Minero, C.; Musci, M.; Casale, C.; Brunotte, M.; Tratzky, S.; Dischinger, N.; Funken, K. H.; Sattler, C.; Vincent, M.; Collares-Pereira, M.; Mendes, J. F.; Rangel, C. M. *Solar Energy* **1999**, *67*, 317–330.
2. Serpone, N.; Sauvè, G.; Koch, R.; Tahiri, H.; Pichat, P.; Piccinini, P.; Pelizzetti, E.; Hidaka, H. *J. Photochem. Photobiol. A: Chem.* **1996**, *94*, 191–193.
3. Pelizzetti, E.; Minero, C. *Electrochim. Acta* **1993**, *38*, 47–55.
4. Mills, A.; Le Hunte, S. *J. Photochem. Photobiol. A: Chem.* **1997**, *108*, 1–35.
5. Hoffmann, M. R.; Martin, S. T.; Choi, W.; Bahnemann, D. W. *Chem. Rev.* **1995**, *95*, 69–96.
6. Pelizzetti, E.; Minero, C. *Comments Inorg. Chem.* **1994**, *15*, 297–337.
7. Pelizzetti, E.; Minero, C. *Colloid Surf. A.* **1999**, *151*, 321–327.
8. Piccinini, P.; Minero, C.; Vincenti, M.; Pelizzetti, E. *J. Chem. Soc. Faraday Trans.* **1997**, *93*, 1993–2000.
9. Minero, C.; Maurino, V.; Pelizzetti, E. *Res. Chem. Interm.* **1997**, *23*, 291–310.
10. Minero, C.; Pelizzetti, E.; Pichat, P.; Sega, M.; Vincenti, M. *Environ. Sci. Technol.* **1995**, *29*, 2226–2234.
11. Calza, P.; Minero, C.; Pelizzetti, E. *Environ. Sci. Technol.* **1997**, *31*, 2198–2203.
12. Minero, C.; Catozzo, F.; Pelizzetti, E. *Langmuir* **1992**, *8*, 481–486.

13. Chen, H. Y.; Zahraa, O.; Bouchy, M. *J. Photochem. Photobiol A: Chem.* **1997**, *108*, 37–44.
14. Peill, N. J.; Hoffmann, M. R. *Environ. Sci. Technol.* **1996**, *30*, 2806–2812.
15. Pozzo, R. L.; Giombi, J. L.; Baltanas M. A.; Cassano, A. E. *Catal. Today.* **2000**, *62*, 175–187.
16. Brandi, R. J.; Alfano, O. M.; Cassano, A. E. *Environ. Sci. Technol.* **2000**, *34*, 2623–2630.
17. Peill, N. J.; Hoffmann, M. R. *Environ. Sci. Technol.* **1998**, *32*, 398–404.
18. Stafford, U.; Gray, K. A.; Kamat, P. V. *Res. Chem. Interm.* **1997**, *23*, 355–388.
19. Gerischer, H.; Heller, A. *J. Electrochem. Soc.* **1991**, *138*, 113–118; Wang, C. M.; Heller, A.; Gerischer, H. *J. Am. Chem. Soc.* **1992**, *114* 5230–5234.
20. Pelizzetti, E.; Carlin, V.; Minero, C.; Graetzel, M. *New J. Chem.* **1991**, *15*, 351–359.
21. Bahnemann, D. W.; Cunningham, J.; Fox, M. A.; Pelizzetti, E.; Pichat, P.; Serpone, N. In *Aquatic and Surface Photochemistry*; Helz, G. R.; Zepp, R. G.; Crosby, D. G.; Eds. Lewis Publishing: Boca Raton, FL, pp. 261–316.
22. Emeline, A. V.; Ryabchuk, V.; Serpone, N. *J. Photochem. Photobiol. A: Chem.* **2000**, *133*, 89–97.
23. Pelizzetti, E.; Minero, C.; Maurino, V.; Sclafani, A.; Hidaka, H.; Serpone, N. *Environ. Sci. Technol.* **1989**, *23*, 1380–1385.
24. Kormann, C.; Bahnemann, D. W.; Hoffmann, M. R. *Environ. Sci. Technol.* **1991**, *25*, 494–500.
25. Minero, C.; Pelizzetti, E.; Malato, S.; Blanco, J. *Solar Energy* **1996**, *56*, 421–428.
26. Cunningham, J.; Srijaranai, S. *J. Photochem. Photobiol. A: Chem.* **1991**, *58*, 361–371.
27. Cunningham, J.; Al-Sayyed, G. *J. Chem. Soc. Faraday Trans.* **1990**, *86*, 3935–3941.
28. Cunningham, J.; Al-Sayyed, G.; Srijaranai, S. In *Aquatic and Surface Photochemistry*; Helz, G. R.; Zepp, R. G.; Crosby, D. G., Eds. Lewis Publishing CRC Press; Boca Raton, FL, 1994, p. 317.
29. Regazzoni, A. E.; Mandelbaum, P.; Matsuyoshi, M.; Schiller, S.; Bilmes, S. A.; Blesa, M. A. *Langmuir* **1998**, *14*, 868–874.
30. Serpone, N.; Salinaro, A.; Emeline, A.; Ryabchuk, V. *J. Photochem. Photobiol. A: Chem.* **2000**, *130*, 83–94.
31. Bahnemann, D.; Henglein, E.; Lilie, J.; Spanhel, L. *J. Phys. Chem.* **1984**, *88*, 709–711.
32. Turchi, C. S.; Ollis, D. F. *J. Catal.* **1990**, *122*, 178–192.
33. Kesselmann, J. M.; Lewis, N. S.; Hoffmann, M. R. *Environ. Sci. Technol.* **1997**, *31*, 2298–2302.
34. Minero, C. *Solar Energy Mater. SOLAR Cells* **1995**, *38*, 421–430.
35. Cunningham, J.; Al-Sayyed, G.; Sedlak, P.; Caffrey, J. *Catal. Today* **1999**, *53*, 145–158.
36. Minero, C.; Mariella, G.; Maurino, V.; Pelizzetti, E. *Langmuir* **2000**, *16*, 2632–2641.
37. Minero, C. *Catal. Today* **1999**, *54*, 205–216.
38. Mandelbaum, P.; Bilmes, S. A.; Regazzoni, A.; Blesa, M. A. *Solar Energy* **1999**, *65*, 75–80.
39. Lawless, D.; Serpone, N.; Meisel, D. *J. Phys. Chem.* **1991**, *95*, 5166–5170.

40. Kesselman, J. M.; Shreve, G. A.; Hoffmann, M. R.; Lewis, N. S. *J. Phys. Chem.* **1994**, *98*, 13,385–13,395.

41. Upadhya, S.; Ollis, D. F. *J. Phys. Chem.* **1997**, *191*, 2625–2631.

42. Alexiadis, A.; Baldi, G.; Mazzarino, I. *Catal. Today* **2001**, *66*, 467–474.

43. Mills, A.; Wang, J. S. *Phys. Chem.* **1999**, *213* (Pt. 1), 49–58.

44. Calza, P.; Pazè, C.; Pelizzetti, E.; Zecchina, A. *Chem. Commun.* **2001**, *20*, 2130–2131.

45. Hoffmann, M. R.; Martin, S. T.; Choi, W.; Bahnemann, D. W. *Chem. Rev.* **1995**, *95*, 69–96.

46. Kesselman, J. M.; Weres, O.; Lewis, N. S.; Hoffmann, M. R. *J. Phys. Chem. B.* **1997**, *101*, 2637–2643.

47. Bahnemann, D. W.; Hilgendorff, M.; Memming, R. *J. Phys. Chem. B.* **1997**, *101*, 4265–4275.

48. Krautler, B.; Bard, A. J. *J. Am. Chem. Soc.* **1978**, *100*, 5985–5989.

49. Carraway, E. R.; Hoffman, A. J.; Hoffmann, M. R. *Environ. Sci. Technol.* **1994**, *28*, 786–793.

50. Mao, Y.; Schoneich, C.; Asmus, K. D. *J. Phys. Chem.* **1991**, *95*, 10,080–10,089.

51. Kormann, C.; Bahnemann, D. W.; Hoffmann, M. R. *Environ. Sci. Technol.* **1988**, *22*, 798–804.

52. Draper, R. B.; Fox, M. A. *Langmuir* **1990**, *6*, 1396–1401.

53. Maurino, V.; Calza, P.; Minero, C.; Pelizzetti, E.; Vincenti, M. *Chemosphere* **1997**, *35*, 2675–2688.

54. Serpone, N.; Terzian, R.; Hidaka, H.; Pelizzetti, E. *J. Phys. Chem.* **1994**, *98*, 2634–2640.

55. Arnold, S. M.; Hickey, W. J.; Harris, R. F. *Environ. Sci. Technol.* **1995**, *29*, 2083–2089.

56. Fun, Y. S.; Pignatello, J. J. *Environ. Sci. Technol.* **1995**, *29*, 2065–2072.

57. Guillard, C. *J. Photochem. Photobiol. A: Chem.* **2000**, *135*, 65–75.

58. Li, X.; Cubbage, J. W.; Tetzlaff, T. A.; Jenks, W. S. *J. Org. Chem.* **1999**, *64*, 8509–8524.

59. Li, X.; Cubbage, J. W.; Tetzlaff, T. A.; Jenks, W. S. *J. Org. Chem.* **1999**, *64*, 8525–8536.

60. Li, X.; Cubbage, J. W.; Jenks, W. S. *J. Photochem. Photobiol. A: Chem.* **2001**, *143*, 69–85.

61. Minero, C.; Mariella, G.; Maurino, V.; Pelizzetti, E. *Langmuir* **2000**, *16*, 2632–2641.

62. Minero, C.; Mariella, G.; Maurino, V.; Vione, D.; Pelizzetti, E. *Langmuir* **2000**, *16*, 8964–8972.

63. Fox, M. A. In *Photocatalysis, Fundamental and Applications*; Serpone, N.; Pelizzetti, E., Eds. Wiley: New York, 1989, pp. 421–455.

64. Cermenati, L.; Pichat, P.; Guillard, C.; Albini, A. *J. Phys Chem. B.* **1997**, *101*, 2650–2658.

65. Cermenati, L.; Albini, A.; Pichat, P.; Guillard, C. *Res. Chem. Intermed.* **2000**, *26*, 221–234.

66. Cannon, R. D. *Electron Transfer Reactions*; Butterworths: London, 1980, pp. 223–266.

67. Minero, C.; Maurino, V.; Bernardez-Cordero, P.; Calza, P.; Vione, D.; Pelizzetti, E. unpublished.
68. Mills, A.; Morris, S. *J. Photochem. Photobiol. A: Chem.* **1993**, *71*, 75–83.
69. Rideh, L.; Weher, A.; Ronze, D.; Zoulalian, A. *Catal. Today* **1999**, *48*, 357–362.
70. Minero, C.; Piccinini, P.; Calza, P.; Pelizzetti, E. *New J. Chem.* **1996**, *20*, 1159–1164.
71. Calza, P.; Minero, C.; Pelizzetti, E. *J. Chem. Soc. Faraday Trans.* **1997**, *93*, 3565–3771.

7

Titanium Dioxide-Photocatalyzed Reactions of Organophosphorus Compounds in Aqueous Media

Kevin E. O'Shea

Florida International University, Miami, Florida, U.S.A.

I. INTRODUCTION

Organophosphorus compounds present a significant threat to the environment because of their extreme toxicity and widespread industrial use. Esters and thioesters of phosphoric acid and thiophosphoric acid are widely used for agriculture purposes as nonsystemic pesticides. Although the organophosphorus compounds, replacements for organochlorine compounds, are less persistent and do not undergo appreciable bioaccumulation, they have come under scrutiny because of their toxicity and potency as cholinesterase inhibitors [1,2]. In addition, large quantities of organophosphorus chemical warfare agents stored at military facilities must be destroyed [3]. While governing agencies initially considered incineration as a viable process for the treatment of chemical warfare agents, public outcry regarding the potential dangers associated with air pollution produced in this process have forced a re-evaluation of the treatment options [4].

Model compounds (simulants) are often used to estimate the reactivity of organophosphorus pesticides and warfare agents toward potential remediation technologies because of their availability and ease of handling [5–18]. The structures of select organophosphorus pesticides, chemical warfare agents, and model compounds (simulants) are presented in Fig. 1.

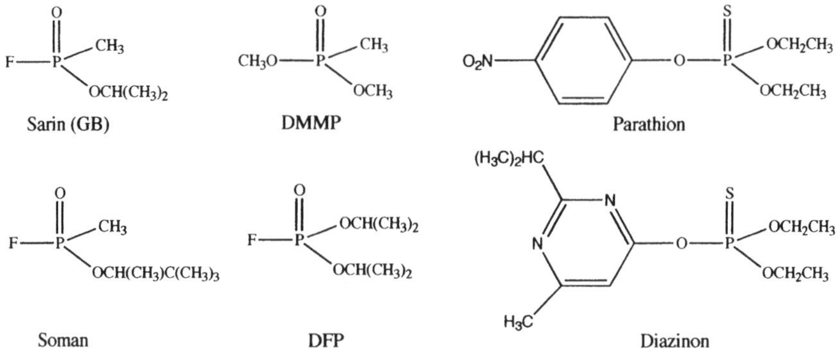

Figure 1 Structures of chemical warfare agents (sarin and soman), simulants (dimethyl methylphosphonate and diisoproyl fluorophosphate), and pesticides (parathion and diazinon).

A number of processes may be applicable for the remediation of hazardous organophosphorus compounds, but photochemical and advanced oxidation processes (AOPs) provide attractive nonthermal treatment alternatives [19]. AOPs generally involve the generation of reactive oxygen species, which, subsequently, can react to degrade a variety of organic compounds [20]. Among the AOPs, titanium dioxide photocatalysis has shown tremendous promise as a mild mineralization process for hundreds of hazardous organic substrates. This technology involves the use of oxygen, light, and TiO_2 semiconductor particles to oxidize organic compounds. Over the past several decades, vigorous research has demonstrated that essentially all chlorinated aliphatics (except carbon tetrachloride), chlorinated aromatics, several pesticides, herbicides, surfactants, and dyes are completely oxidized to innocuous products such as water, carbon dioxide, and mineral acids [21–38]. Hundreds of articles are published each year on research and development involving TiO_2-photocatalyzed water and air treatment. Given the considerable volume of literature, the emphasis of this chapter is directed toward TiO_2 photocatalysis of organophosphorus compounds [6,7,10,11,16–18, 39–48]. Following a brief introduction to the principles of photocatalysis the emphasis of the chapter will be on TiO_2 photocatalyzed degradation of organophosphorus substrates.

II. TITANIUM DIOXIDE PHOTOCATALYSIS

Photoexcitation of TiO_2 with wavelengths ≤380 nm generates an electron–hole pair [Eq. (1)], creating the potential for reduction and oxidation processes to occur at the surface of the semiconductor.

$$TiO_2 \xrightarrow{\text{light}} h_{vb}^+ + e_{cb}^- \tag{1}$$

The positions of the valence and conduction bands of anatase TiO$_2$ at pH $= 1$ relative to silver calomel electrode (SCE) are -0.1 V and $+3.1$ V, respectively [49].

The direct reaction between a substrate adsorbed on the surface of the semiconductor and the h_{vb}^+ or e_{cb}^- via electron transfer is typically not considered to be a significant reaction pathway in dilute oxygenated aqueous media.

Oxygen is generally used to scavenge the electron-forming superoxide anion radical:

$$O_2 + e_{cb}^- \rightarrow O_2^{\cdot -} \tag{2}$$

This prolongs the lifetime of the hole thus increasing the probability to oxidize surface hydroxyl groups or adsorbed water to yield a hydroxyl radical:

$$OH_{ads}(\text{or } H_2O) + h_{vb}^+ \rightarrow \cdot OH \, (H^+) \tag{3}$$

The reactive oxidizing species is likely associated with or bound to the surface of the TiO$_2$ and is often defined as a trapped hole, but for the purpose of this chapter, the oxidizing species will be called a hydroxyl radical [50]. The superoxide anion radical formed during photocatalysis can yield hydrogen peroxide in aqueous solution [51]. Hoffmann et al. used labeling experiments to demonstrate that photochemically produced H$_2$O$_2$ arises from reduction of molecular oxygen by conduction-band electrons during photocatalysis [52]. The hydrogen peroxide can then undergo reduction by e_{cb}^-, also producing a hydroxyl radical but in relatively low yields [53]:

$$H_2O_2 + e_{cb}^- \rightarrow \cdot OH + {-}OH \tag{4}$$

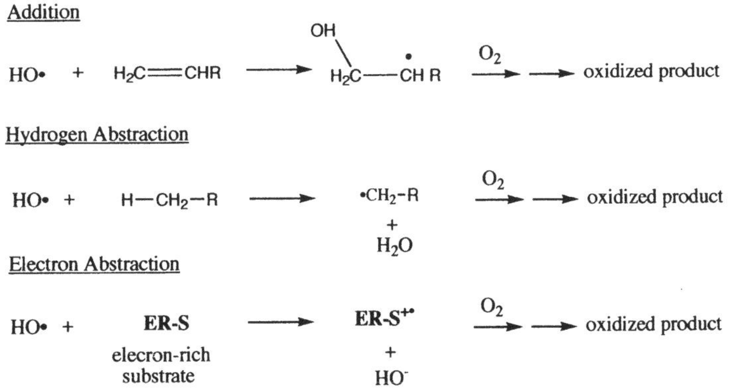

Figure 2   Common reaction pathways of hydroxyl radicals with organic compounds.

Once formed, the hydroxyl radical, a very powerful oxidant, can react with a variety of substrates [54]. Hydroxyl radicals are highly reactive toward organic compounds and the most common reaction pathways are addition, hydrogen and electron abstraction represented in Fig. 2. The reaction pathways will depend on the functionality and electronic characteristics of the organic compound. In general, these reaction pathways lead to radical species, which can form peroxyl radicals upon addition of molecular oxygen [55]. Subsequent and competing reaction pathways yield a variety and often-complex mixture of oxidized products [56]. In most instances, continued photocatalysis leads to complete degradation, giving the corresponding mineral acids, water, and carbon dioxide.

III. CATALYSTS

A. Suspensions

TiO_2 exists in a number of crystalline forms, but the anatase form is most commonly used for photocatalysis. Different composition and photoelectrochemical characteristics can result from different manufacturing processes of anatase forms. Degussa P25 is the most commonly used and is generally considered the most photoactive commercially available form of TiO_2. Dayte et al. suggested that Degussa P25 TiO_2 is largely anatase with a coating or islands of rutile on the surface [57]. The specific mix of these two forms and lattice imperfections help prolong the lifetime of the electron–hole pair, which increases the likelihood that oxidative and reductive processes will occur, improving the photocatalytic activity.

We reported the photoactivity of several different TiO_2 catalysts in the photocatalytic degradation of the chemical warfare simulant, dimethyl methylphosphonate (DMMP) [11]. The results of these studies are illustrated in Fig. 3. The photoactivity increases with surface area for the Degussa P25 TiO_2 catalysts. This observation is consistent with the rationalization that the increased surface area translates into more active sites/reactive species for a given type of catalyst, yielding higher reaction rates. Although the Hombikat UV 100 TiO_2 catalyst has the highest surface area among those studied, it exhibited intermediate photoactivity. Unlike the Degussa P25 catalysts, the Hombikat UV 100 TiO_2 is 100% anatase made by a sulfate processes.* The P25 catalysts are produced via hydrolysis of $TiCl_4$ in hydrogen flame, resulting in a mixture of anatase and rutile with an anatase content of 60–80%. The commercially available 2% iron oxide-doped Degussa P25 TiO_2 (VP F44O) exhibited no photocatalytic activity in the degradation of DMMP. A variety of factors, including surface area, lattice composition,

---

* The specifications and a report of the photoactivity of Hombikat UV 100 is available from Sachtelben Chemie GmbH. P. O. Box 170454. D-47184 Duisburg, Germany.

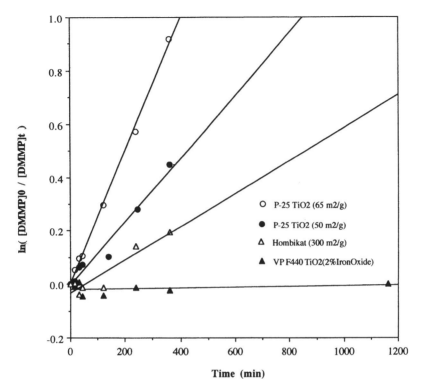

Figure 3   Apparent first-order rates for the photodecomposition of 4.0 mM DMMP in air-saturated aqueous suspensions using different types of TiO₂ (0.1 g/L).

porosity, and adsorption characteristics, play key roles in the observed photoactivity of TiO₂ catalysts.

B.   Immobilized Titanium Dioxide

Lu et al. reported the photocatalytic degradation of an organophosphorus insecticide on glass-supported and suspensions of titanium dioxide [42,45]. The observed rates for the TiO₂ suspensions were faster than the fixed TiO₂. The apparent first-order rate constants determined in a continuous-recirculation reactor with fixed catalyst increase with the flow rate, but reach a plateau at a flow rate of 500 mL/min. With immobilized TiO₂, the requirements for mixing during photolysis and the filtration or settling process required for TiO₂ suspensions are eliminated. Disadvantages of immobilized catalysts include reduced surface areas and mass-transfer limitations, typically yielding slower degradation rates.

C. Doped Titanium Dioxide

The use of doped or modified TiO_2 has a number of potential advantages [36]. By adding small amounts of dopants/metals to the TiO_2, the band gap may decrease such that longer wavelengths of irradiation (solar) become applicable. In addition, the photoactivity may increase through inhibition of the electron–hole pair recombination. Metal ions can also react with H_2O_2 by Fenton or photo-Fenton-type chemistry to form hydroxyl radicals [58].

Platinum-doped TiO_2 can be prepared easily and has exhinited enhanced photocatalytic activity [59]. Several organophosphorus compounds were wholly degraded in the presence of Pt-loaded TiO_2 upon exposure to sunlight [40]. In the presence of suspended Pt-loaded TiO_2, the observed degradation rates were enhancing by approximately five fold relative to nondoped TiO_2. Pt loading may accelerate the formation of $O^{\bullet -}_2$ and the enhanced reactivity is suggested to be the result of retarding electron–hole pair recombination through better scavenging of the electron [60]. Gratzel et al. reported TiO_2-photocatalyzed degradation of p-nitrophenyl diethylphosphonate by irradiation of niobium-doped TiO_2 [61].

D. Catalyst Loading

The effect of catalyst loading has been reported in the TiO_2 photocatalytic decomposition of organophosphorus compounds [11,45]. In general, for a TiO_2 suspension, the rates of degradation increase with increasing catalyst concentrations up to a maximum of ~1 g/L, followed by a gradual decrease at higher catalyst concentrations. At high catalysts concentrations, it is difficult to maintain a homogenous suspension of the TiO_2 in solution and settling can occur. In addition, agglomeration of TiO_2 particles is more pronounced at a high catalyst concentration under photocatalysis conditions [62]. At lower concentrations, light may be transmitted through the sample; thus, not all of the incident light is used. Regardless, light scattering leads to inefficient use of incident photons and is a major drawback in TiO_2 photocatalysis.

IV. MINERALIZATION STUDIES

A. Pesticides

Harada et al. studied the TiO_2 photocatalytic mineralization of trimethyl phosphate, trimethyl phosphite, and O, O-dimethyl ammonium phosphodithioate and reported excellent mass balance based on phosphate, sulfate, and carbon dioxide produced after prolonged illumination [39]. Subsequent studies found the organophosphorus insecticides dimethyl-2,2-dichlorovinyl phosphate [Eq. (5)] and dimethyl-2,2,2-trichloro-1-hydroxyethyl phosphonate were mineralized by solar irradiation in the presence of suspended TiO_2 [40]:

$$(CH_3O)_2POOCHCCl_2 + 9/2\ O_2 \rightarrow PO_4^{3-} + 2\ Cl^- + 4\ CO_2$$
$$+ 5\ H^+ + H_2O \quad (5)$$

Gratzel et al. demonstrated a wide variety of organophosphorus pesticides, including paraoxon, parathion, and malathion, which possess aromatic side chains and/or sulfur atoms can be totally decomposed into carbon dioxide and the corresponding mineral acids by irradiation in a solar simulator in the presence of anatase TiO$_2$ suspensions [41]. The balanced equation based on carbon dioxide evolution for the mineralization of parathion is represented by Eq. (6):

$$p-O_2N-Ph-O-P(S)(OC_2H_5)_2 + 15\ O_2 \rightarrow 10\ CO_2 + H_3PO_4$$
$$+ H_2SO_4 + HNO_3 + 4\ H_2O \quad (6)$$

Mineralization of the pesticide dichlorvos has been achieved by TiO$_2$ suspensions and immobilized forms of TiO$_2$ [42,43,45]. The mineralization was monitored via production of chloride ion measured potentiometrically. For every mole of dimethyl-2,2-dichlorovinyl phosphate stoichometric, 2 mol of chloride ions are simultaneously released [43]. Although the investigators suggest that the initial degradation involves release of the chloride ion, large discrepancies between the rate of degradation of organophosphorus compounds and the formation of the mineralized products are normally observed, indicating that the substrate conversion to phosphate likely requires a number of oxidation steps.

B. Chemical Warfare Simulants

The conversion of DMMP to phosphate under extended irradiation of aqueous TiO$_2$ suspension gives an excellent phosphorus mass balance [10], as illustrated in Fig. 4. The lag in the formation of phosphate and the incomplete phosphorus mass balance during the intermediate reaction times clearly indicate that mineralization to phosphate involves intermediate phosphorus-atom-containing compounds.

C. Surfactants

The mineralization of phosphorus-containing surfactants dodecylpoly(oxyethylene)-phosphates and poly(dodecyldecaoxyethylene) phosphates by irradiated TiO$_2$ suspensions was monitored by the formation of carbon dioxide and inorganic phosphate [46]. The phosphate is formed rapidly during the initial stages of the reaction, with a maximum yield of 94%. The photonic efficiency for the process is wavelength dependent with the highest yield at the most energetic wavelength and, as expected, at wavelengths less energetic than the band-gap energy of the catalyst, no significant degradation occurs.

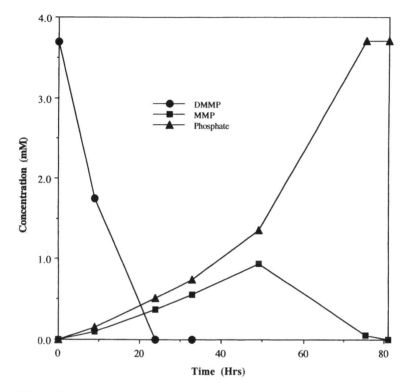

Figure 4   TiO$_2$ photocatalysis of DMMP and the formation and disappearance of phosphorus-containing products as a function of irradiation time.

V. ENVIRONMENTAL FACTORS

A. Temperature Effects

The Arrhenius activation energy for TiO$_2$ photocatalysis of the organophosphorus pesticide dichlorvos is 28.4 kcal/mol [42]. Arrhenius behavior is also reported between 35°C and 65°C for TiO$_2$ photocatalysis of DMMP, yielding an activation energy of 19.2 kJ/mol [11]. The relatively low activation energies are consistent with processes where thermal reactions of the reactant and the effect of temperature on the reaction rate are minimal. The Arrhenius activation energies for the TiO$_2$ photocatalysis have been reported for a variety of organic compounds in the range from 10 to 31 kJ/mol [63–67]. A number of explanations have been presented to explain the modest temperature effect, including rationalization based on the increasing collision frequency of molecules, which increases rate with increasing temperature [42]. The increased activity with temperature may be due to the red shift of the band gap [68] or a temperature-dependent shift in the adsorption–desorption equilibrium influencing the Arrhenius behavior [11].

B. Dissolved Oxygen

In general, oxygen has a pronounced effect on the degradation of substrates. A series of experiments using different oxygen-to-nitrogen ratios demonstrates that the oxidation rate of the dichlorvos increases dramatically with increasing oxygen concentration [42]. A comparison of TiO_2 photocatalyzed degradation rates of a variety of organophosphorus insecticides demonstrated no degradation under argon saturation, modest degradation under air saturation, and a three fold to sixfold relative increase in the degradation rate under oxygen saturation [41]. In the absence of oxygen, no appreciable TiO_2 photocatalysis was observed for DMMP [10].

The primary role of oxygen is that of an electron trap which helps prolong the lifetime of the hole, hence promoting oxidation processes. Heller and colleagues have suggested that the electron transfer to oxygen may be the rate-limiting step in TiO_2 photocatalysis [69–71]. A secondary role of oxygen is to react with carbon-centered radicals to yield the peroxyl radicals which subsequently undergo a variety of oxidative processes.

C. Additives

Several groups have reported enhanced TiO_2 photocatalysis of organophosphorus insecticides upon addition of H_2O_2 [11,41,45–47]. Harada et al. reported a 2–10-fold increase in the degradation of organophosphorus insecticides with the addition of H_2O_2 [40]. We observed a significant increase in the first-order rate constant for the TiO_2 photocatalysis of DMMP with the addition of H_2O_2 [11]. The enhancement reaches a plateau, followed by a gradual decrease at higher H_2O_2 concentrations. It is generally accepted that the increased rate of degradation is due to the formation of additional hydroxyl radicals through reduction of H_2O_2. Although increased rates have been reported for a variety of substrates, a decrease in the rate is observed at high concentrations owing to the competition between H_2O_2 for the active site (competitive adsorption), reaction between the H_2O_2, and the reactive species and recombination of hydroxyl radicals [72].

Gratzel et al. reported that the addition of strong oxidants $K_2S_2O_8$, H_2O_2, $NaIO_4$, and $KBrO_3$ enhances dramatically the rate of TiO_2 photocatalytic activity by inhibiting recombination of the electron–hole pair in addition to direct oxidation of the substrate by the strong oxidants [41].

D. Solution pH

The surface of a TiO_2 particle (Degussa P25) is completely hydroxylated and has zero overall charge at pH ~6.6. As the pH is lowered, the number of protonated sites increases, as does the overall positive charge on the particle. At high pH, the number of oxyanion sites increases and the particle becomes negatively charged. These surface changes will directly affect surface charge, substrate adsorption, and subsequent reactivity. The solution pH drastically affects the TiO_2 surface and consequently, the TiO_2-catalyzed photolysis of substrates in aqueous solutions [73,74].

We have shown that the solution pH drastically influences the rate of degradation of DMMP [10]. The rate of disappearance increases at high pH and decreases at low pH. The increase in the rate of disappearance at high pH may be the result of stronger adsorption and a significant contribution from direct or surface-catalyzed hydrolysis [75]. At low pH (pH<2), the decomposition is negligible. This may be the result of a highly protonated surface which prevents significant adsorption of the simulant onto the surface of the TiO_2 particles.

VI. KINETIC PARAMETERS

A. Rates

TiO_2 photocatalysis of organophosphorus compounds at low concentrations has been reported to follow apparent first-order kinetics [10,42,45]. Although degradation rates at a given concentration may be consistent with first-order kinetics, changes in the apparent rate constants have been observed over wide concentration ranges [11]. Also, a number of reports have appeared in which a reduction in the reaction order is observed in TiO_2 photocatalysis, suggesting the reaction kinetics are more complex than simple first order [11,44,76]. At high substrate concentrations, saturation of the TiO_2 surface can lead to zero-order kinetics, whereas at a low substrate concentration, the number active sites is not rate limiting and the degradation rate becomes proportional to the substrate concentration, in accordance with apparent first-order kinetics. Reaction orders smaller than unity in the TiO_2 photocatalysis of phosphonic acids are rationalized as a result of slow diffusion of the product from the surface [44].

B. Models

A variety of models have been derived to describe the kinetics of semiconductor photocatalysis, but the most commonly used model is the Langmuir–Hinshelwood (LH) model [77–79]. The LH model relates the rate of surface-catalyzed reactions to the surface covered by the substrate. The simplest representation of the LH model [Eq. (7)] assumes no competition with reaction by-products and is normally applied to the initial stages of photocatalysis under air- or oxygen-saturated conditions. Assuming that the surface coverage is related to initial concentration of the substrate and to the adsorption equilibrium constant, K, the initial degradation rate of the substrate can be expressed in a LH kinetic expression as follows:

$$r = \frac{dC}{dt} = \frac{kKC}{1+KC} \tag{7}$$

where C is the initial substrate concentration, r is the observed reaction rate, k is the intrinsic reactivity constant, and K is the equilibrium adsorption constant. The standard application of the LH model involves a plot of $1/r$ versus $1/C$. The linearity of the relationship indicates the validity of the LH model. The adsorption equilibrium rate constants and intrinsic reactivities are calculated from the slope and intercept of the LH plot:

$$\frac{1}{r} = \frac{1}{k} + \frac{1}{kKC} \tag{8}$$

The LH model has proven useful in the TiO$_2$ photocatalysis of phosphonic acids [44] dimethyl methylphosphonate [10], and dichlorvos [42]. Individual differences in LH kinetic parameters for the phosphonic acids are rationalized based on expected differences in the relative stabilities of the organic radicals generated at the surface and the fact that the products may not diffuse from the surface as is assumed by the LH model. The LH kinetic parameters for dichlorvos were determined using a fixed-catalyst reactor and differ significantly from the studies of the organophosphorus compounds determined using TiO$_2$ suspensions. Given the different reaction conditions employed in these studies, it is not meaningful to compare the LH kinetic parameters of these different substrates.

Although it has been reported that adsorption is critical for the generation of reactive intermediates, there is still uncertainty as to the role adsorption plays and how it ultimately affects the reaction pathways [50,80]. These reports provide convincing evidence that LH and related kinetic models are good for predictive purposes, only indicative of apparent kinetics, and may be consistent with several different mechanisms.

VII. REACTION PATHWAYS AND MECHANISMS

A. Product Studies

The majority of studies on the TiO$_2$ photocatalysis of organophosphorus compounds rely primarily on the disappearance of the initial substrate coupled with the monitoring of the mineralization products. Because of the complex mixtures formed in the TiO$_2$ photocatalysis of these compounds, there is limited information on the reaction pathways and mechanisms involved in the degradation processes. A recent report by Konstantinou et al. identified a number of intermediate products in the degradation of dichlofenthion and bromophos methyl by TiO$_2$

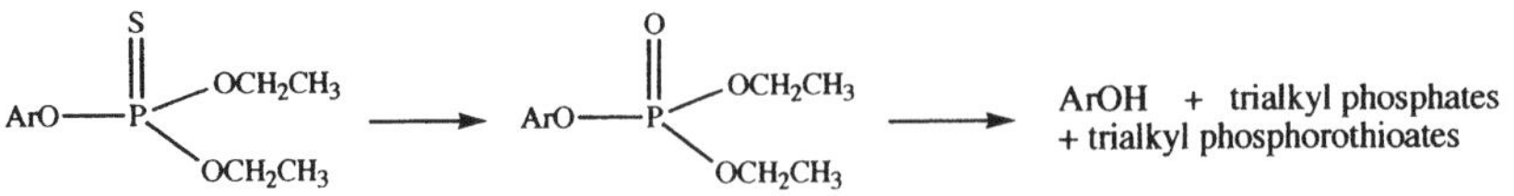

Figure 5 Reaction pathways proposed for the TiO_2 photocatalysis of dichlofenthion and bromophos methyl in aqueous media. (From Ref. 48.)

and simulated solar light ($\geq$290 nm) [48]. They propose a HO· attack on the P=S bond, resulting in the formation of oxon derivative (P=O), followed by the rupture of the P—O ester bond (Fig. 5).

Surface-catalyzed dark and photoassisted phosphate ester hydrolysis (P—O bond rupture) in aqueous suspensions of TiO_2 has been proposed [6,7,41]. A number of products indicative of radical cationic pathways have been reported in the TiO_2 photocatalysis of benzyl phosphonic acid [44].

B. Mechanistic Insight

Valuable mechanistic information is often inferred by comparison to other hydroxyl-radical-generating techniques. A similar reaction pathway is observed from TiO_2 photocatalysis [10], ultrasonic [12], or ionizing radiation [14] of DMMP, as illustrated in Fig. 6. Each of these processes involves the generation of hydroxyl radicals. The transformation of DMMP to methyl methylphosphonate (MMP) to methyl phosphonic acid (MPA) could occur by hydroxyl radical addition to the phosphorus atom, as proposed by Pignatello in the hydroxyl-radical-mediated degradation of methyl parathion [81]. Alternatively, such transformations could occur by electron abstraction to form the DMMP radical cation, followed by the addition of water or by abstraction of hydrogen from the alkyl group of the ester, which are expected to yield MMP.

C. Pulse Radiolysis

Radiolytic studies have been used to provide insight into the mechanisms involved in the TiO_2-catalyzed photo-oxidation of a variety of organic substrates

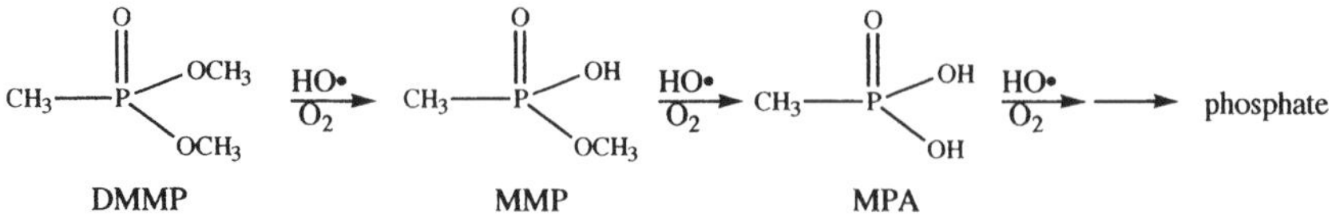

Figure 6 Hydroxyl-radical-media-degradation pathway for DMMP.

[50,82–85]. Because a number of reasonable mechanisms can be proposed to explain the observations discussed earlier, we used pulse radiolysis studies to develop a better understanding of the reactions between organophosphorus compounds and hydroxyl radicals [15]. First, we generated the strong one-electron oxidant Cl$_2^{\cdot-}$, but observed no significant reaction with the DMMP, thus providing convincing evidence that the hydroxyl radical, a weaker one-electron oxidant, does not react with DMMP to form the radical cation. The addition of hydroxyl radical to the phosphorus atom of DMMP is expected to yield MMP independent of oxygen; however, in the absence of oxygen, MMP is not formed. Product studies and the reactions of tetranitromethane during pulse radiolysis provide strong evidence that hydroxyl radical reactions yield a carbon-centered radical likely formed through hydrogen abstraction at the alkyl group of the ester chain.

We propose the following mechanism to explain the hydroxyl-radical-mediated formation of the methyl methylphosphonate (Fig. 7). The abstraction of hydrogen from the methyl ester leads to a radical intermediate, which will rapidly add oxygen to form a peroxyl radical. The dimerization of two peroxy radicals forms the tetroxide, which subsequently reacts to produce a mole of oxygen or hydrogen peroxide molecule, and produces the corresponding formate and/or acetal, both of which will hydrolyze to MMP. The remaining alkyl group can be oxidized in a similar fashion to yield the MPA.

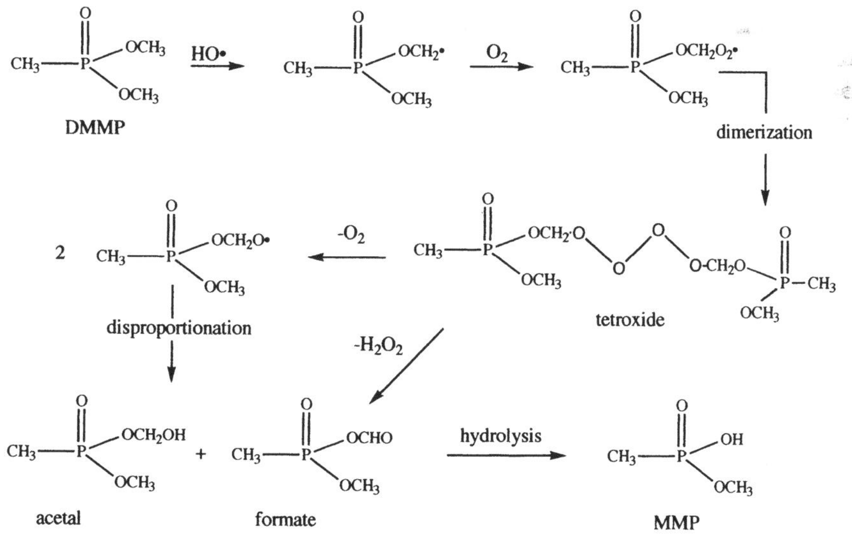

Figure 7 Proposed mechanism for the conversion of DMMP to MMP by hydroxyl-mediated reactions in oxygenated solutions.

Our radiolysis studies also indicate that phosphonates react quite slowly with the superoxide anion radical. Although our studies do not support the formation of radical cations as an initial oxidation step, we cannot rule out the possibility that radical cations are not involved in the oxidation of the C—P bond, as previously proposed [44]. It is also possible that more electron-rich organphosphorus compounds or organophosphorus compounds in the adsorbed state may exhibit different redox and hydroxyl radical chemistries than what is observed under pulse radiolysis employing homogeneous conditions.

VIII. CONCLUSIONS

In this chapter, an attempt has been made to address fundamental mechanistic and kinetic aspects of TiO_2 photocatalysis of organophosphorus compounds. Comparisons between homogeneous (radiolysis) and heterogeneous (photocatalysis) hydroxyl-generating processes have helped to elucidate the reaction pathways and led to number of important mechanistic conclusions. From the various kinetic parameters, the overall rates and efficiencies for the degradation of organophosphorus compounds can be predicted and may find direct application in evaluation and implementation of semiconductor photocatalysis.

Titanium dioxide photocatalysis appears to be a promising technology for the environmental cleanup of dilute organophosphorus compounds. Optimization of treatment conditions will depend on a number of factors, including the catalyst type and size, agglomeration, light scattering, surface area, irradiating volume, and so forth. Because reactor types and experimental conditions vary significantly, these factors should be addressed for individual applications. Although mineralization was achieved for a variety of organophosphorus pesticides, cost-effective treatments to complete mineralization are generally not practical. It is therefore imperative that detailed product studies are carried out to identify toxic intermediates and/or toxicological assessments are conducted at intermediate reaction times prior to implementation of this technology.

ACKNOWLEDGMENTS

Financial Support from the National Science Foundation, the Petroleum Research Fund, and the Dreyfus Foundation is gratefully acknowledged.

REFERENCES

1. Bussiere, J. L.; Kendall, R. J.; Lacher, T. E., Jr.; Bennett, R. S. *Environ. Toxicol. Chem.* **1989**, *8*, 1125.
2. Compton, J. A. F. *Military Chemical and Biological Agents, Chemical Toxicological Properties*; The Telford Press, Caldwell, NJ, 1987.

3. Umatilla Study Group, *Govt. Rept. Announ. Index* **1998**, 88(18), 164, Abstract 847.
4. Yang, Y. -C.; Baker, J. A.; Ward, R. *Chem. Rev.* **1992**, *92*, 1729.
5. Moss, R. A.; Kim, K. Y.; Swarup, S. *J. Am. Chem. Soc.* **1986**, *108*, 788.
6. Rose, T. L.; Najundiah, C. In *Proceedings of the 1985 Scientific Conference on Chemical Defense Research*; Rausa, M., Ed., Aberdeen Proving Ground, Maryland, 1986, p. 299.
7. Lewis, T. J.; Aurian-Blajeni, B.; Rose, T. L. In *Proceedings of the 1987 U.S. Army Chemical Research, Development and Engineering Center Scientific Conference on Chemical Defense Research*; Rausa, M., Ed.; Aberdeen Proving Ground, Maryland, 1988, p. 749.
8. Bailin, L. J.; Sibert, M. E.; Jonas, L. A.; Bell, A. T. *Environ. Sci. Technol.* **1975**, *9*, 254.
9. Aurian-Blajeni, B.; Boucher, M. M. *Langmuir* **1989**, *5*, 170.
10. O'Shea, K. E.; Beightol, S.; Garcia, I.; Hernandez, M.; Kalen, D. V.; Cooper, W. J. *J. Photochem. Photobio. A: Chem.* **1997**, *107*, 221.
11. O'Shea, K. E.; Garcia, L.; Aguilar, M. *Res. Chem. Intermed.* **1997**, *23*, 325.
12. O'Shea, K. E.; Aguila, A.; Vinodgopal, K.; Kamat, P. V. *Res. Chem. Intermed.* **1998**, *24*, 695.
13. Aguila, A.; O'Shea, K. E.; Kamat, P. V. *J. Adv. Oxid. Technol.* **1998**, *3*, 37.
14. O'Shea, K. E.; Kalen, D. V.; Cooper, W. J.; Garcia, I.; Aguilar, M. In *Environmental Applications of Ionizing Radiation*; Cooper, W. J.; Curry, R. D.; O'Shea, K. E., Eds.; Wiley: New York, 1998, p. 569.
15. Aguila, A.; O'Shea, K. E.; Tobien, T.; Asmus, K.-D. *J. Phys. Chem.* **2001**, *105*, 7834.
16. Obee, T. N.; Satyapal, S. *J. Photochem. Photobiol. A: Chem.* **1998**, *118*, 45.
17. Rusu, C. N.; Yates, J. T. *J. Phys. Chem. B.* **2000**, *104*, 12,299.
18. Rusu, C. N.; Yates, J. T. *J. Phys. Chem. B.* **2000**, *104*, 12,292.
19. Legrini, O.; Oliveros, E.; Braun, A. M. *Chem. Rev.* **1993**, *93*, 671.
20. Glaze, W. H.; Kang, W. *J. Am. Water Works Assoc.* **1998**, *5*, 57.
21. Schviavello, M., Ed. *Photochemistry, Photocatalysis and Photoreactors: Fundamentals and Development*; Reidel: Dordrecht, 1985.
22. Henglein, A. *Chem. Rev.* **1989**, *89*, 1861.
23. Serpone, N.; Pelizzetti, E. Eds. *Photocatalsysis: Fundamentals and Applications*; Wiley: New York, 1989.
24. Ollis, D. F.; Pelizzetti, E.; Serpone, N. *Environ. Sci. Technol.* **1991**, *25*, 1523.
25. Pelizzetti, E.; Schviavello, M. Eds. *Photochemical Conversion and Storage of Solar Energy*; Kluwer: Dordrecht, 1991.
26. Fox, M. A. *Chemtech.* **1992**, *92*, 680.
27. Matthews, R. W. *Pure Appl. Chem.* **1992**, *64*, 1285.
28. Fox, M. A.; Dulay, M. T. *Chem. Rev.* **1993**, *93*, 341.
29. Herrmann, J.-M.; Guillard, C.; Pichat, P. *Catal. Today* **1993**, *17*, 7.
30. Kamat, P. V. *Chem. Rev.* **1993**, *93*, 267.
31. Ollis, D. F.; Al-Ekabi, H. Eds, *Photocatalytic Purification and Treatment of Water and Air*; Elsevier: Amsterdam, **1993**.
32. Wold, A. *Chem. Mater.* **1993**, *5*, 280.

33. Helz, G. R.; Zepp. R. G.; Crosby, D. G. Eds. *Aquatic and Surface Photochemistry*; Lewis: Boca Raton, FL, **1994**.

34. Zeltner, W. A.; Hill, C. G., Jr.; Anderson, M. A. *Chemtech* **1994**, *94*, 21.

35. Bahnemann, P.; Cunningham, J.; Fox, M. A.; Pelizzetti, E.; Pichat, R.; Serpone, N. In *Aquatic and Surface Photochemistry*; G. R. Helz; Zepp, R. G.; Crosby, D. G., Eds; Lewis: Boca Raton, FL, **1994**, p. 261.

36. Hoffmann, M. R.; Martin, S. T.; Choi, W.; Bahnemann, D. W. *Chem. Rev.* **1995**, *95*, 69.

37. Linsebigler, A. L.; Lu, G.; Yates, J. T. Jr. *Chem. Rev.* **1995**, *95*, 735.

38. Stafford, U.; Gray, K. A.; Kamat, P. V. *Heterogeneous Chem. Rev.* **1996**, *3*, 77.

39. Harada, K.; Hisanaga, T.; Tanaka, K. *New J. Chem.* **1987**, *11*, 987.

40. Harada, K.; Hisanaga, T.; Tanaka, K. *Water Res.* **1990**, *24*, 1415.

41. Gratzel, C. K.; Jirousek, M.; Gratzel, M. *J. Mol. Catal.* **1990**, *60*, 375.

42. Lu, M.-C.; Roam, G.-D.; Chen, J. N.; Huang, C. P. *J. Photochem. Photobiol. A: Chem.* **1993**, *76*, 103.

43. Hung, S. T.; Mack, M. K. *Environ. Tech.* **1993**, *14*, 265.

44. Krosley, K. W.; Collard, D. M.; Adamson, J.; Fox, M. A. *J. Photochem. Photobiol. A: Chem.* **1993**, *69*, 357.

45. Lu, M.-C.; Roam, G. -D.; Chen, J. -N.; Huang, C. P. *Water Sci. Tech.* **1994**, *30*, 29.

46. Hidaka, H.; Zhao, J.; Satoh, Y.; Nohara, K.; Pelizzetti, E.; Serpone, N. *J. Mol. Catal.* **1994**, *88*, 239.

47. Doong, R.; Chang, W. *J. Photochem. Photobiol. A: Chem.* **1997**, *107*, 239.

48. Konstantinou, J. K.; Sakellarides, T. M.; Sakkas, V. A.; Albanis, T. A. *Environ. Sci. Technol.* **2001**, *35*, 398.

49. Gratzel, M. *Heterogeneous Photochemical Electron Transfer*; CRC Press: Boca Raton, FL, 1989.

50. Lawless, D.; Serpone, N.; D. Meisel, *J. Phys. Chem.* **1991**, *95*, 5166.

51. Carraway, E. R.; Hoffmann, A. J.; Hoffmann, M. R. *Environ. Sci. Technol.* **1991**, *28*, 786.

52. Hoffmann, A. J.; Carraway, E. R.; Hoffmann, M. R. *Environ. Sci. Technol.* **1994**, *28*, 776.

53. Matthews, R. W. *J. Chem. Soc. Faraday Trans.* **1984**, *180*, 457.

54. Buxton, G. V.; Greenstock, C. L.; Hermann, W. P.; Ross, A. B. *J. Phys. Chem. Ref. Data.* **1988**, *17*, 513.

55. Von Sonntag, C.; Schuchmann, H. P. *Angew. Chem.* **1991**, *30*, 1229.

56. Von Sonntag, C.; Schuchmann, H. P. In *The Chemistry of Free Radicals: Peroxyl Radicals*; Alfassi, Z. B.; Ed.; Wiley: New York, 1997, p. 173.

57. Dayte, A. K.; Riegel, G.; Bolton, J. R.; Huang, M.; Prairie, M. R. *J. Solid State Chem.* **1995**, *115*, 236.

58. Fenton, H. J. H. *J. Chem. Soc.* **1894**, *65*, 899.

59. Kraeutler, B.; Bard, A. J. *J. Am. Chem. Soc.* **1978**, *78*, 4317.

60. Tanaka, K.; Harada, K.; Murata, S. *Solar Energy* **1986**, *36*, 159.

61. Gratzel, C. K.; Jirousek, M.; Gratzel, M. *J. Mol. Catal.* **1987**, *39*, 347.

62. O'Shea, K. E.; Pernas, E.; Saiers, J. *Langmuir* **1999**, *15*, 2071.

63. Matthews, R. W. *J. Phys. Chem.* **1987**, *91*, 3328.

64. Okamoto, K. I.; Yamamoto, Y.; Tanaka, H.; Itaya, A. *Bull. Chem. Soc. Jpn.*, **1985**, *58*, 2015.
65. Herrmann, J. -M.; Mozzanega, M. N.; Pichat, P. *J. Photochem.* **1983**, *22*, 333.
66. Bideua, M.; Claudel, B,; Otterbein, M. *J. Photochem.* **1980**, *14*, 291.
67. Harvey, P. R.; Rudham, R., Ward, S. *J. Chem. Soc. Faraday Trans.* **1983**, *II*, *79*, 1381.
68. Wilkinson, F.; Willsher, C. J.; Uhl. S.; Honnen, W.; Oelkrug, D. *J. Photochem.* **1986**, *33*, 273.
69. Gerischer, H.; Heller, A. *J. Electrochem. Soc.* **1992**, *113*, 139.
70. Wang, C. M.; Heller, A. *J. Chem. Soc.* **1992**, *114*, 5230.
71. Gerischer, H.; Heller, A. *J. Phys. Chem.* **1991**, *95*, 5261.
72. Peterson, M. W.; Turner, J. H.; Nozik, A. J. *J. Phys. Chem.* **1991**, *95*, 221.
73. Wei, T.-Y.; Wan, C. -C. *J. Photochem. Photobiol. A. Chem.* **1992**, *69*, 241.
74. D'Oliveria, J. -C.; Al-Sayyed, G.; Pichat, P. *Environ. Sci. Technol.* **1990**, *24*, 990.
75. Baldwin, D. S.; Beattie, J. K.; Jones, D. R. *Water Res.* **1996**, *30*, 1123.
76. Sabin, F.; Turk T.; Vogler, A. *J. Photochem. Photobiol. A: Chem.* **1992**, *63*, 99.
77. Adamson, A. W. In *Physical Chemistry of Surfaces*, 4th ed.; Wiley: New York, 1982, p. 374.
78. Parfitt, G. D.; Rochester, C. H. In *Adsorption from Solution at Solid-Liquid Interface*; Parfitt, G. D.; Rochester, C. H.; Eds.; Academic Press: London, 1983. p. 4.
79. Al-Ekabi, H.; De Mayo, P. *J. Phys. Chem.* **1986**, *90*, 4075.
80. Turchi, C. S.; Ollis, D. F. *J. Catal.* **1990**, *122*, 178.
81. Pignatello, J. J.; Sun, Y. *Water Res.* **1995**, *29*, 1837.
82. Terzian, R.; Serpone, N.; Draper, R. B.; Fox, M. A.; Pelizzetti, E. *Langmuir* **1991**, *7*, 3081.
83. Stafford, U.; Gray, K. A.; Kamat, P. V. *J. Phys. Chem.* **1994**, *98*, 6343.
84. Mao, Y.; Schoneich, C.; Asmus, K. -D. *J. Phys. Chem.* **1991**, *95*, 10,080.
85. Goldstein, S.; Czapski, G.; Rabani, J. *J. Phys. Chem.* **1994**, *98*, 6586.

8

Photocatalytic Oxidation of Gas-Phase Aromatic Contaminants

**Michael Lewandowski and
David F. Ollis**

North Carolina State University, Raleigh, North Carolina,
U.S.A.

I. INTRODUCTION

The photocatalyzed oxidation of gas-phase contaminants in air has been demonstrated for a wide variety of organic compounds, including common aromatics like benzene, toluene, and xylenes. For gas-phase aromatic concentrations in the sub-100-ppm range, typical of common air contaminants in enclosed spaces (office buildings, factories, aircraft, and automobiles), photocatalytic treatment leads typically to complete oxidation to CO_2 and H_2O. This generality of total destruction of aromatic contaminants at ambient temperatures is attractive as a potential air purification and remediation technology.

However, the photocatalytic reactivity of aromatics, especially benzene, is lower than some of the more promising pollutants for this emerging technology. Also, at aromatic concentrations of as little as 10 ppm, photocatalysts may exhibit apparent deactivation. Here, we summarize achievements to date and discuss methods for increasing aromatic reactivity, minimizing photocatalyst deactivation, and periodically regenerating used protocatalysts.

II. RANGE OF PHOTOCATALYTIC EXPERIMENTS

A. Aromatic Contaminants of Interest

1. Simple Aromatics

Benzene (C_6H_6), toluene ($C_6H_5CH_3$), ethylbenzene ($C_6H_5C_2H_5$), and xylenes [o-xylene, m-xylene, and p-xylene, $C_6H_4(CH_3)_2$] (known as the BTEX compounds) make up the simplest and most widely used group of aromatic compounds. They are often collectively referred to as the BTEX group, from the first letters of the four compounds' names. The BTEX compounds normally occur in petroleum deposits, particularly in coal tars. They are used for a variety of industrial purposes, such as solvents and feedstocks for the manufacture of other chemicals. Aromatic compounds, particularly the branched aromatics, are found in some types of paint, finish, and adhesive. All of the BTEX compounds may be found in gasoline, particularly toluene, which is sometimes added to gasoline to increase the octane rating [1].

  Airstreams may become contaminated with aromatic contaminants through the evaporation of aromatic solvents or gasoline products, or through volatilization during the drying of paints, finishes, and adhesives. Automotive exhausts may contain traces of BTEX compounds as well. Spills or leaking chemical storage tanks might produce soil contamination, which can be transferred to the gas phase through air-stripping of the contaminated soils. Indoor air in buildings and vehicles may also contain low levels of aromatic contaminants, usually produced by outgassing of certain types of polymer, paint, and finish.

  All four BTEX compounds are classified as hazardous air pollutants under the 1990 U.S. Clean Air Act. Benzene is classified as a carcinogenic compound as well [1]. The Occupational Safety and Health Administration (OSHA) imposes time-weighted average permissible exposure levels (TWA-PEL) for industrial workers for the BTEX compounds. The OSHA permissible exposure limit for benzene is the lowest at 3.26 mg/m^3 (1 ppm), due predominantly to the known carcinogenic nature of benzene. The PEL for toluene is somewhat higher at 75 mg/m^3 (~20 ppm). The exposure limits for ethylbenzene and xylenes are significantly higher at 435 mg/m^3 (100 ppm) and 655 mg/m^3 (~150 ppm), respectively [1]. Aromatic contaminants may produce odor problems as well as health effects. The odor threshold for benzene, for example, is 4.7 ppm, whereas the odor threshold for toluene is even lower than the OSHA PEL limit at 2.14 ppm [2]. In addition to health effects and potential odor problems, gas-phase aromatic contaminants may play a role in the formation of photochemical smog.

  Photocatalytic research has explored benzene destruction [3–7] because of its carcinogenic properties and low permissible exposure limits. However, these properties also impose limits on the industrial use of benzene, so other aromatics, particularly toluene, are used more often and might be present at higher levels in industrial settings. This circumstance has led research efforts on the photocatalytic oxidation of aromatic contaminants in air to focus on the branched aromatics: toluene [8–16] ethylbenzene [17], and xylenes [18,19]. Because the branched aromatic compounds tend to behave similarly under photocatalytic conditions, toluene is typically employed as the model branched aromatic.

2. Nitroaromatics

Although most research conducted on the photocatalytic oxidation of aromatic contaminants has focused on the BTEX compounds, several other classes of aromatic compounds, including nitro-aromatics and chlorinated aromatics, are also of special interest. Nitro-aromatics, particularly 2,4,6-trinitrotoluene (TNT) and its associated by-products, represent an industrially significant class of aromatic contaminants. TNT has been used extensively as an explosive for both military and civilian purposes for nearly a century [20]. Although other explosive compounds have replaced TNT in many modern applications, considerable quantities of TNT are still produced in numerous locations. Particulate, vapor, and contaminated water emissions may be a concern anywhere TNT is manufactured, as well as in explosives fabrication and storage facilities. In production facilities, in particular, additional aromatic contaminants may be associated with TNT emissions. These may include significant levels of toluene, which is used as a feedstock for TNT production, along with traces of benzene and xylene. By-products generated during TNT production, including mononitrotoulenes, dinitrotoluenes, and nitrobenzenes, may be expected to be present as well.

Because TNT is an explosive, safety considerations may impose constraints on potential treatment options. Photocatalytic oxidation offers several significant advantages in this regard. Oxidative destruction of the contaminants, rather than separation and concentration, helps prevent the buildup of potentially explosive residues. Low (ambient)–temperature operation circumvents potential problems associated with ignition sources located in close proximity to large quantities of explosive mixtures. These advantages have prompted several examinations of photocatalytic oxidation as potential treatment technology for use in association with TNT, as well as other explosive materials [21–28].

3. Chlorinated Aromatics

Chlorinated aromatics are primarily of interest in the liquid, rather than gas, phase and will not be covered in detail. However, chlorinated aromatic contaminants, particularly various chlorophenols, represent a significant class of water contaminants, and considerable research into the photocatalytic degradation of these materials has been conducted in recent years [29–35]. Many chlorinated aromatics are resistant to biological degradation, allowing them to accumulate in the environment and persist for long periods of time. Some are known to have significant, detrimental effects on the activity of microbiological agents, which may hinder traditional biological water treatment methods [36].

These factors have led researchers to examine novel methods for the degradation of these contaminants, including liquid-phase photocatalytic oxidation. The treatment of wastewaters contaminated with agents like chlorinated aromatics has also prompted research into the optimum methods of integrating chemical oxidation techniques with existing water treatment methods, particularly biological treatment and ozonation techniques [37].

B.   Types of Photoreactors

1.   Batch Reactors: Recirculation Systems

Batch reactors, in numerous configurations, have been used to study various aspects of photocatalytic reactions, particularly reaction kinetics. Gas-phase batch reactors are generally not well suited for use in air remediation in industrial settings, where continuous-flow waste streams are often targeted for treatment. However, recirculating batch reactors, incorporating various photocatalyst arrangements (including fluidized beds, packed beds, flat plates, annular tubes, and monoliths) are of interest in indoor air-quality maintenance and odor control.

In these potential applications, a fixed volume of air (inside an enclosed building, vehicle, aircraft, etc.) may be repeatedly cycled through a photocatalytic reactor, which may be either free standing or part of a larger ventilation system (Fig. 1). Normally, such a system would be required to maintain air quality by continuously removing routinely generated contaminants (e.g., generated by occupants or by outgassing of materials). However, such systems should also be capable of rapidly reducing the concentration of a contaminant spontaneously released into the air volume as well. The magnitude of such a spontaneous release may vary considerably depending on the potential application; examples may range from brief bursts of cigarette smoke to chemical spills.

2.   Continuous-Flow Reactors

In many industrial applications, contaminated airstreams may be produced on a continuous or near-continuous basis, and short residence time flow reactors are needed to treat these streams continuously as they are produced.

a.   Powder-Layer Reactors.   Powder-layer photoreactors have been used extensively for bench-scale experimental studies of the photocatalytic oxidation of continuous contaminant streams. Powder-layer reactors rely on a simple bed of TiO_2 catalyst powder, usually supported on a porous glass frit or filter (Fig. 2). Reactors of this type are relatively simple to construct and operate. Catalyst preparation is generally straightforward, and the quantity of catalyst employed is easily controlled. Measurement of the illumination depth into the powder layer

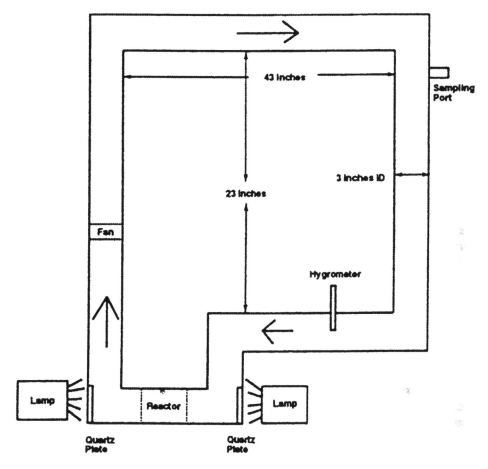

Figure 1   Schematic diagram of a recirculating photoreactor system. (From Ref. 38.)

is also relatively simple [40], allowing the catalyst load to be adjusted to provide total absorption of incident radiation or complete illumination of the entire catalyst load as the experiment requires.

Powder-layer reactors are not appropriate for full-scale commercial applications, however. Loose catalyst powders are generally not compatible with high-flow-rate situations, as the powder may become compressed or dispersed into the airstream, depending on the direction of the airflow. Powder-layers are also a poor choice in situations where the reactor may be subjected to vibrations or sudden shocks, as might be expected in reactors associated with certain mobile

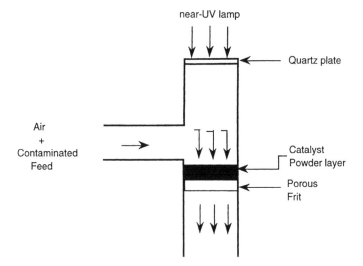

Figure 2 Schematic diagram of a continuous-flow, powder-layer photoreactor. (From Ref. 39.)

emission sources. The powder layer may become unevenly distributed across the support under such conditions, allowing channeling of contaminated air through the reactor with little or no contact with the catalyst powder.

b. Coated Surfaces, Foams, and Monoliths. Supported titanium dioxide catalysts are expected to be of more commercial significance than unsupported powders. Supported catalysts are much easier to transport and handle, less susceptible to shock and vibration, and certain configurations (monoliths, reticulated foams) can accommodate higher air flow rates with lower pressure drops than powder layers. However, achieving consistent preparation and accurately determining the catalyst load, surface area, and extent of illumination is not yet routine.

A variety of photocatalyst supports has been examined experimentally. Dip-coated glass slides or plates have been used in many experimental systems as a simple lab-scale supported photocatalyst system. Coated glass offers many of the important features of a supported photocatalyst while still offering relatively simple preparation. Honeycomb monoliths, widely used as commercial catalyst supports for a variety of gas-phase applications, have also been examined as photocatalyst supports (Fig. 3). Although these monoliths offer good stability and excellent throughput, providing illumination for the photocatalyst inside the monolith channels can be problematic [41,42]. Randomly structured support materials, like fiber-based filters, reticulated foams, and similar materials, have been used

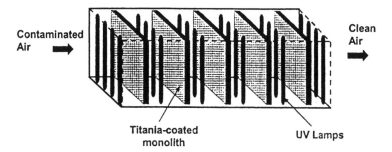

Contaminated Air →

Clean Air →

Titania-coated monolith

UV Lamps

Figure 3 Schematic diagram of a continuous-flow, monolith photoreactor. (From Ref. 41.)

in some experimental studies. These materials offer many of the advantages of honeycomb-type monolith structures, but may also suffer light penetration problems [41,42].

III. EXPERIMENTAL FINDINGS

A. Dark Adsorption of Aromatic Contaminants onto TiO$_2$

1. Trends Within the BTEX Family

The simple aromatic (BTEX) contaminants display relatively weak binding to TiO$_2$ surfaces. Larson and Falconer examined the binding of benzene, toluene, xylene, and mesitylene (1,3,5-trimethylbenzene) on Degussa P25 TiO$_2$ using temperature-programmed desorption (TPD) techniques [43]. Desorption of these aromatic contaminants was observed to begin just above room temperature (Fig. 4). Benzene desorbed at the lowest temperatures, followed by toluene, p-xylene, and mesitylene, respectively. Their TPD data may be used to estimate binding energies, using the (presumed) first-order desorption expression

$$\frac{E_b}{RT_M^2} = \frac{k_0}{\beta} \exp \frac{-E_b}{RT_M}$$

where β is the heating rate (1 K/sec) and T_M is the temperature at which the desorption rate is at its maximum [44]. From the TPD data provided by Larson and Falconer, the binding energies for benzene, toluene, p-xylene, and mestiylene are calculated to be 23.0, 24.2, 27.5, and 28.6 kcal/mol, respectively. These values are consistent with relatively weak chemisorption.

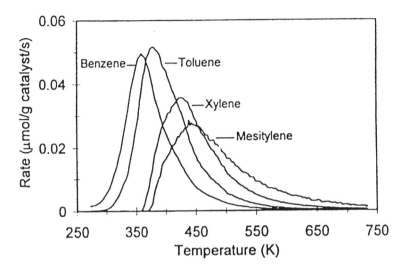

Figure 4   Temperature-programmed desorption/oxidation spectra for benzene, toluene, *p*-xylene, and mesitylene adsorbed on TiO$_2$. (From Ref. 43.)

It has been suggested that the aromatic contaminants bind to TiO$_2$ surfaces via the π-bonds of the aromatic ring structure [45,46]. The presence of methyl and ethyl side groups on the aromatic ring appear to further stabilize this π-bonding, increasing the binding energy somewhat. The aromatic contaminants are believed to π-bond with exposed titanium cations on dehydroxylated TiO$_2$ surfaces or to surface hydroxyl groups on hydroxylated titanium dioxide.

The quantity of aromatic contaminants that adsorb onto TiO$_2$ surfaces is also relatively low. d'Hennezel and Ollis [47] measured the dark adsorption of the BTEX compounds at a gas-phase concentration of 50 mg/m^3. Benzene displays the lowest dark adsorption, followed by ethylbenzene. Higher dark adsorption was observed for toluene and xylenes. At 50 mg/m^3, the dark adsorption of *m*-xylene was nearly 10 times that of benzene (Table 1).

2.   Aromatics Relative to Other Contaminants of Interest

Some hydrocarbons display modest dark adsorption similar to that seen with aromatic contaminants. However, oxygenated compounds, particularly aldehydes and alcohols, display stronger binding and significantly higher dark adsorption quantities. For example, temperature-programmed desorption/oxidation studies conducted by Muggli et al. [48] reveal noticeably higher desorption temperatures

Table 1 Dark Adsorption of Various Compounds onto TiO_2

Pollutants	Molecular weight	Corrected dark adsorption (molecule/cm^2)
1,4-dioxane	88	5.36×10^{13}
Vinyl acetate	86	3.34×10^{14}
Methyl acrylate	86	8.76×10^{13}
Chloroform	118	9.83×10^{11}
Methylene chloride	84	3.57×10^{12}
1,1,1-Trichloroethane	132	9.09×10^{10}
2-Butanone	72	2.53×10^{13}
Acetone	58	3.09×10^{13}
Butyraldehyde	72	2.65×10^{14}
Acetaldehyde	44	2.48×10^{14}
1-Butanol	74	3.68×10^{14}
Methanol	32	1.61×10^{14}
Benzene	77	4.44×10^{12}
M-Xylene	105	1.69×10^{13}
Toluene	91	9.68×10^{12}
Ethyl benzene	106	5.87×10^{12}
TCE	130	2.31×10^{13}
Hexane	86	1.60×10^{12}
MTBE	88	1.86×10^{13}

Source: Ref. 47.

for ethanol and acetaldehyde than those seen for aromatic contaminants (Fig. 5). CO and CO_2 normally bind only weakly to titania; consequently, the desorption of significant quantities of CO and CO_2 at quite high temperatures (650–700 K) indicated that a portion of the adsorbed organic material present was very strongly bound and underwent oxidation before temperatures sufficient to initiate desorption were even reached. The multiple desorption peaks suggested that there may be several distinct adsorption configurations for these compounds, whereas, aromatic contaminants displayed only single peaks, suggesting a single type of adsorption.

Dark adsorption measurements made for a variety of compounds by d'Hennezel and Ollis [47] displayed a considerable range in the amount of material adsorbed onto TiO_2 catalyst surfaces. For gas-phase contaminant concentrations of 50 mg/m^3 and a relative humidity of 7%, dark adsorption measurements for aldehydes and alcohols show approximately 10 times more material adsorbing than do aromatic contaminants (Table 1). Similarly, adsorption isotherms gener-

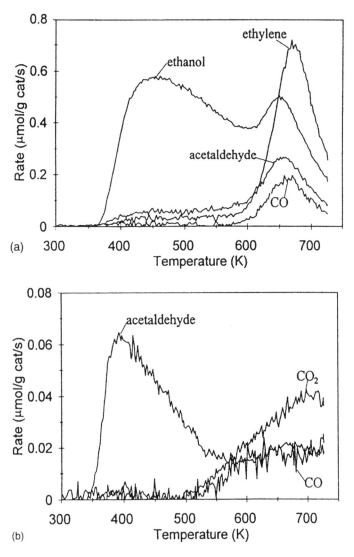

Figure 5 Temperature-programmed desorption/oxidation for (a) ethanol and (b) acetaldehyde on TiO_2. (From Ref. 48.)

ated for ethanol and acetaldehyde by Nimlos et al. [49] show much greater dark adsorption than that predicted by a toluene adsorption isotherm [50]. Isotherms for acetaldehyde and ethanol show several hundred times the dark adsorption seen in the toluene isotherm at low gas-phase concentrations, although humidity influences may play some role in the observed difference (Fig. 6).

The higher binding strengths and adsorption quantities associated with oxygenated materials when compared with aromatics may play a significant role in gas-phase photocatalytic oxidation. In particular, intermediates formed during oxidation of aromatics may include hydroxyl, carbonyl, and carboxylate groups and would be expected to bind strongly. This circumstance will affect the generation of gas-phase reaction by-products and influence apparent photocatalyst deactivation phenomena, both of which are discussed in more detail in later sections.

B. Photocatalytic Reaction Rate Trends

1. Trends Within the BTEX Family

d'Hennezel and Ollis [17] evaluated the continuous photocatalytic oxidation of BTEX compounds at a gas-phase concentration of 50 mg/m^3 in a powder-layer photoreactor (7% relative humidity). The photocatalytic reaction rates for the BTEX contaminants follow the same general trend as the dark adsorption of these materials. The pseudo-steady-state photocatalytic reaction rate is lowest for benzene and higher for the branched aromatics. Ethylbenzene is reported to be slightly more reactive than toluene (the reverse of the reported dark adsorption trend for these two compounds), whereas xylenes are noticeably more reactive than the single-branch aromatics.

In all four cases, the initial reaction rates at the start of illumination in the continuous-feed photoreactor were higher than the pseudo-steady-state reaction rates; the reaction rates declined over time until pseudo-steady-state operation was achieved. This apparent deactivation phenomenon, often observed with aromatic contaminants, is discussed in Sec. III.E. In a transient reaction system, the time required to reach pseudo-steady-state operation also appears to increase in the same order as the reaction rates. For example, for the continuous photocatalytic oxidation of aromatic contaminants at 50 mg/m^3 in a powder-layer photoreactor, the time required for pseudo-steady-state operation to be achieved was reported to be approximately 90 min for benzene, 120 min for toluene, and as long as 6 hr for m-xylene [50,51]. Under such conditions, the difference in reaction rates between the aromatic contaminants is magnified by the fact that the more reactive aromatics retain their higher transient reaction rates for longer periods (Fig. 7).

2. Aromatics Relative to Other Contaminants of Interest

d'Hennezel and Ollis considered the continuous photocatalytic oxidation of a variety of air contaminants under identical reaction conditions [17]. In addition

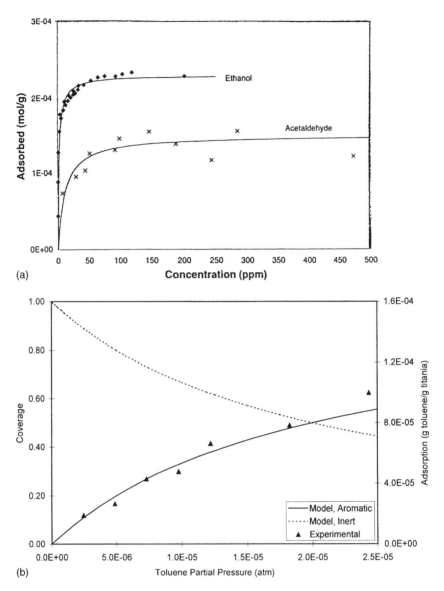

Figure 6 Adsorption isotherms for (a) ethanol and acetaldehyde (from Ref. 49), and (b) toluene (From Ref. 50) on TiO$_2$.

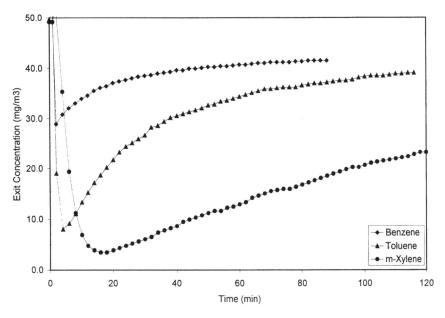

Figure 7 Apparent deactivation during the continuous photocatalytic oxidation of benzene, toluene, and m-xylene. (From Ref. 50.)

to the BTEX compounds, representative aldehydes, alcohols, ketones, and chloroalkanes were examined individually at gas-phase concentrations of 50 mg/m^3 with 7% relative humidity. The contaminants were allowed to adsorb onto the TiO$_2$ catalyst powder in the dark until equilibrium was reached prior to ultraviolet (UV) illumination. The aromatic contaminants displayed initial conversions of approximately 40% for benzene and roughly 80–90% for the branched aromatics (toluene, ethylbenzene, and m-xylene). Conversion of all four aromatic contaminants was observed to decline toward pseudo-steady-state conversion levels in the 20–40% range.

Under the same reaction conditions, acetaldehyde and butyraldehyde displayed near-complete conversion (greater than 95%). The photocatalytic oxidation of the alcohol 1-butanol displayed similarly high conversion levels, although conversion of methanol was somewhat lower. The oxygenated compounds methyl-t-butyl ether (MTBE), methyl acrylate, 1,4 dioxane, and vinyl acetate displayed conversion levels ranging from 92% to 100%. The lowest conversion levels of the oxygenated compounds studied were seen with the ketones used [acetone and 2-butanone (methylethylketone)], which displayed conversions of approximately 80%. The initial conversion levels seen with n-hexane were similar to those of the aromatics (62%), although the conversion remained slightly higher than that of most aromatics as the experiment progressed (~50%). Of all of the compounds considered, only the chloroalkanes displayed lower conversion levels than the aromatic contaminants. Chloroform, methylene chloride, and 1,1,1-tri-chloroethane were reported to reach, at most, only 15% conversion at pseudo-steady-state.

C. Gas-Phase Reaction Products

1. Low Aromatic Concentrations (<25 ppm)

Most photocatalytic studies conducted at low aromatic concentrations report no detectable concentrations of gas-phase intermediates [12,17,18]. Traces of inter-mediates may be present in the gas phase, but at levels below the detection limits of the analytical instruments employed in these studies. There is evidence, however, for either reaction intermediates or reaction by-products on the catalyst surface, even at these low concentrations. Catalyst discoloration, typically a yel-lowish or brownish color, is often reported following the photocatalytic oxidation of aromatic contaminants at low to moderate gas-phase concentrations [3,4,7,17,52]. These intermediates or reaction by-products may be largely trapped on the catalyst surface by the higher affinity of oxygenated species, like alcohols and aldehydes, for TiO_2 surfaces when compared to the aromatic parent com-pounds.

2. Moderate Aromatic Concentrations (25–500 ppm)

As the feed concentrations of aromatic contaminants are increased, low concentra-tions of gas-phase intermediate products are occasionally reported. During the photocatalytic oxidation of aromatic contaminants in the 25–500-ppm range, some researchers have observed up to several ppm of gas-phase intermediate products. For example, Ibusuki and Takeuchi [8] detected 1–2 ppm of benzalde-hyde in the gas phase following photocatalytic oxidation of 80 ppm of toluene in a batch-reactor system. Even at moderate aromatic feed concentrations, gas-phase intermediate concentrations may be too low to be detected reliably.

3. High Aromatic Concentrations (~1% in air)

Although photocatalytic oxidation is generally applied to only low concentrations of gas-phase contaminants, typically in the ppm range, several studies have ad-dressed higher concentrations as well. Augugliaro et al. [14] and Martra et al. [16] evaluated the photocatalytic oxidation of toluene at concentrations of up to 1.3 mol% in air. At these high concentrations, significant levels of gas-phase intermediate products were observed. Of the toluene removed from the gas phase in the photocatalytic system, nearly 20% was converted into benzaldehyde, as detected in the reactor off-gas. In addition to benzaldehyde, lower levels of benzene, benzyl alcohol, benzoic acid, and phenol were also detected.

D. Humidity Influences

Humidity has a significant influence on the photocatalytic oxidation of aromatic contaminants in the gas phase. This is of particular interest, because commercial photocatalytic systems will be required to operate under a broad range of relative-humidity levels. The specific influence of relative humidity on the photocatalytic reaction has generally proved rather difficult to quantify because water has a dual role: It may compete with contaminants for surface adsorption sites (a negative influence) and it plays a role in the regeneration of surface hydroxyl groups during photocatalysis (a positive influence).

1. Zero Relative Humidity

In the absence of water, aromatic contaminants experience reduced competition for catalyst-binding sites, which may initially lead to increased levels of adsorption. This heightened aromatic loading may produce a brief, enhanced transient period of photocatalytic oxidation. However, the lack of water generally leads to a rapid loss in catalyst activity, particularly at higher aromatic concentrations, possibly due to the irreversible consumption of surface-bound water and hydroxyl groups. Photocatalysis in the absence of water may produce higher concentrations of partial oxidation products, both in the gas phase [8] and on the catalyst surface, and a pronounced discoloration of the photocatalyst is often reported [7]. Humidification of the gas phase appears vital to the long-term operation of photocatalytic reactor systems.

2. Modest Relative Humidity

The presence of gas-phase water is generally beneficial to the photocatalytic oxidation of aromatic contaminants. In continuous photoreactors, humidity appears to prolong catalyst activity and delay or prevent catalyst deactivation. The effects of humidity on reaction rates, however, appear to vary, depending on the aromatic contaminant concentration and the humidity level. For example, Peral and Ollis [18] examined the continuous photocatalytic oxidation of m-xylene in a powder-layer photoreactor at several different relative humidity levels. The m-xylene photo-oxidation reaction rate was observed to increase for gas-phase water concentrations up to 1000 mg/m^3 ($\sim$7% relative humidity). Increasing the humidity level further (up to 5500 mg/m^3) produced a gradual decrease in the observed reaction rate, possibly due to increased adsorption-site competition between xylene and water. The reported xylene removal rate for a water concentration of 5500 mg/m^3 was approximately half that seen at 1000 mg/m^3.

Obee and Brown [9] observed a similar removal rate maximum at moderate humidity levels in the photocatalytic oxidation of low concentrations of toluene

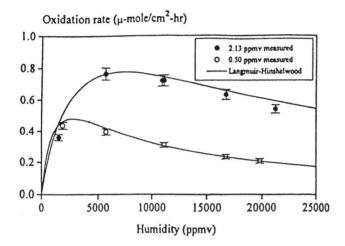

Figure 8   Influence of moderate relative humidity levels on the photocatalytic oxidation of toluene. (From Ref. 9.)

(Fig. 8). The location of this rate maximum was observed to depend on the gas-phase toluene concentration considered. For toluene concentrations of 0.5 ppmv, the maximum oxidation rate was reported to occur at approximately 2500 ppmv of water. When the toluene concentration was increased to 2.13 ppmv, however, the maximum oxidation rate was observed to shift to a higher humidity level, occurring at 7500 ppmv of water.

3.   High Relative Humidity

High humidity levels are not reported to produce decreasing aromatic photo-oxidation rates in all cases, however. Under some conditions, generally when higher aromatic loadings are used, the photocatalytic oxidation rate for aromatic contaminants has been found to continue to rise with increasing humidity levels. For example, Ameen and Raupp [19] examined the continuous photocatalytic oxidation of 25 ppmv of o-xylene at relative humidity levels ranging from 30% to 90%. Pseudo-steady-state o-xylene removal rates were observed to increase with increasing relative humidity, with the 90% relative humidity case producing the highest observed removal rates (Fig. 9).

E.   Apparent Deactivation Phenomena

1.   Experimental Evidence

During the continuous photocatalytic oxidation of organic contaminants, the rate of contaminant removal has been commonly observed to decline over time, some-

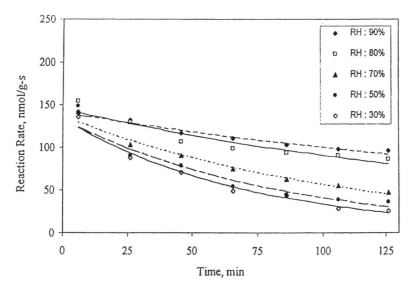

Figure 9 Influence of humidity on the photocatalytic oxidation of *o*-xylene. (From Ref. 19.)

times significantly. The photocatalytic oxidation of aromatics, even at comparatively low gas-phase concentrations, appears particularly prone to this apparent deactivation (Fig. 7). In some cases, the rate of removal of aromatic contaminants may decline by a factor of 4 over the course of several hours of operation [50].

One explanation for this phenomenon claims that this loss of activity is not, in fact, a deactivation of the catalyst, but a gradual approach toward a steady-state level of conversion [53]. As the degradation of pollutant molecules proceeds, reaction intermediates are formed on the catalyst surface. Some of these intermediate compounds may bind strongly to the surface of the catalyst particles. Calculations indicate that, at the ambient temperatures usually employed in photocatalytic systems, even a modest 30 kcal/mol binding energy can lead to a substantial residence time (up to several days) of these compounds on the catalyst surface, unless they are removed by further reaction [53]. This strong binding of reaction intermediates appears consistent with the generally low levels of observed gas-phase intermediates, as well as the appearance of catalyst discoloration during continuous photocatalysis. These strongly bound intermediate compounds can, therefore, be expected to accumulate gradually on the catalyst surface, occupying surface sites and preventing additional pollutant molecules from adsorbing onto the catalyst. This behavior could lead to the observed decrease in the rate of pollutant consumption and a decrease in overall conversion.

The possible accumulation of intermediates during the photocatalytic oxidation of aromatic contaminants has been studied with a variety of techniques, including temperature-programmed desorption (TPD), oxidation (TPO), and hydrogenation (TPH), Fourier transform infrared analysis (FTIR), and extraction of adsorbed species with a variety of solvents.

2. Accumulated Reaction Intermediates

a. TPD/TPO/TPH Analysis.   Larson and Falconer [43] applied TPD and TPO techniques to titanium dioxide catalysts used for the photo-oxidation of aromatic contaminants. As described previously, unreacted aromatic contaminants desorb readily from the titanium dioxide surface at low temperatures (Fig. 4). The TPO of catalyst samples previously used for the photo-oxidation of benzene, toluene, or xylene revealed the presence of residual, unreacted aromatic contaminants, as well as some additional surface species formed du_ing photocatalysis. These new compounds did not desorb intact, but oxidized to CO and CO_2 at higher temperatures, indicating much stronger binding than seen with the parent aromatic contaminants. Off-gas monitoring during photocatalysis was reported to have shown higher O_2 uptake than CO_2 production, suggesting that the strongly bound material observed with TPO consisted of partially oxidized reaction intermediates.

Blount and Falconer [54] further examined the photocatalytic oxidation of toluene using TPH. During TPH analysis of used catalyst samples, the strongly bound intermediates observed by Larson and Falconer [43] were reported to be hydrogenated and desorbed predominantly as toluene, along with smaller quantities of benzene. This indicated that the intermediate species responsible for apparent catalyst deactivation during toluene photooxidation retained an aromatic ring structure.

Neither Larson and Falconer [43] nor Blount and Falconer [54] definitely identified the compound or compounds responsible for the observed drop in catalyst activity during the photocatalytic oxidation of aromatic contaminants. Some possible intermediates, including benzaldehyde and benzoic acid, were considered, but were ruled out as being the species responsible for the apparent deactivation of the photocatalysts.

b. FTIR Analysis.   Ameen and Raupp [19] examined the photocatalytic oxidation of *o*-xylene in a powder-layer reactor. In situ infrared spectroscopy was used to monitor surface species generated during the photocatalytic oxidation of *o*-xylene. A number of IR bands was reported to develop during UV illumination of the photocatalyst, indicating that intermediate species were being formed and were accumulating on the catalyst surface. Several of the observed bands were assigned to *o*-tolualdehyde, *o*-toluic acid, and benzoate ions. Under low-humidity conditions, the relative concentrations of these surface species were reported to be higher than under high-humidity (80–90% relative humidity) conditions. From the infrared data, Ameen and Raupp concluded that apparent deactivation was occurring through the accumulation of recalcitrant, partially oxidized intermediates. However, they were unable to identify conclusively which compound(s) were most directly responsible for this deactivation.

Mendez-Roman and Cardona-Martinez [55] examined titanium dioxide catalysts with FTIR spectroscopy during the photocatalytic oxidation of toluene. Reaction intermediates, believed to be benzaldehyde and benzoic acid, were reported to accumulate on catalyst samples. This accumulation of intermediates was found to be reduced in the presence of gas-phase water. Mendez-Roman and Cardona-Martinez concluded that toluene appeared to be converted to benzaldehyde, which was then oxidized further to form benzoic acid. They suggested that the accumulation of benzoic acid led to the observed apparent catalyst deactivation. Other researchers, however, have argued that benzoic acid is unlikely to be the compound responsible for apparent deactivation in the photocatalytic oxidation of aromatics. For example, Larson and Falconer [43] concluded, based on higher CO_2 evolution rates for benzoic acid relative to toluene during photooxidation, that benzoic acid was not sufficiently recalcitrant to be responsible for the deactivation seen with aromatic contaminants.

c. Liquid Extraction Analysis. Sitkiewitz and Heller [4] examined the photocatalytic oxidation of benzene in a recirculating reaction system. In experiments with no gas-phase humidification, catalyst deactivation was observed, and discoloration of the catalyst was reported. Used catalysts were exposed to acetonitrile solvent with extended agitation in order to extract the compound(s) responsible for the discoloration. FTIR was employed to examine the extracted material. The spectra obtained suggested the presence of both aldehyde and hydroxyl functionalities, as well as C—H stretching vibrations consistent with both aromatic and aliphatic structures. However, no precise identification of the compounds present was made.

d'Hennezel et al. [56] considered the continuous photocatalytic oxidation of both toluene and benzene (50 mg/m^3) in slightly humidified air. Pronounced catalyst discoloration was observed following both benzene and toluene photooxidation. Surface-bound intermediates were extracted from used catalysts using either diethyl ether, followed by gas chromatography–mass spectroscopy (GC/MS) analysis of the extract, or water, followed by high-performance liquid chromatography (HPLC) and UV adsorbance detection. Diethyl ether extracts from catalysts used for toluene photocatalytic oxidation revealed the presence of benzaldehyde and benzyl alcohol, but did not appear to remove the intermediates responsible for catalyst discoloration. Water extractions from catalyst samples used with toluene were reported to remove benzoic acid, benzyl alcohol, and benzaldehyde, as well as smaller quantities of 4-hydroxybenzyl alcohol, 4-hy-

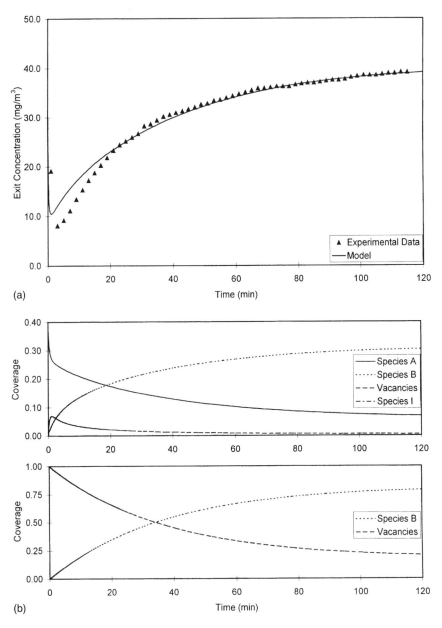

Figure 10 Two-site kinetic modeling results for the continuous photocatalytic oxidation of toluene (50 mg/m^3, 7% relative humidity): (a) flame ionization detector (FID) traces; (b) predicted surface coverages. (From Ref. 50.)

droxybenzaldehyde, and 3-hydroxybenzaldehyde. Water extracts from catalysts used for benzene photocatalytic oxidation were reported to contain hydroquinone, 1,4-benzoquinone, and phenol. In both cases, the water extraction did appear to release the agent(s) responsible for catalyst discoloration from the titania, but the compound(s) did not dissolve readily in the water and were not identified.

3. Kinetic Modeling

Lewandowski and Ollis have proposed a simple kinetic model describing the transient photocatalytic oxidation of aromatic contaminants [50]. The model considered three chemical species: an aromatic contaminant, preadsorbed onto the catalyst in the dark and refreshed continuously from the gas phase; a strongly bound, recalcitrant reaction intermediate; and final reaction products (CO or CO_2), assumed for simplicity to be strictly gas-phase species. The model also assumed that two types of catalyst site were present on the photocatalyst surface, with the first suitable for the adsorption of aromatic contaminants, as well as reaction intermediates, and the second type considered to be more polar in nature, suitable only for adsorption of partially oxidized reaction intermediates.

This kinetic model was used to simulate the photocatalytic oxidation of benzene and m-xylene at 50 mg/m^3, and toluene over a range of gas-phase concentrations (10–100 mg/m^3) [50,57]. The model system was found to be reasonably capable of reproducing the trends observed in experimental data (Fig. 10).

IV. RATE ENHANCEMENT AND ACTIVITY MAINTENANCE IN THE PHOTO-OXIDATION OF AROMATICS

The relatively low reaction rates and apparent deactivation typically seen with aromatic contaminants may pose a significant barrier to the commercialization of photocatalytic systems for the destruction of aromatics. Early economic estimates developed by Miller and Fox [58] suggested that photocatalytic systems may not be capable of offering cost advantages over the traditional treatment technologies of carbon adsorption and catalytic incineration when the compounds to be treated display low photocatalytic reaction rates, as do aromatics. These limitations have prompted researchers to examine methods that may increase the effectiveness of photocatalytic systems toward aromatic contaminants. Several techniques for increasing the photocatalytic reaction rate for aromatic contaminants and prolonging higher catalyst activity levels are discussed in the following subsections.

A. Chlorine Promotion

Chlorinated olefins, particularly trichloroethylene (TCE), appear uniquely suited to destruction by photocatalytic oxidation. The photocatalytic oxidation of TCE typically displays a high reaction rate and high conversion per pass with little or no apparent deactivation [59–63]. Often, total conversion of the contaminant may be achieved. This phenomenon was first attributed by Nimlos et al. [61] to the generation of chlorine radicals during the degradation of the TCE parent molecules. Once generated, these chlorine radicals may react with additional TCE, promoting further TCE degradation and generating additional chlorine radicals. This chlorine radical chain reaction was argued, by analogy to the known gas-phase mechanism [64], to allow for rapid conversion of the TCE with excellent photocatalytic efficiencies [61]:

$$CCl_2 = CHCl + Cl\cdot \rightarrow CHCl_2CCl_2\cdot$$
$$CHCl_2CCl_2\cdot + O_2 \rightarrow CHCl_2CCl_2OO\cdot$$
$$2\ CHCl_2CCl_2OO\cdot \rightarrow 2\ CHCl_2CCl_2O\cdot + O_2$$
$$CHCl_2CCl_2O\cdot \rightarrow CHCl_2CCl(O) + Cl\cdot$$

The presence of chlorine radical chain reactions in a photocatalytic reaction system may significantly increase reaction rates and photocatalytic efficiencies. These enhancements would appear to have the potential to overcome the shortcomings typically associated with the photocatalytic oxidation of aromatic contaminants if a chlorine radical chain reaction could be initiated in conjunction with an aromatic photocatalytic reaction and if the chlorine radicals were capable of reacting with (and thus accelerating the conversion of) the aromatic contaminant of interest. Two potential configurations for combining chlorine radical promotion with the photocatalytic oxidation of aromatic contaminants have been examined in some detail: mixed contaminant feeds and prechlorinated catalysts.

1. Cofed Contaminants

One method for combining chlorine radical chain reactions with photocatalytic oxidation involves cofeeding a reactive compound like TCE along with a less-reactive contaminant into the photoreactor. Berman and Dong [65] first demonstrated that this TCE cofeeding arrangement could lead to enhanced conversion of the less reactive contaminants isooctane, methylene chloride, and chloroform (Table 2). Furthermore, in two of these situations (isooctane/TCE and methylene chloride/TCE), complete conversion of the TCE was also maintained. This phenomenon runs counter to the general observation in catalysis that the presence of two reactants competing for catalyst sites will reduce reaction rates compared to single-component conditions.

Luo and Ollis have shown that this cofeeding strategy may also be used to enhance the photocatalytic oxidation of toluene [12]. Near-complete photocatalytic oxidation of toluene at less than ~ 200 mg/m^3 in the gas-phase was achieved in the presence of TCE (either 226 or 753 mg/m^3). Based on the homogenous gas-phase TCE degradation mechanism proposed by Sanhueza et al. [64], which

Table 2 Influence of Cofed TCE on the Photocatalytic
Oxidation of Isooctane, Methylene Chloride, and Chloroform

Compound	Time for 90% conversion (sec)
Isooctane	22.5
Isooctane + 400 ppm TCE	7
Methylene chloride	550
Methylene chloride + 699 ppm TCE	300
Methylene chloride + 0.373% TCE	270
Methylene chloride + 0.752% TCE	155
Chloroform	720
Chloroform + 0.206% TCE	380

Source: Ref. 65.

was applied to TCE photocatalytic oxidation by Nimlos et al. [61], Luo and Ollis
proposed a photocatalytic chain-transfer mechanism to describe how the chlorine
radicals generated through simultaneous TCE degradation could react with adsorbed toluene molecules via a hydrogen abstraction reaction [12]:

$$TCE + TiO_2 + h\nu \rightarrow Cl\cdot + \ldots$$
$$Cl\cdot + C_6H_5—CH_3 \rightarrow C_6H_5—CH_2\cdot + HCl$$

Such a reaction mechanism could initiate toluene oxidation and increase the observed removal rates and overall conversion of the toluene.

These findings were further extended by studies conducted by Sauer et
al., who demonstrated that other chlorinated olefins, including perchloroethylene
(PCE) and trichloropropene (TCP), could produce enhanced toluene conversion
in the same manner as the TCE cofeed used previously [11]. d'Hennezel and
Ollis extended the scope of the previous TCE cofeed studies to consider benzene,
ethylbenzene, and *m*-xylene in addition to toluene [17]. Although this study found
enhanced photocatalytic oxidation of the branched aromatics (toluene, ethylbenzene, and *m*-xylene) in the presence of TCE cofeeds (Fig. 11), no increase in
benzene conversion was reported. Benzene conversion appeared to be slightly
depressed by the presence of TCE in the gas phase, suggesting that the chlorine
radicals produced by TCE degradation could not effectively initiate oxidation of
the adsorbed benzene molecules (Fig. 12).

Lewandowski and Ollis carried out thermodynamic calculations for the
reaction of chlorine radicals with branched (toluene, xylenes) and unbranched
(benzene) aromatic contaminants [51]. On a branched aromatic contaminant, abstraction of the hydrogen atom on a methyl side group by a chlorine radical,

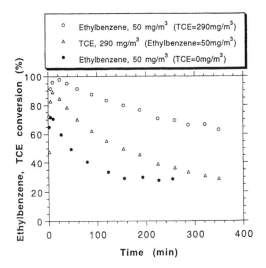

Figure 11   Photocatalytic oxidation of ethylbenzene cofed with TCE. (From Ref. 17.)

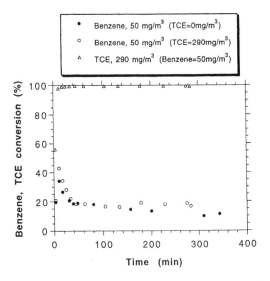

Figure 12   Photocatalytic oxidation of benzene cofed with TCE. (From Ref. 17.)

$$Cl\cdot + C_6H_5{-}CH_3 \rightarrow HCl + C_6H_5{-}CH_2\cdot$$

the enthalpy change on reaction, ΔH_{rxn}, was calculated to be negative (exothermic) (Table 3). This reaction would be expected to occur spontaneously, suggesting that chlorine radicals could initiate oxidation of adsorbed toluene in photocatalytic systems. In the case of hydrogen abstraction from an aromatic ring by the chlorine radicals, as may occur in the case of benzene oxidation,

$$Cl\cdot + C_6H_6 \rightarrow HCl + C_6H_5\cdot$$

the enthalpy change was calculated to be positive (endothermic), suggesting that this reaction would be thermodynamically unfavorable. This prediction of a thermodynamically unfavorable, chlorine radical-initiated ring hydrogen abstraction may explain why chlorinated olefin cofeeds are effective at enhancing photocatalytic oxidation for branched aromatics, but not for benzene.

Although TCE cofeeds are unable to increase the photocatalytic oxidation of benzene, the effects on the oxidation of branched aromatics are significant. However, practical considerations will hinder the use of TCE cofeeds. First, TCE and branched aromatic contaminants are not often present in the same airstream. Adding TCE or similar chlorinated olefins to an environment or airstream containing branched aromatic contaminants is not practical, because TCE is, itself, classi-

Table 3 Calculated ΔH_{rxn} Values for Halide Radicals Abstracting Hydrogens from Aromatic Ring of Benzene or Toluene Methyl Group

	ΔH_{rxn} (kcal/mol)	ΔH_{rxn} (kJ/mol)
Aromatic ring of benzene		
$OH\cdot + C_6H_6 \rightarrow HOH + C_6H_5\cdot$	-8.33	-34.84
$F\cdot + C_6H_6 \rightarrow HF + C_6H_5\cdot$	-24.90	-104.18
$Cl\cdot + C_6H_6 \rightarrow HCI + C_6H_5\cdot$	7.64	31.95
$Br\cdot + C_6H_6 \rightarrow HBr + C_6H_5\cdot$	23.34	97.65
$I\cdot + C_6H_6 \rightarrow HI + C_6H_5\cdot$	39.58	165.58
Toulene methyl group		
$OH\cdot + C_6H_5CH_3 \rightarrow HOH + C_6H_5CH_2\cdot$	-31.26	-130.77
$F\cdot + C_6H_5CH_3 \rightarrow HF + C_6H_5CH_2\cdot$	-47.83	-200.12
$Cl\cdot + C_6H_5CH_3 \rightarrow HCl + C_6H_5CH_2\cdot$	-15.29	-63.99
$Br\cdot + C_6H_5CH_3 \rightarrow HBr + C_6H_5CH_2\cdot$	0.41	1.71
$I\cdot + C_6H_5CH_3 \rightarrow HI + C_6H_5CH_2\cdot$	16.65	69.64

Source: Ref. 51.

fied as a hazardous air pollutant under the U.S. Clean Air Act (as is PCE). Further, even in situations where TCE and branched aromatic contaminants may normally occur together in an airstream, the studies conducted by Luo and Ollis [12] and by Sauer et al. [11] suggest that the relative concentrations of the TCE and the aromatic contaminant must be maintained within a specific range, with the chlorinated compound present at the higher concentration. Luo and Ollis demonstrated that a TCE to toluene ratio of approximately 3 to 1 is required to maintain enhanced toluene conversion [12]. When the toluene concentration was increased beyond this limit, toluene conversion was observed to decrease to levels similar to those observed for the photocatalytic oxidation of toluene alone, and TCE conversion was suppressed. Similar results were obtained by Sauer et al. for reactions employing PCE as a chlorine radical source [11]. This effect was attributed to a quenching of the chlorine radical chain reaction by the higher toluene surface concentrations.

In order to circumvent the addition of a hazardous air pollutant like TCE or PCE to generate enhanced removal of branched aromatics, d'Hennezel and Ollis explored a catalyst pretreatment intended to introduce chlorine atoms into the surface structure of TiO_2 photocatalysts.

2. Catalyst Pretreatment

Previous studies with alumina and titania have demonstrated that chlorine may be introduced into the surface structure of these materials through reactions with various chlorine-containing compounds. Chlorination of titania has been shown using gas-phase treatments of carbon tetrachloride or hydrochloric acid [66,67]. For chlorination with gaseous carbon tetrachloride, the proposed reaction mechanism involved an exchange of four chlorine groups for two oxygens, producing chlorinated titania and carbon dioxide:

$$CCl_{4(g)} + 2O_{(s)} \rightarrow CO_{2(g)} + 4Cl_{(s)}$$

For titanium dioxide chlorination by HCl, two reactions were proposed:

$$-Ti-O-Ti-_{(s)} + HCl_{(g)} \rightarrow -Ti-Cl_{(s)} + -Ti-OH_{(s)}$$
$$-TiOH_{(s)} + HCl_{(g)} \rightarrow -TiCl_{(s)} + H_2O_{(nds)}$$

The first reaction was claimed by the authors to operate primarily on dehydroxylated titania, and the second was believed to occur on titania that was partially or completely hydroxylated [66].

A variation of this process, using an aqueous-phase treatment of TiO_2 with 3 N hydrochloric acid solution, was used by d'Hennezel and Ollis for prechlorination of a titanium dioxide photocatalyst [68]:

$$-TiOH_{(s)} + Cl^-_{(aq)} \rightarrow -TiCl_{(s)} + OH^-_{(aq)}$$

Upon illumination, photogenerated holes could convert surface chloride groups into chlorine radicals, which could then initiate radical reactions with adsorbed organic material. If sufficient coverage of chloride groups could be obtained through the pretreatment process, chlorine radical chain reaction of the sort seen with TCE cofeeds might occur.

Experiments conducted using aqueous chloride salts as pretreatment agents prompted Lewandowski and Ollis to adjust this mechanism to reflect the possible role of aqueous protons (H^+) in the pretreatment process [69]. Based on kinetic studies of the halogenation of alcohols, they suggested that acidic conditions were required to displace surface hydroxyl groups prior to chloride addition to the catalyst:

$$—TiOH_{(s)} + H_{(aq)}^+ \rightarrow —TiOH_{2(s)}^+$$
$$—TiOH_{2(s)} \rightarrow —Ti_{(s)}^+ + H_2O_{(aq)}$$
$$—Ti_{(s)}^+ + Cl_{(aq)}^- \rightarrow —TiCl_{(s)}$$

As in the mechanism proposed by d'Hennezel and Ollis, UV illumination was predicted to initiate the conversion of surface chloride groups into chlorine radicals.

Alternately, Lewandowski and Ollis have also suggested that molecular HCl may bind directly to the catalyst surface, rather than having chlorine become incorporated into the catalyst surface structure [51]. In this case, the interaction of the HCl molecules with the surface may weaken the hydrogen–chlorine bond sufficiently to allow for direct photolysis of the bond by incident long-wave UV.

Regardless of the exact mechanism at work, HCl catalyst pretreatment have been demonstrated to enhance the photocatalytic oxidation of toluene at low concentrations [68,69]. The apparent deactivation of the photocatalyst is noticeably delayed over HCl–pretreated catalyst samples in a manner similar to that seen with cofed toluene and TCE (Fig. 13). However, the pseudo-steady-state level of conversion appears to be nearly identical on both untreated and HCl-pretreated catalysts. Because the batch HCl pretreatment process incorporates a limited quantity of HCl into the catalyst surface structure, this similarity in long-term activity may be the result of surface chlorine depletion.

Alternatively, some of the reaction intermediates generated during the photocatalytic oxidation of aromatic contaminants may be rather recalcitrant to chlorine radicals, leading to the apparent deactivation phenomenon. Thermodynamically, intermediate species that are recalcitrant toward hydroxyl radicals may be recalcitrant toward chlorine radicals as well (Table 3). The accumulation of these recalcitrant reaction intermediates on the catalyst surface may gradually reduce the effectiveness of chlorine radicals in the photocatalytic system, decreasing the chlorine-promoted enhancement in toluene oxidation to negligible levels [51].

Benzene photocatalysis, as suggested by the thermodynamic calculations of Lewandowski and Ollis [51], is not enhanced when conducted on HCl-pretreated TiO$_2$. As in previous studies using benzene/TCE cofeeds, benzene conver-

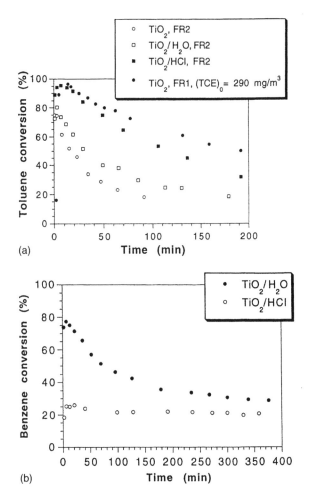

Figure 13 Photocatalytic oxidation of (a) toluene (from Ref. 68) and (b) benzene (from Ref. 56) on HCl-pretreated TiO_2 catalysts.

sion is slightly depressed on HCl-pretreated catalyst samples compared with un-treated TiO_2 powders.

The photocatalytic oxidation of m-xylene on an HCl-pretreated catalyst showed little or no enhancement [51]. This result is unusual, because enhanced photocatalytic oxidation of m-xylene has been demonstrated with cofed TCE [17]. Lewandowski and Ollis suggested that xylene, which adsorbs onto TiO_2 in larger quantities than other aromatic contaminants, may be present at levels sufficient to hinder the chlorine radical chain reaction, possibly by displacing chloride and/ or HCl from the catalyst surface or by hindering the readsorption of HCl formed following hydrogen abstractions [51]. This could lead to a quenching effect similar to that reported by Luo and Ollis for cofed toluene and TCE at higher toluene to TCE ratios [12]. Because catalyst pretreatments with aqueous HCl solutions appear to provide a limited quantity of chlorine on the catalyst surface, the range of concentrations at which HCl pretreatments are likely to enhance the conversion of branched aromatics, particularly the xylenes, may be somewhat limited [51,69].

B. Catalyst Regeneration

The decline in catalyst activity seen in some continuous photocatalytic systems has prompted researchers to examine methods of restoring activity to used photocatalysts. Because the decline in catalyst activity is often attributed to the accumulation of recalcitrant intermediates or by-products on the catalyst surface, most catalyst regeneration techniques focus on the removal of these presumed species. Two such methods, thermal regeneration and photocatalytic regeneration, have been examined for use in association with the photocatalytic oxidation of aromatic contaminants.

1. Thermal Regeneration

Heating a used photcatalyst to a sufficiently high temperature, generally several hundred degrees Celsius, can allow for the desorption or thermal oxidation of accumulated surface intermediates. Such a thermal regeneration scheme was examined by Cao et. al. [52] for the treatment of titanium dioxide catalyst samples used for the photocatalytic oxidation of toluene. Following 100 min or more of toluene photocatalytic oxidation, during which toluene oxidation rates were observed to decline to pseudo-steady-state levels and a yellowish catalyst discoloration was noted, the used photocatalysts were heated to 350°C or 420°C in air to remove accumulated reaction intermediates. A 2-hr thermal regeneration at 350°C was reported to reduce the catalyst discoloration and provide a partial recovery of catalyst activity. The 420°C thermal regeneration produced a complete recovery of the photocatalytic activity and returned the catalyst to its original white color.

Cao et. al. also examined titanium dioxide photocatalysts doped with 0.5% platinum. These doped catalysts displayed complete activity recovery following thermal regeneration at 350°C. Presumably, the addition of platinum, which may act as a thermal oxidation catalyst, allows for the destruction of accumulated intermediates generated during the photocatalytic oxidation at lower temperatures than untreated titanium dioxide.

A thermal regeneration technique is most appropriate for use in an industrial setting, where high-temperature operations are feasible and high regeneration rates are desirable. However, in other photocatalytic applications, such as deodorization and indoor air-quality maintenance, such high-temperature regeneration operations are unsuitable. In these situations, photocatalytic regeneration techniques, which may be carried out at ambient temperatures, appear more appropriate.

2. Photocatalytic Regeneration

In photocatalytic regeneration, recalcitrant surface species are gradually removed from the photocatalyst through continuous exposure to UV illumination in clean (uncontaminated) flowing air. In general, humidified air is preferred for photocatalytic regeneration because complete oxidation reaction may be significantly hindered in the complete absence of water. Although photocatalytic regeneration offers the advantage of low-temperature operation, the regeneration period required to restore a photocatalyst following photo-oxidation of aromatic contaminants may be rather long compared with the operating time available before apparent catalyst deactivation sets in.

Lewandowski and Ollis examined the effects of increasing regeneration times on the recovery of catalyst activity in the continuous photocatalytic oxidation of toluene at 50 mg/m^3 and 7% relative humidity [57]. After approximately 2 hr of use, fresh catalysts displayed a decline in activity (to 20% toluene removal from a maximum initial level of ~80%) and developed a dark yellowish discoloration. Photocatalytic regenerations of 2, 4, and 6 hr were observed to provide gradually increasing recovery of the previously observed (fresh) catalyst activity (Fig. 14). After 6 hr of regeneration, titanium dioxide photocatalyst was seen to provide roughly 80% of its previous initial activity, whereas the catalyst was almost completely restored to its original color.

Other researchers have examined the effects of longer photocatalytic regeneration periods, of 12 hr or more, following apparent deactivation of fresh photocatalysts during the photo-oxidation of various aromatic contaminants. Jacoby et al. reported the photocatalytic regeneration of titanium dioxide catalysts used for the photo-oxidation of benzene [5]. The evolution of CO and CO$_2$ was observed when used catalysts were exposed to clean flowing air and UV illumination, which they attributed to the continuing photocatalytic oxidation of residual benzene and strongly bound reaction intermediates. Following photocatalytic regeneration periods of 20 hr or more, complete recovery of catalyst activity was noted.

d'Hennezel et al. considered photocatalytic regeneration following photocatalytic oxidation of benzene and toluene at 50 mg/m^3 with 7% relative humidity [56]. A 16-hr photocatalytic regeneration was observed to restore previous catalyst activity on titanium dioxide samples used for benzene photo-oxidation. This regeneration period also nearly restored prior toluene removal rates and appeared to fully restore the photocatalyst to its original color.

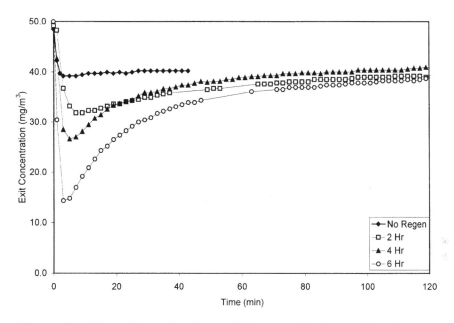

Figure 14 Effects of increasing photocatalytic regeneration time on the subsequent photocatalytic oxidation of toluene. (From Ref. 57.)

Ameen and Raupp examined photocatalytic regeneration following photocatalytic oxidation of *o*-xylene at 25 ppmv and 80% relative humidity [19]. Exposure to clean flowing air and UV illumination for 12 hr was found to restore the titanium dioxide photocatalyst activity to levels observed with fresh catalysts.

V. CONCLUSIONS

Photocatalytic oxidation over illuminated titanium dioxide has been demonstrated to be effective at removing low concentrations of a variety of hazardous aromatic contaminants from air at ambient temperatures. At low contaminant concentration levels and modest humidity levels, complete or nearly complete oxidation of aromatic contaminants can be obtained in photocatalytic systems. Although aromatic contaminants are less reactive than many other potential air pollutants, and apparent catalyst deactivation may occur in situations where recalcitrant reaction intermediates build up on the catalyst surface, several approaches have already been developed to counter these potential problems. The introduction of a chlorine source, either in the form of a reactive chloro-olefin cofeed or an HCl-pretreated catalyst, has been demonstrated to promote the photocatalytic oxidation of branched aromatic contaminants. Apparent deactivation has been shown to be a reversible phenomenon, which may be countered using several different catalyst regeneration schemes, including thermal regeneration and photocatalytic regeneration.

REFERENCES

1. EPA Health Effects Notebook for Hazardous Air Pollutants—Draft, EPA-452/D-95-00, PB95-503579, 1994.
2. Heinsohn, R. J. *Industrial Ventilation: Engineering Principles*; Wiley: New York, 1991.
3. Fu, X. Z.; Zeltner, W. A.; Anderson, M. A. *Appl. Catal. B.* **1995**, *6*, 209.
4. Sitkiewitz, S.; Heller, A. *New J. Chem.* **1996**, *20*, 233.
5. Jacoby, W. A.; Blake, D. M.; Fennell, J. A.; Boulter, J. E.; Vargo, L. M.; George M. C.; Dolberg, S. K. *J. Air Waste Manag. Assoc.* **1996**, *46*, 891.
6. Lichtin, N. N.; Sadeghi, M. *J. Photochem. Photobiol. A.* **1998**, *113*, 81.
7. Einaga, H.; Futamura, S.; Ibusuki, T. *Phys. Chem. Chem. Phys.* **1999**, *1*, 4903.
8. Ibusuki, T.; Takeuchi, K. *Atmos. Environ.* **1986**, *20*, 1711.
9. Obee, T. N.; Brown, R. T. *Environ. Sci. Technol.* **1995**, *29*, 1223.
10. Obee, T. N. *Environ. Sci. Technol.* **1996**, *30*, 3578.
11. Sauer, M. L.; Hale, M. A.; Ollis, D. F. *J. Photochem. Photobiol. A.* **1995**, *88*, 169.
12. Luo, Y.; Ollis, D. F. *J. Catal.* **1996**, *163*, 1.
13. Alberici, R. M.; Jardim, W. E. *Appl. Catal. B.* **1997**, *14*, 55.
14. Augugliaro, V.; Coluccia, S.; Loddo, V.; Marchese, L.; Martra, G.; Palmisano, L.; Schiavello, M. *Appl. Catal. B.* **1999**, *20*, 15.
15. Hager, S.; Bauer, R. *Chemosphere* **1999**, *38*, 1549.
16. Martra, G.; Coluccia, S.; Marchese, L.; Auguliaro, V.; Loddo, V.; Palmisano, L.; Schiavello, M. *Catal. Today* **1999**, *53*, 695.
17. d'Hennezel, O.; Ollis, D. F. *J. Catal.* **1997**, *167*, 118.
18. Peral, J.; Ollis, D. F. *J. Catal.* **1992**, *136*, 554.
19. Ameen, M. M.; Raupp, G. B. *J. Catal.* **1999**, *184*, 112.
20. Akhavan, J. *The Chemistry of Explosives*, The Royal Society of Chemistry: Cambridge, 1998.
21. Wang, Z. K.; Kutal, C. *Chemosphere* **1995**, *30*, 1125.
22. Dillert, R.; Brandt, M.; Fornefett, I.; Siebers, U.; Bahnemann, D. *Chemosphere* **1995**, *30*, 2333.
23. Schmelling, D. C.; Gray, K. A. *Water Res.* **1995**, *29*, 2651.
24. Schmelling, D. C.; Gray, K. A.; Kamat, P. V. *Water Res.* **1997**, *31*, 1439.
25. Nahen, M.; Bahnemann, D.; Dillert, R.; Fels, G. *J. Photochem. Photobiol. A.* **1997**, *110*, 191.
26. Kumar, S.; Davis, A. P. *Water Environ. Res.* **1997**, *69*, 1238.
27. Schmidt, A.; Butte W. *Chemosphere* **1999**, *38*, 1293.
28. Vohra, M. S.; Tanaka, K. *Water Res.* **2002**, *36*, 59.
29. Al-Ekabi, H.; Serpone, N. *J. Phys. Chem.* **1988**, *92*, 5726.
30. Stafford, U.; Gray, K. A.; Kamat, P. V. *J. Phys. Chem.* **1994**, *98*, 6343.

31. Vinodgopal, K.; Stafford, U.; Gray, K. A.; Kamat, P. V. *J. Phys. Chem.* **1994**, *98*, 6797.
32. Alberici, R. M.; Jardim, W. F. *Water Res.* **1994**, *28*, 1845.
33. Mills, A.; Davies, R. *J. Photochem. Photobiol. A.* **1995**, *85*, 173.
34. Ku, Y.; Leu, R. M.; Lee, K. C. *Water Res.* **1996**, *30*, 2578.
35. Jardim, W. F.; Moraes, S. G.; Takiyama, M. M. K. *Water Res.* **1997**, *31*, 1728.
36. Gliman, A. P.; Douglas, V. M.; Newhook, R. C.; Arbuckle, T. E. *Chlorophenols and Their Impurities: A Health Hazard Evaluation*; Department of National Health and Welfare: Ottawa, Ontario, 1989.
37. Scott, J. P.; Ollis, D. F. *Environ. Prog.* **1995**, *14*, 88.
38. Sauer, M. L.; Ollis, D. F. *J. Catal.* **1994**, *149*, 81.
39. d'Hennezel, O. Thesis, North Carolina State University, 1998, p. 144.
40. Formenti, M.; Juillet, F.; Meriaude, P.; Teichner, S. *J. Chem. Technol.* **1971**, *1*, 680.
41. Bendfeldt, P.; Hall, R. J.; Obee, T. N.; Hay, S. O.; Sangiovanni, J. J. *A. Feasibility Study of Photocatalytic Air Purification for Commercial Passenger Airliners*; United Technologies Research Center: Hartford, CT, 1997.
42. Raupp, G. B.; Alexiadis, A.; Hossain, Md. M.; Changrani, R. *Catal. Today* **2001**, *69*, 41.
43. Larson, S. A.; Falconer, J. L. *Catal. Lett.* **1997**, *44*, 57.
44. Boudart, M.; Djega-Mariadassou, G. *Kinetics of Heterogeneous Catalytic Reactions*; Princeton University Press: Princeton, NJ, 1984.
45. Suda, Y. *Langmuir* **1988**, *4*, 147.
46. Nagao, M.; Suda, Y. *Langmuir* **1989**, *5*, 42.
47. d'Hennezel, Thesis, North Carolina State University, 1998, p. 66.
48. Muggli, D. S.; McCue, J. Y.; Falconer, J. L. *J. Catal.* **1998**, *173*, 470.
49. Nimlos, M. R.; Wolfrum, E. J.; Brewer, M. L.; Fennell, J. A.; Bintner, G. *Environ. Sci. Technol.* **1996**, *30*, 3102.
50. Lewandowski, M; Thesis, North Carolina State University, **2002**, p. 69.
51. Lewandowski, M; Thesis, North Carolina State University, **2002**, p. 34.
52. Cao, L. X.; Gao, Z.; Suib, S. L.; Obee, T. N.; Hay, S. O.; Freihaut, J. D. *J. Catal.* **2000**, *196*, 253.
53. Ollis, D. F. The 3rd International Conference on TiO₂ Photocatalytic Purification and Treatment of Water and Air, 1997.
54. Blount, M. C.; Falconer, J. L. *J. Catal.* **2001**, *200*, 21.
55. Mendez-Roman, R.; Cardona-Martinez, N.; *Catal. Today* **1998**, *40*, 353.
56. d'Hennezel, O.; Pichat, P.; Ollis, D. F. *J. Photochem. Photobio. A.* **1998**, *118*, 197.
57. Lewandowski, M; Thesis, North Carolina State University, **2002**, p. 115.
58. Miller, R.; Fox, R. In *Photocatalytic Purification of Water and Air*; Ollis, D. F.; Al-Ekabi, H., Eds. Elsevier: Amsterdam, 1993, p. 573.
59. Dibble, L. A., Raupp, G. B. *Catal. Lett.* **1990**, *4*, 345.
60. Dibble, L. A., Raupp, G. B. *Environ. Sci. Technol.* **1992**, *26*, 492.
61. Nimlos, M. R.; Jacoby, W. A.; Blake, D. M.; Milne, T. A. *Environ. Sci. Technol.* **1993**, *27*, 732.
62. Jacoby, W. A.; Nimlos, M. R., Blake, D. M. *Environ. Sci. Technol.* **1994**, *28*, 1661.
63. Larson, S. A.; Falconer, J. L. *Appl. Catal. B.* **1994**, *4*, 325.

64. Sanhueza, E.; Hisatsune, J.; Heicklen, J. *Chem. Rev.* **1976**, *76*, 801.
65. Berman, E.; Dong, J. In *The Third International Symposium on Chemical Oxidation: Technologies for the Nineties*, Volume 3; Eckenfelder, W. W.; Bowers, A. R.; Roth, J. A., Eds. Technomic Publishing: Lancaster, PA, 1993, p. 183.
66. Primet, M.; Basset, J.; Mathieu, M. V.; Prettre, M. *J. Phys. Chem.* **1970**, *74*, 2868.
67. Boonstra, A. H.; Mutsaers, C. A. H. A. *J. Phys. Chem.* **1975**, *79*, 1694.
68. d'Hennezel, O.; Thesis, North Carolina State University, 1998, p. 68.
69. Lewandowski, M.; Ollis, D. F. *J. Adv. Oxid. Technol.* **2002**, *5*, 33.

9

Design and Development of New Titanium Dioxide Semiconductor Photocatalysts

Masakazu Anpo, Masato Takeuchi, Hiromi Yamashita, and Satoru Kishiguchi
Osaka Prefecture University, Osaka, Japan

Ann Davidson and Michel Che
Université Pierre et Marie Curie, Paris, France

I. INTRODUCTION

Environmental pollution and destruction on a global scale have drawn attention to the vital need for totally new, safe, and clean chemical technologies and processes, the most important challenge facing chemical scientists for the 21st century. Strong contenders as environmentally harmonious catalysts are titanium oxide photocatalysts which can operate at room temperature in a clean and safe manner. Such photocatalytic systems are urgently needed for the purification of polluted water, the decomposition of offensive atmospheric odors as well as toxins, the fixation of CO_2, and the decomposition of NO_x and chlorofluorocarbons on a global scale [1–8]. Furthermore, transparent titanium oxide thin-film photocatalysts prepared on glass, tile, and various architectural materials have been actively investigated as promising antibacterial, self-cleaning, and deodorization systems [6,7].

However, unlike photosynthesis in green plants, the titanium oxide photocatalyst does not absorb visible light and, therefore, it can make use of only 3–4% of solar photons that reach the Earth. Therefore, to address such enormous tasks, photocatalytic systems which are able to operate effectively and efficiently not only under ultraviolet (UV) but also under sunlight must be established. To this end, it is vital to design and develop unique titanium oxide photocatalysts which can absorb and operate with high efficiency under solar and/or visible-light irradiation [9–16].

This chapter deals with the design and development of such unique second-generation titanium oxide photocatalysts which absorb UV–visible light and operate effectively under visible and/or solar irradiation by applying an advanced metal ion-implantation method.

II. PRACTICAL APPLICATIONS OF TITANIUM OXIDE PHOTOCATALYSTS

When titanium oxides are irradiated with UV light that is greater than the band-gap energy of the catalyst (about $\lambda < 380$ nm), electrons (e^-) and holes (h^+) are produced in the conduction and valence bands, respectively. These electrons and holes have a high reductive potential and oxidative potential, respectively, which, together, cause catalytic reactions on the surfaces; namely photocatalytic reactions are induced. Because of its similarity with the mechanism observed with photosynthesis in green plants, photocatalysis may also be referred to as artificial photosynthesis [1–4]. As will be introduced in a later section, there are no limits to the possibilities and applications of titanium oxide photocatalysts as "environmentally harmonious catalysts" and/or "sustainable green chemical systems."

In the presence of O_2 and H_2O, the photo-formed e^- and h^+ react with these molecules on the titanium oxide surfaces to produce O_2^- and OH radicals, respectively. These O_2^- and OH radicals have a very high oxidation potential, inducing the complete oxidation reaction of various organic compounds such as toxic halocarbons, as shown in

$$C_nH_mO_zCl_y \xrightarrow[\text{TiO}_2]{h_2} nCO_2 + yHCl + wH_2O \tag{1}$$

Such high photocatalytic reactivities of photo-formed e^- and h^+ can be expected to induce various catalytic reactions to remove toxic compounds and can actually be applied for the reduction or elimination of polluted compounds in air such as NO_x, cigarette smoke, as well as volatile compounds arising from various construction materials, oxidizing them into CO_2. In water, such toxins as chloroalkenes, specifically trichloroethylene and tetrachloroethene, as well as dioxins can be completely degraded into CO_2 and H_2O. Such highly photocatalytically reactive systems are also applicable in protecting lamp covers and walls in tunnels from becoming dark and sooty by emission gases. Soundproof highway walls coated with titanium oxide photocatalysts have been constructed on heavily congested roads for the elimination of NO_x (Fig. 1). The reactivity of photo-formed O_2^- and OH radicals is high enough to decompose or kill bacteria so that new cements and tiles mixed or coated with titanium oxides have been commercialized and are already in use in the operation rooms of hospitals to maintain a sterile and bacteria-free environment.

Furthermore, titanium oxide thin films have been found to exhibit a unique and useful function (i.e., a superhydrophilic property). Usually, the contact angle of a water droplet on a surface is $50°–70°$; therefore, metal oxide surfaces become cloudy when water is dropped on them or if there is moisture in the atmosphere. However, under UV light irradiation of the titanium oxide surfaces, this contact angle of water droplets becomes smaller, even reaching zero (superhydrophilicity), its extent depending on the UV irradiation time and irradiation intensity. Thus, under UV light irradiation, titanium oxide thin-film surfaces never become

Figure 1 View of the soundproof-highway walls coated with titanium oxide photocatalysts for the elimination of NO_x (the walls were constructed in Osaka, in April 1999) (20 m^3 of the polluted air can be purified at the rate of 1 m^2 photocatalyst/hr).

Figure 2 Antifogging effect of titanium oxide thin-film-coated surface. The glass mirror, whose right side was coated with titanium oxide thin film, exhibits a clear image even in a high-water-moisture room like in a bathroom. Decrease in the contact angle of a water droplet under UV irradiation of the titanium oxide thin film surface, leading to a photoinduced superhydrophilic property of the mirror. (Supplied by TOTO.)

cloudy, even in the rain. This remarkable function can also be applied for the production of new mirrors which can be used even in bathrooms and side mirrors for cars to keep visibility against rain (Fig. 2).

Some practical applications of titanium oxide photocatalysts in Japan are as follows:

1. Air cleaners containing titanium oxide photocatalysts
2. White paper containing titanium oxide photocatalysts
3. Antibacterial textile fibers containing titanium oxide photocatalysts
4. Systems for the purification of polluted air (e.g., the elimination of NO_x)
5. Superhydrophilic, self-cleaning systems and coating materials for cars
6. Soundproof highway walls covered with titanium oxide photocatalysts
7. Lamp covers coated with titanium oxide thin-film photocatalysts

8. Cements containing titanium oxide photocatalyst powders
9. Architectural materials using titanium oxide photocatalysts
10. Coating materials using titanium oxides for architectural walls
11. Self-cleaning tents
12. Glass tablewares
13. Outdoor antennas coated with titanium oxide thin films

III. PREPARATION AND CHARACTERIZATION OF NEW TITANIUM OXIDE PHOTOCATALYSTS

The main characteristics of the various titanium oxide catalysts used in this chapter are summarized in Table 1. Titanium oxide thin film photocatalysts were prepared using an ionized cluster beam (ICB) deposition method [13–16]. In ICB deposition method, the titanium metal target was heated to 2200 K in a crucible and Ti vapor was introduced into the high vacuum chamber to produce Ti clusters. These clusters then reacted with O_2 in the chamber and stoichiometric titanium oxide clusters were formed. The ionized titanium oxide clusters formed by electron beam irradiation were accelerated by a high electric field and bombarded onto the glass substrate to form titanium oxide thin films.

The metal ion implantation of the catalysts was carried out using an ion implanter consisting of a metal ion source, a mass analyzer, a high-voltage ion accelerator (50–200 keV), and a high vacuum pump (Fig. 3) [7]. The metal ions were expected to be injected into the deep bulk of the catalyst when high acceleration energy was applied to the metal ions. In fact, as expected, SIMS analyses using a Shimadzu/Kratos SIMS1030 clearly showed that the metal ions implanted into the titanium oxide catalyst exist in a highly dispersed state and are injected into the deep bulk of the catalyst, exhibiting a distribution maximum at around 1000–3000 Å from the surface and zero distribution at the surface [10–12]. Although such a distribution depends on the acceleration energy and the kind of

Table 1 Characteristics of the Titanium Oxides Used in the Present Study

Catalyst	Anatase (%)	BET surface area (m²/g)	Particle size (nm)	Purity as TiO_2 (%)	Band-gap energy (eV)
F-2	72.3	27.1	23.4	99.97	3.25
F-4	87.5	54.2	15.0	99.97	3.251
F-6	81.0	102	9.30	99.99	3.262
P-25	70.9	50.2	18.6	99.54	3.250
S-1	86.1	30.6	30.2	99.90	3.252

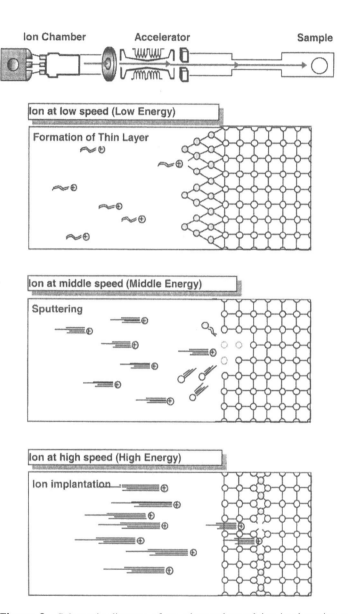

Figure 3 Schematic diagram of an advanced metal ion-implantation method. High-energy implantation (bottom) was used in the present study.

catalyst, one of the most significant advantages of using the metal ion-implantation method is to modify the bulk electronic properties of a catalyst.

The metal ion-implanted titanium oxide catalysts were calcined in O_2 at around 725–823 K for 5 hr. Prior to various spectroscopic measurements such as UV-vis diffuse reflectance, SIMS, XRD, EXAFS, ESR, and ESCA, as well as investigations on the photocatalytic reactions, both the metal ion-implanted and unimplanted original pure titanium oxide photocatalysts were heated in O_2 at 750 K and then degassed in cells at 725 K for 2 h, heated in O_2 at the same temperature for 2 h, and, finally, outgassed at 473 K to 10^{-6} torr [12–15].

Light irradiation of the photocatalysts in the presence of reactant molecules such as NO_x and a mixture of CH_3CCH and H_2O was carried out using a high-pressure Hg lamp (Toshiba SHL-100UV) through water and color filters (i.e., $\lambda > 450$ nm for visible light irradiation and $\lambda < 380$ nm for UV irradiation, respectively, at 275–295 K). The reaction products were analyzed by GC and GC-MASS. The UV-vis diffuse reflectance spectra were measured using a Shimadzu UV-2200A spectrophotometer at 295 K. The ESR spectra were recorded at 77 K with a Bruker ESP300E and a JEOL RE-2X spectrometer (X-band). The binding energies and the elemental composition of the catalysts were measured using a Shimadzu ECSA-3200 electron spectrometer. The XAFS (XANES and FT-EXAFS) spectra were measured at the BL-7C facility of the Photon Factory at the National Laboratory for High-Energy Physics, Tsukuba, Japan.

IV. RESULTS AND DISCUSSION

The metal ion-implantation method was applied to modify the electronic properties of titanium oxide photocatalysts by bombarding them with high-energy metal ions, and it was discovered that metal ion implantation with various transition metal ions such as V, Cr, Mn, Fe, and Ni accelerated by a high-voltage enables a large shift in the absorption band of the titanium oxide catalysts towards the visible-light region, with differing levels of effectiveness. However, Ar, Mg, or Ti ion-implanted titanium oxides exhibited no shift, showing that such a shift is not caused by the high-energy implantation process itself, but to some interaction of the transition metal ions with the titanium oxide catalyst. As can be seen in Figs. 4b–4d, the absorption band of the Cr ion-implanted titanium oxide shifts smoothly to the visible-light region, the extent of the red shift depending on the amount and type of metal ions implanted, with the absorption maximum and minimum values always remaining constant. The order of the effectiveness in the red shift was found to be V > Cr > Mn > Fe > Ni ions. Such a shift allows the metal ion-implanted titanium oxide to use solar irradiation more effectively with efficiencies in the range of 20–30% [4,12].

Furthermore, as shown in Fig. 5, such red shifts in the absorption band of the metal ion-implanted titanium oxide photocatalysts can be observed for any

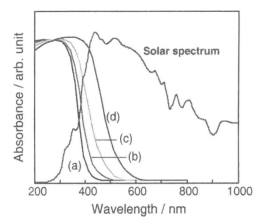

Figure 4 UV-vis absorption spectra (diffuse reflectance) of the original unimplanted pure TiO_2 (a) and the Cr ion-implanted TiO_2 (b–d), and the solar spectrum which reaches the earth (amounts of Cr ion implanted in 10^{-7} mol/g; b: 2.2; 6.6; d: 13.0).

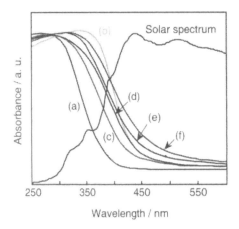

Figure 5 Shifts in the absorption spectra of various types of TiO_2 photocatalysts (shown in Table 1) implanted with the same amounts of V ions. (a) original unimplanted pure P-25, (b) V/F-6, (c) V/F-4, (d) V/P-25, (e) V/F-2, (f) V/F-1. Amount of V ion implanted was 6.6×10^{-7} mol/g (3.4×10^{-3} wt%).

kind of titanium oxide except amorphous types, the extent of the shift changing from sample to sample. It was also found that such shifts in the absorption band can be observed only after calcination of the metal ion-implanted titanium oxide samples in O_2 at around 723–823 K. Therefore, calcination in O_2 in combination with metal ion implantation was found to be instrumental in the shift of the absorption spectrum toward the visible-light region. These results clearly show that such a shift in the absorption band of the titanium oxides by metal ion implantation is a general phenomenon, not a special feature of a certain kind of titanium oxide catalyst.

Figure 6 shows the absorption bands of the titanium oxide photocatalysts impregnated or chemically doped with Cr ions in large amounts as compared with those for Cr-ion-implanted samples. The Cr-ion-doped catalysts show no shift in the absorption band; however, a new absorption shoulder appears at around 420 nm due to the formation of the impurity energy level within the band gap, its intensity increasing with the amount of Cr ions chemically doped. Such results indicate that the method of doping causes the electronic properties of the titanium oxides to be modified in completely different ways, thus confirming that only metal ion-implanted titanium oxide catalysts show shifts in the absorption band toward the visible-light region.

With unimplanted or chemically doped titanium oxide photocatalysts, the photocatalytic reaction does not proceed under visible-light irradiation ($\lambda > 450$ nm). However, we have found that visible-light irradiation of metal ion-implanted

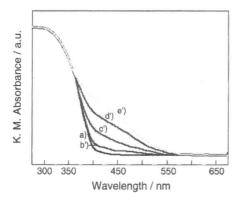

Figure 6 UV-vis absorption spectra (diffuse reflectance) of the original undoped pure TiO$_2$ (a) and TiO$_2$ chemically doped with Cr ions (b′–e′). Cr ions chemically doped in 10^{-7} mol/g: (a) undoped original pure TiO$_2$ (P-25), (b′) 16, (c′) 200, (d′) 1000, (e′) 2000. The TiO$_2$ photocatalysts chemically doped with Cr ions did not exhibit any photocatalytic reactivity.

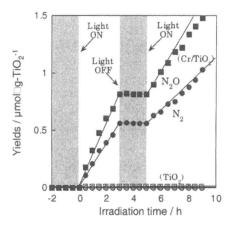

Figure 7   Reaction-time profiles of the photocatalytic decomposition of NO on the Cr-ion-implanted TiO₂ photocatalyst under visible-light (λ> 450 nm) irradiation at 295 K. The unimplanted original pure TiO₂ photocatalyst did not show any photocatalytic reactivity under the same reaction conditions.

titanium oxide photocatalysts can initiate various significant photocatalytic reactions. As shown in Fig. 7, visible-light irradiation ($\lambda > 450$ nm) of the Cr-ion-implanted titanium oxide in the presence of NO at 275 K leads to the decomposition of NO into N_2, O_2, and N_2O with a good linearity against the irradiation time. Under the same conditions of visible-light irradiation, the unimplanted original pure titanium oxide photocatalyst did not exhibit any photocatalytic reactivity. The action spectrum for the reaction on the metal ion-implanted titanium oxide was in good agreement with the absorption spectrum of the photocatalyst shown in Fig. 4, indicating that only metal ion-implanted titanium oxide photocatalysts were effective for the photocatalytic decomposition reaction of NO. Thus, metal ion-implanted titanium oxide photocatalysts were found to enable the absorption of visible light up to a wavelength of 600 nm and were also able to operate effectively as photocatalysts [4,11,15].

It is important to emphasize that the photocatalytic reactivity of the metal ion-implanted titanium oxides under UV light ($\lambda < 380$ nm) retained the same photocatalytic efficiency as the unimplanted original pure titanium oxides under the same UV light irradiation conditions. When metal ions were chemically doped into the titanium oxide photocatalyst, the photocatalytic efficiency decreased dramatically under UV irradiation due to the effective recombination of the photo-formed electrons and holes through the impurity energy levels formed by the doped metal ions within the band gap of the photocatalyst (in the case of Fig. 6)

[17]. These results clearly suggest that metal ions physically implanted do not work as electron and hole recombination centers but only work to modify the electronic property of the catalyst [11,12,14,15].

We have conducted various field work experiments to test the photocatalytic reactivity of the newly developed titanium oxide photocatalysts under solar irradiation. As can be seen in Fig. 8, under outdoor solar light irradiation at ordinary temperatures, the Cr-and V-ion-implanted titanium oxide photocatalysts showed several times higher photocatalytic reactivity for the photocatalytic decomposition of NO. As Fig. 9 shows, it is also found that under solar light irradiation at ordinary temperatures, the V-ion-implanted titanium oxide photocatalysts showed several times higher photocatalytic reactivity for the photocatalytic hydrogenation of CH_3CCH with H_2O than the unimplanted original pure titanium oxide photocatalysts. These results, together with the results shown in Fig. 4, clearly show that by using second-generation titanium oxide photocatalysts developed by applying the metal ion-implantation method, we are able to utilize visible-light and solar light energy more efficiently.

The relationship between the depth profiles of the metal ions of the metal ion-implanted titanium oxide photocatalysts having the same number of metal

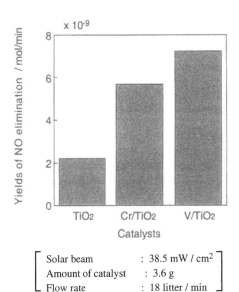

Figure 8 Effect of the Cr- and V-ion-implantation on the photocatalytic reactivity of TiO_2 under outdoor solar beam irradiation for the photocatalytic decomposition of NO at 295 K (solar beam: 38.5 mW/cm^2).

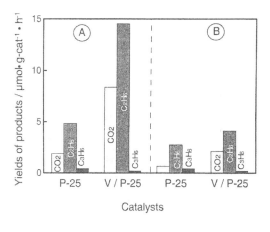

	(A)	(B)
Weather:	fine	cloudy
Intensity of light ($\mu W/cm^2$):		
250 nm :	250	160
360 nm :	2600	1600
400 nm :	10100	6500

Figure 9 Effect of the Cr- and V-ion-implantation on the photocatalytic reactivity of TiO_2 under outdoor solar beam irradiation for the photocatalytic reaction of CH_3CCH with H_2O leading a hydogenolysis reaction of CH_3CCH with H_2O at 295 K. A: fine, B: cloudy weather (solar intensity during fine weather: 12 mW/cm^2; amount of photocatalyst: 6.0 g).

ions, such as V or Cr ions, and their photocatalytic efficiency under visible-light irradiation were investigated. Figure 10 shows the relationship between the depth of V ions in the V-ion-implanted titanium oxide photocatalysts and their photocatalytic reactivity for NO decomposition under visible-light irradiation. As can be seen in Fig. 10, when the metal ions were implanted in the same amounts into the deep bulk of the catalyst by applying high-voltage acceleration energy, the photocatalyst exhibited a high photocatalytic efficiency under visible-light irradiation. This is also one of the most significant advantages to use the high-energy metal ion-implantation techniques to achieve such distribution effect of the metal ions. On the other hand, when a low voltage was applied, the photocatalyst exhibited a low efficiency under the same conditions of visible-light irradiation [15,16].

It was also found that increasing the number (or amounts) of metal ion implanted into the deep bulk of the titanium oxides caused the photocatalytic efficiency of these photocatalysts to increase under visible-light irradiation, passing through a maximum at around 6×10^{16} ions/cm^2 of the catalyst, then decreas-

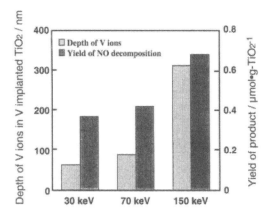

Figure 10 Effect of the depth profile of V ions in the V-ion-implanted titanium oxide photocatalyst on their photocatalytic reactivity for the decomposition of NO_x under visible light ($\lambda > 450$ nm) irradiation at 295 K.

ing with a further increase in the number of metal ions implanted. Only on samples implanted with an increased number of metal ions could the presence of ions at the near surfaces be observed by ESCA measurements. Thus, these results clearly suggest that there are specific conditions in the depth and number of metal ions implanted to achieve optimal photocatalytic reactivity under visible-light irradiation.

The ESR spectra of the V-ion-implanted titanium oxide catalysts were measured before and after calcination of the samples in O_2 at around 723–823 K, respectively (Fig. 11). Distinct and characteristic reticular V^{4+} ions were detected only after calcination at around 723–823 K. It was found that only when a shift in the absorption band toward visible-light regions was observed, the reticular V^{4+} ions could be detected by ESR. No such reticular V ions or shift in the absorption band have ever been observed with titanium oxides chemically doped with V ions [16,18,19].

Figure 12 shows the XANES and FT-EXAFS spectra of the titanium oxide catalysts chemically doped with Cr ions (a and A) and physically implanted with Cr ions (b and B). Analyses of these XANES and FT-EXAFS spectra show that in the titanium oxide catalysts chemically doped with Cr ions by an impregnation or sol–gel method, the ions are present as aggregated Cr oxides having an octahedral coordination similar to Cr_2O_3 and tetrahedral coordination similar to CrO_3, respectively. On the other hand, in the catalysts physically implanted with Cr ions, the ions are present in a highly dispersed and isolated state in octahedral

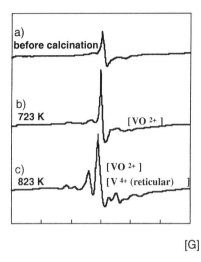

Figure 11 Effect of calcination temperature of the V-ion-implanted titanium oxide sample on the ESR spectra of V^{4+} species in the V-ion-implanted titanium oxide photocatalyst at 77 K.

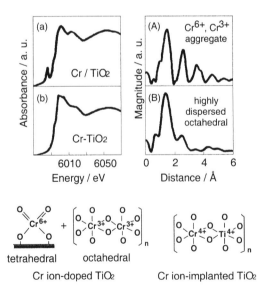

Figure 12 XANES (left) and FT-EXAFS spectra (right) of Cr-ion chemically doped TiO_2 (a, A) and Cr-ion-implanted TiO_2 catalysts (b, B).

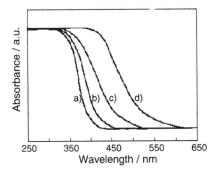

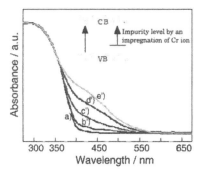

Figure 13 UV-vis absorption spectra (diffuse reflectance) of the Cr-ion-implanted titanium oxide photocatalyst (top) and the chemically doped Cr-ion titanium oxide photocatalyst (bottom) and their modified band-gap energy structures.

coordination, clearly suggesting that the Cr ions are incorporated in the lattice positions of the catalyst in place of the Ti ions.

Our results clearly show that modification of the electronic state of titanium oxide by metal ion implantation is closely associated with the strong and long-distance interaction which arises between the titanium oxide and the metal ions implanted, as shown in Fig. 13, and not by the formation of impurity energy levels within the band gap of the titanium oxides resulting from the formation of impurity oxide clusters which are often observed in the chemical doping of metal ions, as shown in Figs. 6 and 13.

V. CONCLUSIONS

The advanced metal ion-implantation method has been successfully applied to modify the electronic properties of the titanium oxide photocatalysts, enabling the absorption of visible light even longer than 550 nm and initiating the photocatalytic reactions effectively not only under UV but also visible-light irradiation. The results obtained in the photocatalytic reactions and various spectroscopic measurements of the photocatalysts indicate that the implanted metal ions are highly dispersed within the deep bulk of the catalysts and work to modify the electronic nature of the photocatalysts without any changes in the chemical properties of the surfaces. These modifications were found to be closely associated with an improvement in the reactivity and sensitivity of the photocatalyst, thus enabling the titanium oxides to absorb and operate effectively not only under UV but also under visible-light irradiation. As a result, under outdoor solar light irradiation at ordinary temperatures, metal ion-implanted titanium oxide photocatalysts showed several times higher photocatalytic efficiency than the unimplanted original pure titanium oxide photocatalyst.

Thus, the advanced metal ion-implantation method has opened the way to many innovative possibilities, and the design and development of such unique titanium oxide photocatalysts can also be considered an important breakthrough in the utilization of solar light energy, which will advance research in sustainable green chemistry for a better environment [4,5,15,16,20–22].

ACKNOWLEDGMENTS

One of the authors (M. A.) is greatly indebted to the Petroleum Energy (PEC) supported by the New Energy and Industrial Technology Development (NEDO) as well as The Ministry of International Trade and Industry (MITI) and the 1998 Mitsubishi Foundation for their financial support.

REFERENCES

1. Serpone, N.; Pelizzetti, E.; Eds. *Photocatalysis*. Wiley: New York, **1989**.
2. Anpo, M. *Catal. Surveys Jpn.* **1997**, *2*, 167.
3. Anpo, M. Proceedings of the 1st Int. Conf. Protect. the Environ. **1998**, p. 75.
4. Anpo, M. In *Green Chemistry*; Tundo, P.; Anastas, P.; Eds. Oxford University Press: Oxford, **2000**, p. 1.
5. Anpo, M. *Pure Appl. Chem.* **2000**, *72*, 1787.
6. Takeuchi, M.; Yamashita, H.; Matsuoka, M.; Anpo, M. *Catal. Lett.* **2000**, *67*, 135.
7. Negishi, N.; Iyoda, T.; Hashimoto, K.; Fujishima, A. *Chem. Lett.* **1995**, 841.
8. Takami, K.; Sagawa, N.; Uehara, H.; Anpo, M. *Shokubai* **1999**, *41*, 295.
9. Yamashita, H.; Honda, M.; Harada, M.; Ichihashi, Y.; Anpo, M. *J. Phys. Chem. B* **1998**, *102*, 10,707.
10. Anpo, M.; Takeuchi, M. In *Handbook of Ion Engineering*, **2002**, p. 943.
11. Anpo, M.; Yamashita, H.; Ichihashi, Y. *Optronics* **1997**, *186*, 161.
12. Anpo, M.; Ichihashi, Y.; Takeuchi, M.; Yamashita, H. *Res. Chem. Intermed.* **1998**, *24*, 143.

13. Anpo, M.; Ichihashi, Y.; Takeuchi, M.; Yamashita, H. *Studies Surface Sci. Catal.*, **1999**, *121*, 305.
14. Anpo, M.; Takeuchi, M.; Kishiguchi, S.; Yamshita, H. *Surface Sci. Jpn.* **1999**, *20*, 60.
15. Anpo, M.; Yamashita, H.; Kanai, S.; Sato, K.; Fujimoto, T. US Patent 6,077,492, June 20, **2000**.
16. Anpo, M. in: *Studies in Surface Science Catalysis 130 A, 12th Intern. Congr. Catal.* Corma, A.; Melo, F. V.; Mendioroz, S.; Fierro, G., Eds. Elsevier Science: A, Amsterdam **2000**, Part A, P. *157-*.
17. Maruska, H. P.; Ghosh, A. K. *Solar Energy Mater. Solar Cells* **1979**, *1*, 237.
18. Morin, B.; Davidson, A.; Che, M.; Ichihashi, Y.; Anpo, M. Unpublished data.
19. Anpo, M.; Kishiguchi, S.; Ikeue, K.; Takeuchi, M.; Yamashita, H.; Morin, B.; Davidson, A.; Che, M. *Res. Chem. Intermed.* **2001**, *27*, 459.
20. Anpo, M. In: *Photocatalysis—Foundamental and Applications*; Kaneko, M.; Okura, I., Eds. I., Kodanshya, **2002**, p. 175.
21. Anpo, M. *Pure Appl. Chem.* **2001**, *72*, 1265.
22. Takeuchi, M.; Yamashita, H.; Matsuoka, M.; Anpo, M.; Hirao, A.; Itoh, N.; Iwamoto, N. *Catal. Lett.* **2000**, *67*, 135.

10

Dye-Sensitized Solar Cells Based on Mesoscopic Oxide Semiconductor Films

Mohammad K. Nazeeruddin and Michael Grätzel
Swiss Federal Institute of Technology, Lausanne, Switzerland

I. INTRODUCTION

A. Overview

The greenhouse effect is related to carbon dioxide emission in the world's atmosphere and many other pollutants that are released by the extensive use of the *finite* conventional energy supplies [1]. The increasing global need for energy and the depletion of fossil fuel reserves necessitate the development of clean alternative energy sources [2,3]. Although nuclear power was regarded to be a solution for the increasing energy demands, the storage of nuclear waste is too frightening for future generations to tolerate radioactive fission products. Therefore, to maintain our planet habitable, we will have to develop methods for harnessing solar power, which is clean, nonhazardous, and *infinite*.

Photochemistry is expected to make important contributions to identify environmentally friendly and inexpensive alternative solutions to the energy-harnessing problem. One attractive strategy discussed in this chapter is the development of systems that mimic photosynthesis in the conversion and storage of solar energy. A way to successfully trap solar radiation is by a sensitizer molecule anchored to a rough titania surface analogous to the light-absorbing chlorophyll

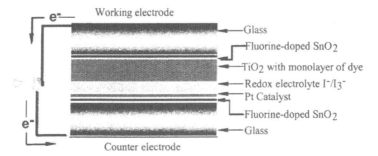

Figure 1   Schematic representation of the cross section of a dye-sensitized solar cell.

molecule found in nature. Sensitization of transparent material such as TiO_2 to the visible spectra is a very old topic and covered by many excellent reviews [4–8]. Nevertheless, the past decade has witnessed a great triumph for dye–sensitized solar cell technology in various laboratories, which have achieved 8–10% efficiencies [9–22]

The dye-sensitized solar cell technology developed at the EPFL contains broadly five components: (1) conductive mechanical support, (2) semiconductor film, (3) sensitizer, (4) redox couple, and (5) counterelectrode with a platinum catalyst. A cross section of the dye-sensitized solar cell is shown in Fig. 1. The total efficiency of the dye-sensitized solar cell depends on optimization and compatibility of each of these components constituents [23]. To a large extent, the nanocrystalline semiconductor film technology along with the dye spectral response are mainly responsible for the high efficiency. The high surface area and the thickness of the semiconductor film yields increased dye optical density resulting in efficient light harvesting [24].

The sensitizer plays a crucial role in the harvesting of the sunlight. To trap solar radiation efficiently in the visible and the near-infrared region of the solar spectrum requires engineering of sensitizers at a molecular level [25]. The electrochemical and photophysical properties of the ground and the excited state of the sensitizer play an important role in the charge-transfer dynamics at the semiconductor interface. The open-circuit potential of the cell depends on the redox couple, which shuttles between the sensitizer and the counterelectrode. The goal of this chapter is to provide the reader with a detailed description of dye-sensitized solar cell and concepts for developing efficient dyes and redox couples for solar cell applications.

B. Dye-Sensitized Solar Cell

Briefly, we describe here various steps involved in the preparation of dye-sensitized solar cell. The nanocrystalline TiO_2 films were prepared by depositing TiO_2

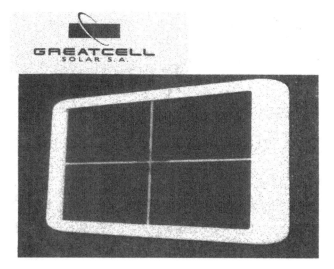

Figure 2 A module of an photoelectrochemical solar cell based on a dye-sensitized TiO$_2$ nanocrystalline film.

colloids on a transparent fluorine-doped tin oxide conducting glass using either screen-printing or the doctor blade technique [26,27]. The films are then dried in air and fired at 450°C. The hot electrodes (~100°C) were immersed into the dye solutions, which were usually prepared in ethanol [(2–5) $\times$ 10^{-4} M]. The dye–deposited film is used as a working electrode. The platinum-coated F-SnO$_2$ glass plate is used as a counterelectrode, which was placed on top of the dye-coated TiO$_2$ film with a thin Surlyn frame. The sandwiched electrodes were held together tightly and heat (130°C) was applied around the Surlyn frame to seal the two electrodes. A thin layer of an electrolyte, containing a I$_3^-$/I$^-$ redox-active couple in methoxyacetonitrile was introduced into the interelectrode space from the counterelectrode side through a pre-drilled hole. The pre-drilled hole was sealed again using Surlyn and a glass cover slip. A module of a photoelectrochemical solar cell based on dye-sensitized TiO$_2$ nanocrystalline film is shown in Fig. 2.

C. Operating Principles of the Dye-Sensitized Solar Cell

The details of the operating principles of the dye-sensitized solar cell are given in Figure 3. The photoexcitation of the metal-to-ligand charge transfer of the adsorbed sensitizer [Eq. (1)] leads to injection of electrons into the conduction band of the oxide [Eq. (2)]. The oxidized dye is subsequently reduced by electron donation from an electrolyte containing the iodide/triiodide redox system [Eq.

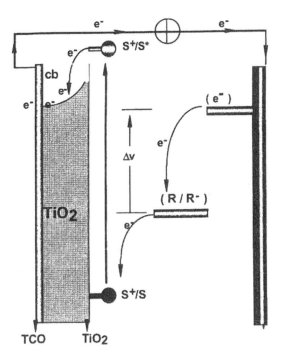

Figure 3 Operating principles and energy-level diagram of dye-sensitized solar cell. S, S^+ and S^* are the Sensitizers in the ground, oxidized, and excited states, respectively. R/ R^- is the redox mediator (I^-/I_3^-).

(3)]. The injected electron flows through the semiconductor network to arrive at the back contact and then through the external load to the counterelectrode. At the counterelectrode, reduction of the triiodide, in turn, regenerates iodide [Eq. (4)], which completes the circuit. With a closed external circuit and under illumination, the device then constitutes a photovoltaic energy-conversion system, which is regenerative and stable. However, there are undesirable reactions, such as the injected electrons recombining either with oxidized sensitizer [Eq. (5)] or with the oxidized redox couple at the TiO_2 surface [Eq. (6)] resulting in losses in cell efficiency.

$$S_{(\text{adsorbed on } TiO_2)} + h\nu \rightarrow S^*_{(\text{adsorbed on } TiO_2)} \tag{1}$$

$$S^*_{(\text{adsorbed on } TiO_2)} \rightarrow S^+_{(\text{adsorbed on } TiO_2)} + e^-_{(\text{injected})} \tag{2}$$

$$S^+_{(\text{adsorbed on } TiO_2)} + \frac{3}{2}I^- \rightarrow S_{(\text{adsorbed on } TiO_2)} + \frac{1}{2}I_3^- \tag{3}$$

$$I_{3\ (cathode)}^{-} + 2e^{-} \rightarrow 3I_{(cathode)}^{-} \tag{4}$$

$$S_{(adsorbed\ on\ TiO_2)}^{+} + e_{(TiO_2)}^{-} \rightarrow S_{(adsorbed\ on\ TiO_2)} \tag{5}$$

$$I_{3\ (cathode)}^{-} + e_{(TiO_2)}^{-} \rightarrow I_{3\ (anode)}^{-} \tag{6}$$

D. Incident Photon to Current Efficiency and Open-Circuit Photovoltage

The incident monochromatic photon-to-current conversion efficiency (IPCE), defined as the number of electrons generated by light in the external circuit divided by the number of incident photons as a function of excitation wavelength is expressed by Eq. (7) [23]. The open-circuit photovoltage is determined by the energy difference between the Fermi level of the solid under illumination and the Nernst potential of the redox couple in the electrolyte (Fig. 3). However, the experimentally observed open-circuit potential for various sensitizers is smaller than the difference between the conduction band and the redox couple, likely due to the competition between electron transfer and charge recombination pathways. Knowledge of the rates and mechanisms of these competing reactions are vital for the design of efficient sensitizers and, thereby, improvement of the solar devices [11,28,29].

$$IPCE = \frac{(1.25 \times 10^3) \times \text{Photocurrent density (mA/cm}^2)}{\text{Wavelength [nm] x Photon flux (W/m}^2)} \tag{7}$$

II. SEMICONDUCTOR FILMS

A. Preparation of Mesoscopic TiO$_2$ Colloids

The favored semiconductor material is titanium dioxide, because of its abundance, low cost, and nontoxicity. This last characteristic promotes its use in health care products and paints. As mentioned in Section I, the efficiency of the solar cell depends to a large extent on the nanocrystalline semiconductor films; the high surface area of these films yields high dye loading and high optical density, resulting in efficient light absorption. The nanoporous structure permits surface coverage of the dye to be sufficiently high for total absorption of the incident light, necessary for efficient solar energy conversion, because the area available for monomolecular distribution of adsorbate is two to three orders of magnitude higher than the geometric area of the substrate.

The original substrate structure used for our early photosensitization experiments was a fractal surface derived by hydrolysis of an organo-titanium compound, but this has since been replaced with a nanostructured layer deposited from a colloidal suspension of TiO$_2$. This evidently provides for a much more reproducible and controlled high-surface-area nanotexture. Further, because it is compatible with screen-printing technology, it anticipates future production requirements. Although commercially available titania powders produced by a pyrolysis route from a chloride precursor have been successfully employed, the present optimized material is the result of a hydrothermal technique, described by Brooks et al. [30]. A specific advantage of the procedure is the ease of control

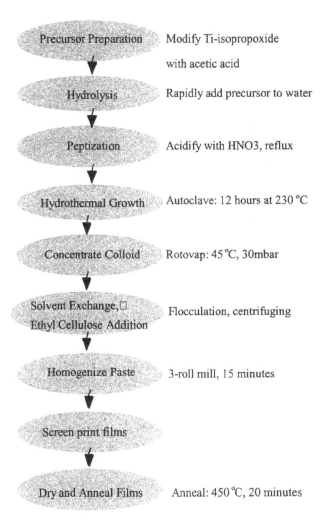

Figure 4   A flow diagram showing the reaction steps involved in the sol-gel preparation of TiO$_2$ colloids and its films.

Figure 5   SEM images of the surface of a mesoporous film prepared from the hydrothermal TiO$_2$ colloid.

of the particle size and, hence, of the nanostructure and porosity of the resultant semiconductor substrate. Figure 4 shows the relevant flow diagram and the reaction steps involved in the sol-gel preparation of the films. The procedure involves the hydrolysis of the titanium alkoxide precursor producing an amorphous precipitate followed by peptization under acid or alkaline water conditions to produce a sol, which is subjected to hyrdrothermal Ostwald ripening in an autoclave. The resulting crystalites of TiO$_2$ consist of anatase or a mixture of anatase and rutile, depending on reaction conditions.

The temperature of the hydrothermal treatment has a decisive influence on the particle size. The standard sol, treated for about 12 hr at 230°C in the autoclave has an average particle diameter of 15 nm. Figure 5 shows scanning microscope pictures of a TiO$_2$ film deposited on transparent conducting glass by screen printing from a concentrated solution of such particles. Each particle is a single crystal exposing predominantly surface planes with (101) direction. The formation of the latter face is favored by its low surface energy. The autoclaving temperature has an effect on pore size distribution. The maximum shift to larger pore diameters with increasing temperature reflects enhanced particle growth. For the 230°C sol, the most frequent pore size is about 10 nm.

B.   Preparation of Films

The TiO$_2$ paste is deposited onto a sheet glass (Nippon Sheet Glass, Hyogo, Japan) which has been coated with a fluorine-doped stannic oxide layer with sheet resistance of 8–10 Ω/cm^2) using a screen-printing technique. The resulting layer is dried in air at 100°C for 15 min, followed by another 15 min at 150°C. For the final processing, the layers are heated using a titanium hot plate (Bioblock Scientific) to 325°C at a rate of 30°C/min and kept at this temperature for 5 min. Then, the temperature is raised to 375°C at a rate of 10°C/min and held there for 5 min. Finally, the layers are fired to 450°C at a rate of 15°C/min under flowing oxygen and left at this temperature for 20 min before cooling to room temperature. The film thickness is 13–18 μm. The heated electrodes are impregnated with a 0.05 M titanium tetrachloride solution (65 μL/cm^2) in a water-saturated desiccator for 30 min at 70°C and washed with distilled water [31]. The TiO_2 layers are again fired at 450°C for 20 min and allowed to cool to 100°C temperature before plunging into a dye solution.

A striking and unexpected behaviour of the mesoporous TiO_2 films prepared by the sol-gel method is that the high surface roughness does not promote charge-carrier loss by recombination. The reason for this behavior is that the electron and the positive charge find themselves, within femtoseconds after light excitation of the dye, on opposite sides of the liquid–solid interface. The carrier loss mechanisms are comparatively slow although conventionally referred to as recombination by analogy with the solid-state process. The loss of a photoexcited electron from the semiconductor should be regarded as a recapture by an oxidized dye species or as a redox capture when the electron reacts directly with an ion in the electrolyte. Either occurs on a millisecond time scale.

III. MOLECULAR SENSITIZERS

A. Requirements of the Sensitizers

The optimal sensitizer for the dye-sensitized solar cell should be panchromatic (i.e., absorb visible light of all colors). Ideally, all photons below a threshold wavelength of about 920 nm should be harvested and converted to electric current. This limit is derived from thermodynamic considerations showing that the conversion efficiency of any single-junction photovoltaic solar converter peaks at approximately 33% near a threshold energy of 1.4 eV [32,33]. In addition, the sensitizer should fulfill several demanding conditions: (1) It must be firmly grafted to the semiconductor oxide surface and inject electrons into the conduction band with a quantum yield of unity, (2) its redox potential should be sufficiently high that it can be regenerated rapidly via electron donation from the electrolyte or a hole conductor, (3) it should be stable enough to sustain at least 10^8 redox turnovers under illumination corresponding to about 20 yr of exposure to natural sunlight.

Molecular engineering of ruthenium complexes that can act as panchromatic charge-transfer sensitizers for TiO_2-based solar cells presents a challenging task, as several requirements which are very difficult to be met simultaneously have to be fulfilled by the dye. The lowest unoccupied molecular orbitals (LUMOs) and the highest occupied molecular orbitals (HOMOs) have to be maintained at levels where photo-induced electron transfer in the TiO_2 conduction band and regeneration of the dye by iodide can take place at practically 100% yield. This restricts greatly the options available to accomplish the desired red shift of the metal-to-ligand charge-transfer transitions (MLCT) to about 900 nm.

The spectral and redox properties of ruthenium polypyridyl complexes can be tuned in two ways: first, by introducing a ligand with a low-lying π^* molecular orbital and, second, by destabilization of the metal t_{2g}-orbital through the introduction of a strong donor ligand. Meyer et al. have used these strategies to tune the MLCT transitions in ruthenium complexes considerably [34]. Heteroleptic complexes containing bidentate ligands with low-lying π^*-orbitals together with others having strong sigma-donating properties show impressive panchromatic absorption properties. However, the extension of the spectral response into the near infrared (IR) was gained at the expense of shifting the LUMO orbital to lower levels from where charge injection into the TiO_2 conduction band can no longer occur [35].

A near-infrared response can also be gained by upward shifting of the Ru t_{2g} (HOMO) levels. However, it turns out that the mere introduction of strong sigma-donor ligands into the complex often does not lead to the desired spectral result, as both the HOMO and LUMO are displaced in the same direction. Furthermore, the HOMO position cannot be varied freely, as the redox potential of the dye must be maintained sufficiently positive to ascertain rapid regeneration of the dye by electron donation from iodide following charge injection into the TiO_2.

Based on extensive screening of hundreds of ruthenium complexes, we discovered that the sensitizer excited-state oxidation potential should be negative and at least -0.9 V versus saturated calomel electrode (SCE), in order to inject electrons efficiently onto the TiO_2 conduction band. The ground-state oxidation potential should be about 0.5 V versus SCE, in order to be regenerated rapidly via electron donation from the electrolyte (iodide/triiodide redox system) or an hole conductor. A significant decrease in electron–injection efficiencies will occur if the excited-and ground-state redox potentials are lower than these values.

B. Tuning of MLCT Transitions

To illustrate the tuning aspects of the MLCT transitions in ruthenium polypyridyl complexes, let us begin by considering the well-known ruthenium *tris*-bipyridine complex (**1**). Complex **1** shows strong visible band at 466 nm, due to charge-transfer transition from metal t_{2g} (HOMO) orbitals to π^* orbitals (LUMO) of the ligand. The Ru(II)/(III) oxidation potential is at 1.3 V, and the ligand-based reduction potential is at -1.5 V versus SCE [36]. From spectro chemical and electrochemical studies of polypyridyl complexes of ruthenium, it has been con-

(1)

(2)

(3)

(4)

(5)

(6)

Structures 1–6

cluded that the oxidation and reduction potentials are the best indicators of the energy levels of the HOMO and LUMO [37]. The energy between the metal t_{2g} orbitals and π^*-orbitals can be reduced either by raising the energy of the t_{2g}-orbitals or by decreasing the energy of the π^*-orbitals with donor–acceptor ligands, respectively (Fig. 6). In the following sections we will be discussing the ways to tune HOMO and LUMO energy levels by introducing various ligands.

C. Spectral Tuning in "Push-Pull" Type of Complexes

The lowest-energy MLCT transition of ruthenium polypyridyl complexes of type **1**, can be lowered so that it absorbs more in the red region of the visible spectrum by substituting one of the bidentate ligands with a donor ligand (pushing ligand). The remaining two electron-withdrawing bidentate ligands act as pulls ligand. Complexes **1–9** show where the absorption maxima of the complexes are tuned from 466 nm to 560 nm by introducing different donor ligands. The oxidation

Structures 7–9

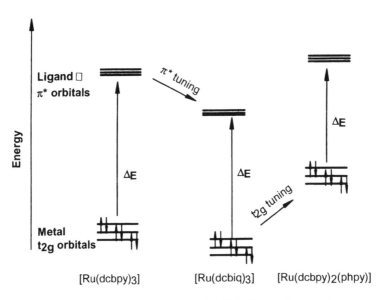

Figure 6   Tuning of HOMO (t_{2g}) and LUMO (π^*) orbital energy in various ruthenium polypyridyl complexes.

potential in these complexes was tuned from 1.3 V to 0.3 V versus SCE, going from **1** to **9**. Thus, the energy of the HOMO is varied by 1 V in this series of complexes[38]. Meyer et al. have used the same concept to synthesize black absorbers by judiciously selecting push and pull ligands [34].

It is interesting to note the magnitude of the spectral shift for the lowest–energy charge-transfer transition (i.e., ~0.5 eV) and the shift in the oxidation potential (~1 eV). This clearly shows that the HOMO tuning is not all translated into the spectral shift of the complex. The apparent 0.5-eV difference is involved in raising the energy of π^*-orbitals of the pulling ligands caused by the pushing ligands [38].

The other interesting class of compounds which is shown as Structures 10–20, contain the ligand 2,6-bis(1-methylbenzimidazol-2′-yl)pyridine. This hybrid ligand, which contains donor (benzimidazol-2′-yl) and acceptor groups, was further tuned by introducing different substituents on the benzimidazol-2′-yl group. The complexes containing these ligands show intense ultraviolet (UV) absorption bands at 362 and 346 due to intraligand π–π^* transition of 2,6-bis(1-methylbenzimidazol-2′-yl)pyridine.

The MLCT bands of these complexes are broad and red-shifted by approximately 140 nm, compared to (**1**). The lowest-energy MLCT transitions within

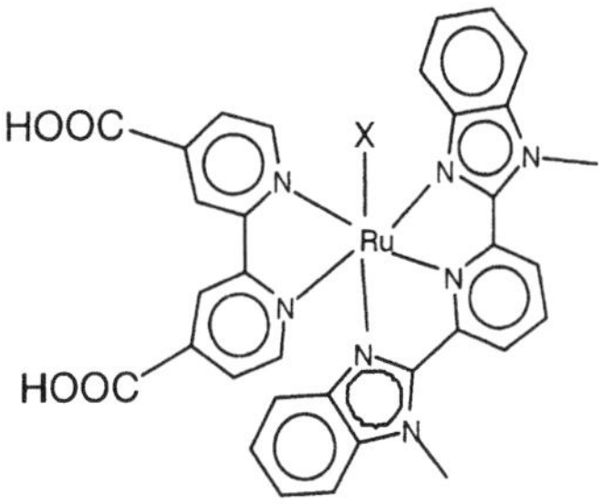

(**10**)	X = Cl
(**11**)	X = H$_2$O
(**12**)	X = OH
(**13**)	X = NCS
(**14**)	X = CN
(**15**)	X = N(CN)$_2$
(**16**)	X = (Cl3-pcd)
(**17**)	X = NCN^{2-}

Structures 10–17

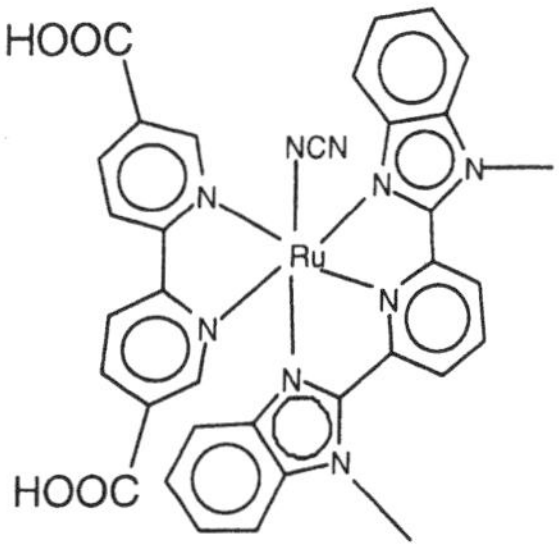

(**18**)

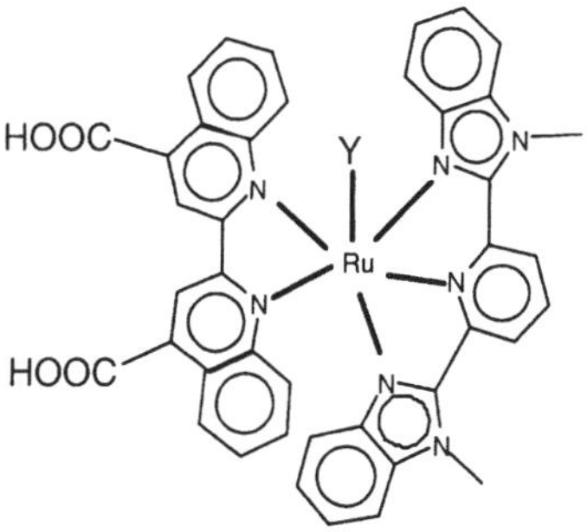

(**19**)	Y = NCS
(**20**)	Y = Cl

Structures 18–20

this series were shifted from 486 to 608 nm, and the HOMO level varied over an extent of 0.45 V versus SCE [39]. The energy of the MLCT transition in these complexes decreases with the decrease in the π acceptor strength of the ancillary ligand (i.e., CN^-, NCS^-, H_2O, NCN^- and Cl^-. The red shift of the absorption maxima, in complexes (**19**) and (**20**) containing 4,4'-dicarboxy-2,2'-biquinoline (dcbiq) compared to the 4,4'-dicarboxy-2,2'-bipyridine (dcbpy) acceptor ligand is due to the low energy π*-orbitals of 4,4'-dicarboxy-2,2'-biquinoline. The resonance Raman spectra of these complexes for excitation at 568 nm show predominantly bands associated with the dcbpy and dcbiq ligands, indicating that the lowest excited state is a metal-to-dcbpy or dcbiq ligand charge-transfer state [39–41].

The spectral properties of ruthenium polypyridyl complexes can also be tuned by introducing donor non-chromophoric ligands such as NCS^-, which destabilize the metal t_{2g}-orbitals. A comparison of the visible absorption spectra of complexes **21–24**, where the number of nonchromophoric ligands are varied from one to three, show the most intense MLCT transition maxima at 500, 535, and 570 nm, respectively. The 70-nm red shift of complex **23** compared to the complex **21** is due to an increase in the energy of the metal t_{2g}-orbitals caused by introducing non-chromophoric ligands. The structure of complex **24** is very similar to complex **23**, except for the three carboxylic groups at the 4, 4', and

Structures 21–24

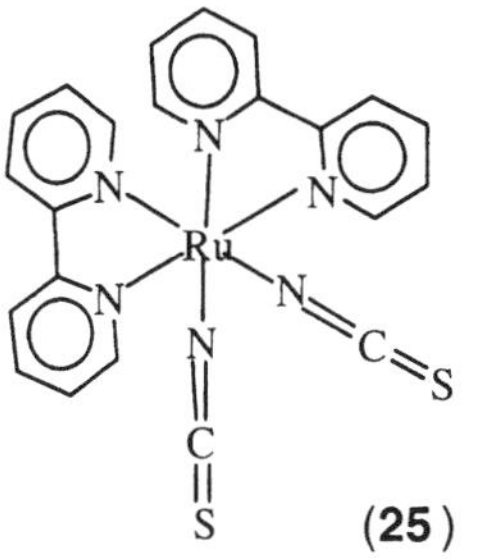

Structures 25 and 26

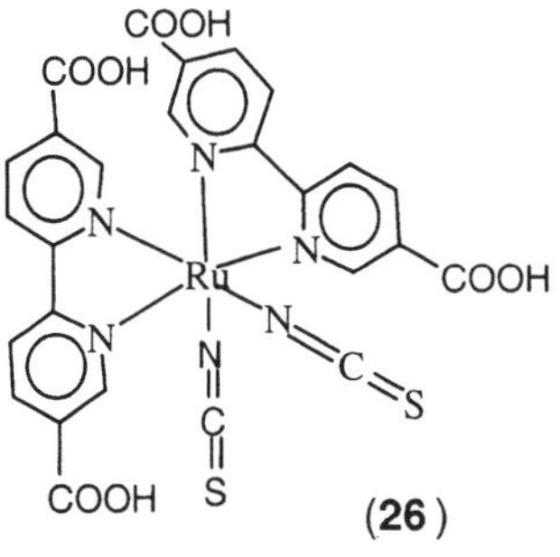

4″ positions. Complex **24** show a maximum at 620 nm and the differences between complexes **24** and **23** reflects the extent of decrease in LUMO energy due to substitution of three carboxylic acid groups at 4,4′,4″-positions of 2,2′:6′,2″-terpyridine compared to the LUMO of 2,2′:6′,2″-terpyridine [9,42,43].

In an effort to further tune the spectral responses of the ruthenium complexes containing polypyridyl ligands, Bignozzi and others have explored the influence of position of the carboxylic acid groups. Comparisons of complexes **25–28** show the extent of LUMO tuning [44–46]. The lowest MLCT absorption maximum of complex **25** is seen at 510 nm in ethanol. By substituting two carboxylic acid groups at the 4,4′-position of the 2,2-bipyridine ligand (**22**), the MLCT maxima red-shifted to 535 nm. However, substitution at the 5,5′-position of 2,2′-bipyridine (**26**), the maxima shifted further into the red (580 nm). On the contrary, the MLCT maxima blue-shifted (500 nm) by substituting at the 6,6′-

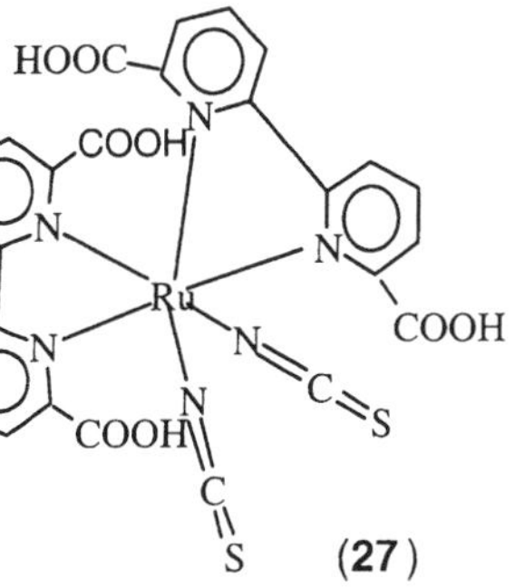

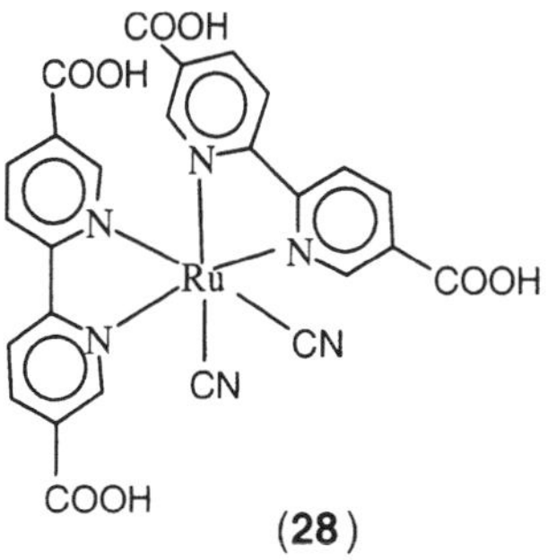

Structures 27 and 28

positions of 2,2′-bipyridine (**27**). The enhanced red response of complexes containing the 5,5′-dicarboxylic acid-2,2′-bipyridine is due to a decrease in the energy of the π*-orbitals, which makes them attractive as sensitizers for nanocrystalline TiO₂ films. Bignozzi and co-workers found that the IPCE of complexes having the 5,5′-dicarboxylic acid-2,2′-bipyridine ligands were lower than the analogous complexes that contain 4,4′-dicarboxylic acid-2,2′-bipyridine [44]. They rationalized the low efficiency of the sensitizers containing 5,5′-dicarboxylic acid-2,2′-bipyridine ligands in terms of low excited-state redox potentials.

In search of new sensitizers that absorb strongly in the visible region of the spectrum, Arakawa et al. have developed several sensitizers based on 1,10-phenanthroline ligands. Among these new classes of compounds, complexes **29** and **30** are noteworthy; they show an intense and a broad MLCT absorption band at 525 nm in ethanol. The energy levels of the LUMO and HOMO for **29** were estimated to be − 1.02 and 0.89 V versus SCE, respectively, which are slightly more positive than the sensitizer **22**. When anchored onto the TiO₂ surface, these sensitizers yield more than 75% photon-to-electron-injection efficiencies [15,47,48].

D. Osmium Complexes

Although there are a large number of osmium polypyridyl complexes, very few have been used as sensitizers in dye-sensitized solar cells. Osmium complexes have several advantages compared to their ruthenium analogs: (1) Osmium has a stronger ligand field splitting compared to ruthenium and (2) the spin orbit coupling leads to excellent response in the red region due to enhanced singlet–triplet charge-transfer transitions [49]. Heimer et al. found that complex **31** is extremely stable under irradiation in a homogeneous aqueous solution compared to the analogous ruthenium complex [50]. The greater photostability for

Structures 29 and 30

Structures 31 and 32

osmium is consistent with a stronger crystal field splitting of metal *d*-orbitals that inhibits efficient population of *d–d* states [51]. Lewis et al. have developed osmium based sensitizers (**31**) and (**32**) and found nearly 80% incident monochromatic photon-to-current conversion efficiencies [20].

E. MLCT Transitions in Geometrical Isomers

Isomerization is another approach for tuning the spectral properties of metal complexes [52–54]. The UV-vis absorption spectrum of the *trans*-dichloro complex (**35**) in DMF solution shows at least three MLCT absorption bands in the visible region at 690, 592, and 440 nm. On the other hand, the *cis*-dichloro complex (**33**) in DMF solution shows only two distinct broad bands in the visible region at 590 and 434 nm, which were assigned to metal-to-ligand charge-transfer transitions. The lowest-energy MLCT band in the trans complexes (**35, 36, 37**) is significantly red-shifted compared to the corresponding cis complexes (**22, 33, 34**) (Fig. 7). This red shift is due to stabilization of the LUMO of the dcbpy ligand in the trans species relative to the cis species. The red shift (108 nm) of the lowest-energy MLCT absorption in the spectrum of the *trans*-dichloro complex compared to the spectrum of the *trans*-dithiocyanato complex is due to the strong σ donor property of the CI$^-$ compared to the NCS$^-$ ligand. The chloride ligands cause destabilization of the metal t_{2g}-orbitals and raises them in energy relatively closer to the ligand π*-orbitals, resulting in a lower-energy MLCT transition.

F. Sensitizers Containing Functionalized Hybrid Tetradentate Ligands

The main drawback of the trans complexes discussed earlier is their thermal and photo-induced isomerization back to the cis configuration. In an effort to stabilize

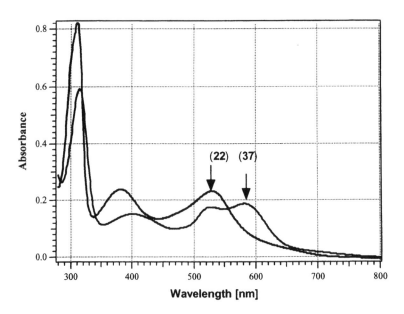

Structures 33–37

(33) X = Cl⁻

(34) X = H₂O

(35) X = Cl

(36) X = H₂O

(37) X = NCS⁻

Figure 7 Ultraviolet-visible absorption spectra of complexes **22** and **37** in ethanol solution at room temperature.

the trans configuration of an octahedral ruthenium complex and integrate the concepts of donor and acceptor (discussed in Sec. III.C) in a single complex, Renouard et al. have developed functionalized hybrid tetradentate ligands and their ruthenium complexes (**38–47**) [55]. In these complexes, the donor units of the tetradentate ligand (benzimidazole in **38** and **39** and *tert*-butylpyridine in **40** and **41**) tune the metal t_{2g}-orbital energies and the acceptor units (methoxycarbonyl) tune the π^* molecular orbitals. The application of a tetradentate ligand

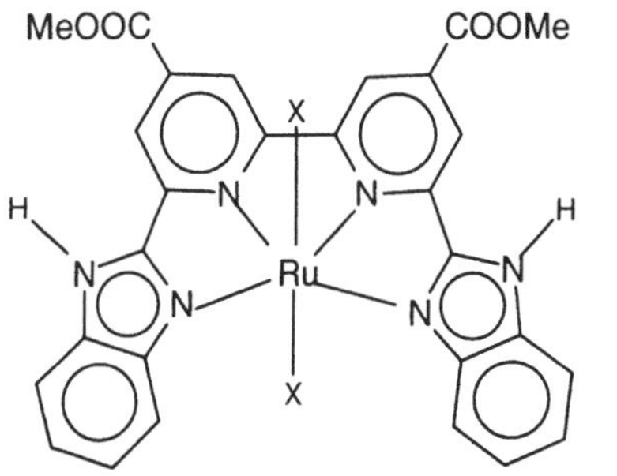

(**38**) X = Cl⁻

(**39**) X = NCS⁻

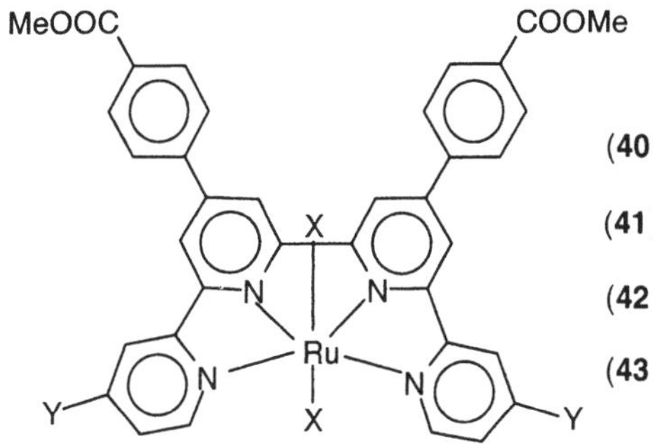

(**40**) X = Cl⁻, Y = tert-butyl

(**41**) X = NCS⁻, Y = tert-butyl

(**42**) X = Cl⁻, Y = H

(**43**) X = NCS⁻, Y = H

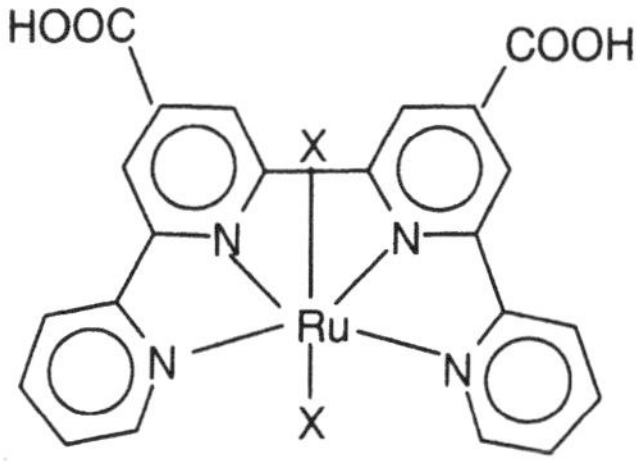

(**44**) X = Cl⁻

(**45**) X = NCS⁻

Structures 38–45

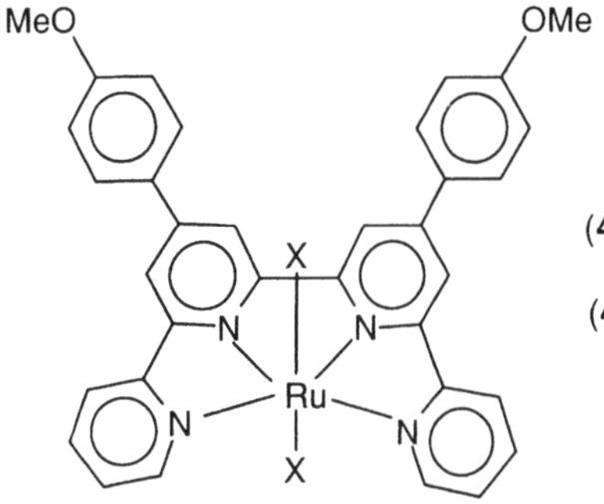

MeO   OMe

(**46**) X = Cl⁻

(**47**) X = NCS⁻

Structures 46 and 47

will inhibit the trans to cis isomerization process. The axial coordination sites are used further to fine-tune the spectral and redox properties and to stabilize the hole that is being generated on the metal, after having injected an electron into the conduction band.

The *trans*-dichloro and dithiocyanate complexes show MLCT transitions in the entire visible and near-IR region. The lowest-energy MLCT transition band of the *trans*-dichloro complexes is around 700 nm in DMF solution and they show weak and broad emission signals with onset at above 950 nm. The absorption and emission maxima of the *trans*-dithiocyanate complexes are blue-shifted compared to its *trans*-dichloro analogs due to the strong π acceptor property of the NCS⁻ compared to the Cl⁻, which is consistent with the elctrochemical data of these complexes.

The $Ru^{III/II}$ redox potentials of the thiocyanate complexes were more positive (by ~350 mV) than its corresponding dichloro complexes and show quasireversible behavior. This is in good agreement with the Ligand Electrochemical Parameters scale, according to which the thiocyanate $Ru^{III/II}$ wave should be ~340 mV more positive than the dichloro species $Ru^{III/II}$ potential [56]. The i_{ox}/i_{red} peak current is substantially greater than unity due to the oxidation of the thiocyanate ligand subsequent to the oxidation of the ruthenium(II) center.

The electronic spectra of **38** were calculated by intermediate neglect of differential overlap (INDO/S) and compared with the experimental data. Geometry optimization of **38** produced a structure with C2 point-group symmetry. The three highest occupied molecular orbitals (HOMO, HOMO-1, and HOMO-2) for complex **38** are mostly formed from $4d$(Ru)-orbitals and their contribution ranges from 52% to 84%. The d_{yz}(Ru)-orbital directed toward the ligand L is coupled to the π-orbitals of the ligand and is thereby delocalized to a considerable degree,

LUMO + 2   LUMO + 1

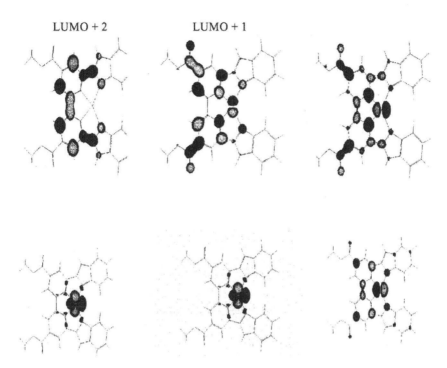

Figure 8   Pictures of the frontier orbitals of complex **38**.

greater than any other $d(t_{2g})$-orbital. The LUMOs are almost entirely localized on the ligand (Fig. 8). Extensive π-back-donation between metal 4d- and ligand π^*-orbitals is observed. Complex **45**, when anchored on TiO$_2$ layers, show 75% IPCE, yielding 18 mA/cm^2 current density under standard AM 1.5 sunlight [55].

G. Hydrophobic Sensitizers

The other important aspect in dye-sensitized solar cells is water-induced desorption of the sensitizer from the surface. Extensive efforts have been made in our laboratory to overcome this problem by introducing hydrophobic properties in the ligands (**48–53**). The photocurrent action spectra of these complexes shows broad features covering a large part of visible spectrum and displays a maxima around 550 nm, where the monochromatic IPCE exceeds 80%. The performance of these hydrophobic complexes as charge-transfer photosensitizers in nanocrys-

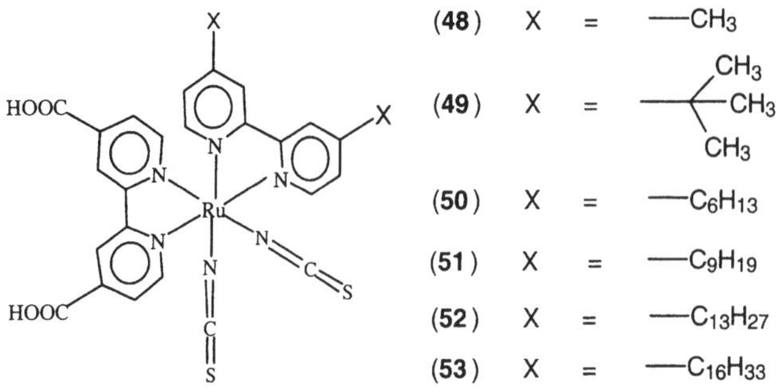

(48) X = —CH₃

(49) X = —C(CH₃)₃

(50) X = —C₆H₁₃

(51) X = —C₉H₁₉

(52) X = —C₁₃H₂₇

(53) X = —C₁₆H₃₃

Structures 48–53

talline TiO$_2$-based solar cell shows excellent stability toward water-induced desorption [57].

The rate of electron transport in a dye-sensitized solar cell is a major element of the overall efficiency of the cells. The injected electrons into the conduction band from optically excited dye can traverse the TiO$_2$ network and be collected at the transparent conducting glass or can react either with the oxidized dye molecule or with the oxidized redox couple (recombination). The reaction of injected electrons into the conduction band with the oxidized redox mediator gives undesirable dark currents, reducing significantly the charge-collection efficiency, and there by decreasing the total efficiency of the cell (Scheme 1).

Several groups have tried to reduce the recombination reaction by using sophisticated device architecture such as composite metal oxides as the semiconductor with different band gaps [58,59]. Gregg et al. have examined surface passivation by deposition of insulating polymers [60]. We have studied the influence of spacer units between the dye and the TiO$_2$ surface with little success [61]. Nevertheless, by using TiO$_2$ films containing hydrophobic sensitizers that contain long aliphatic chains (**50–53**), the recombination reaction is suppressed considerably. The most likely explanation for the reduced dark current is that the long chains of the sensitizer interacts laterally to form an aliphatic network as shown in Scheme 2, thereby preventing triiodide from reaching the TiO$_2$ surface.

H. Near-IR Sensitizers

Phthalocyanines possess intense absorption bands in the near-IR region and are known for their excellent stability, rendering them attractive for photovoltaic applications [62]. They have been repeatedly tested in the past as sensitizers of wide-band-gap oxide semiconductors and obtained poor incident photon-to-electric current conversion yields remaining under 1%, which is insufficient for

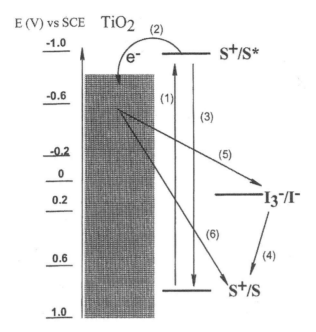

Scheme 1 Illustration of the interfacial charge-transfer processes in nanocrystalline dye sensitized solar cell. S, S^+, and S^* represents the sensitizer in the ground, oxidized, and excited states, respectively. Visible light absorption by the sensitizer (1) leads to an excited state, followed by electron injection (2) onto the conduction band of TiO_2. The oxidized sensitizer (3) is reduced by the I^-/I_3^- redox couple (4). The injected electrons into the conduction band may react either with the oxidized redox couple (5) or with oxidized dye molecule (6).

solar cell applications [63–66]. One of the reasons for such low efficiencies is aggregation of the dye on the TiO_2 surface. This association often leads to undesirable photophysical properties, such as self-quenching and excited-state annihilation. However, the advantage of this class of complexes is the near-IR response, which is very strong, having extinction coefficients of close to $50,000\ M^{-1}\ cm^{-1}$ at 650 nm compared to the polypyridyl complexes, which have small extinction coefficients at this wavelength.

1. Ruthenium Phthalocyanine

The ruthenium phthalocyanine complex (**54**) shows a visible absorption band at 650 nm (ϵ-49,000 $M^{-1}\ cm^{-1}$) and a phosphorescence band located at 895 nm. The triplet-state lifetime is 474 nsec under anaerobic conditions. The emission

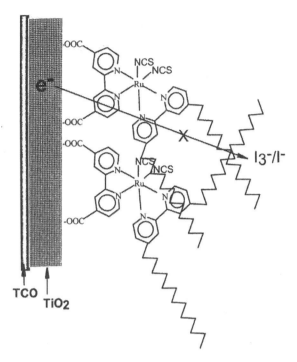

Scheme 2   Pictorial representation of blocking of the oxidized redox couple reaching onto surface of TiO$_2$ for conduction band electrons using hydrophobic sensitizers, which forms an aliphatic net work.

Structure 54

is entirely quenched when complex **54** is adsorbed onto a nanocrystalline TiO_2 film. The very efficient quenching of the emission of **54** was found to be due to electron injection from the excited singlet/triplet state of the phthalocyanine into the conduction band of TiO_2 [67]. The photocurrent action spectrum, where the incident photon-to-current conversion efficiency is plotted as a function of wavelength shows 60% with a maximum at around 660 nm (Fig. 9). These are, by far, the highest conversion efficiencies obtained with the phthalocyanine-type sensitizers.

It is fascinating to note that this class of dyes are injecting efficiently into the conduction band of TiO_2, despite the fact that the pyridyl orbitals do not participate in the π–π* excitation, which is responsible for the 650-nm absorption band. This phenomenon shows that the electronic coupling of the excited state of the dye to the Ti (3*d*) conduction band manifold is strong enough through this

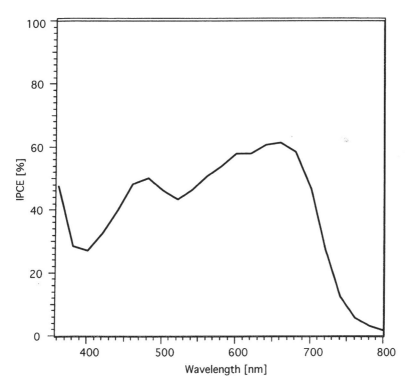

Figure 9 Photocurrent action spectra of nanocrystalline TiO_2 films sensitized by bis(3,4-dicarboxypyridine)Ru(II)1,4,8,11,15,18,22,25-octamethyl-phthalocyanin (**54**). The incident photon-to-current conversion efficiency is plotted as a function of wavelength.

mode of attachment to render charge injection very efficient. These results establish a new pathway for grafting dyes to oxide surfaces through axially attached pyridine ligands.

2. Phthalocyanines Containing 3d Metals

Several zinc(II) and aluminum(III) phthalocyanines substituted by carboxylic acid and sulfonic acid groups were anchored to nanocrystalline TiO_2 films and tested for their photovoltaic behavior [68,69]. Interestingly, zinc(II) 2,9,16,23-tetracarboxyphthalocyanine (**55**) exhibited 45% monochromatic current conversion efficiency at 700 nm. It is shown that electron injection to TiO_2 occurs from the excited singlet state of the phthalocyanine derivatives. The inherent problem of phthalocyanine aggregation is reduced considerably in *3d* metal phthalocyanines by introducing 4-*tert*-butylpyridine and 3α, 7α-dihydroxy-5β-cholic acid (cheno) into the dye solution. The added 4-*tert*-butylpyridine probably coordinates to the metal in the axial position and thereby prevents aggregation of the phthalocyanines.

This type of sensitizer opens up new avenues for improving the near-IR response of the nanocrystalline injection solar cell. In addition, important applications can be foreseen for the development of photovoltaic windows transmitting part of the visible light. Such devices would remain transparent to the eye while

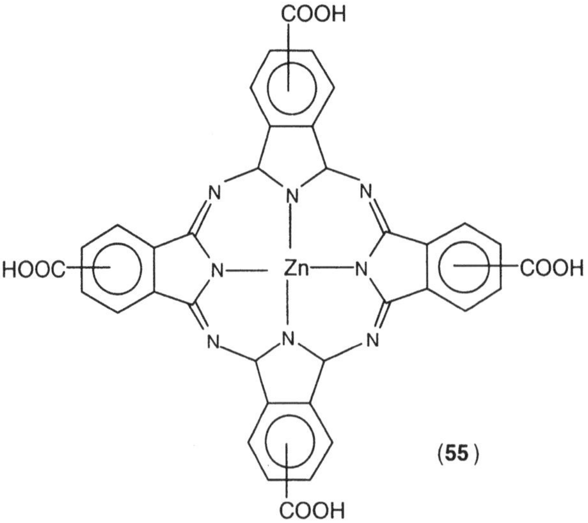

(**55**)

Structure 55

absorbing enough solar energy photons in the near IR to render efficiencies acceptable for practical applications.

I. Surface Chelation of Polypyridyl Complexes onto the TiO$_2$ Oxide Surface

The functional groups serve as grafting agents for the oxide surface of the TiO$_2$ films. The grafting of polypyridyl complexes onto the oxide surface, which allows for electronic communication between the complex and the substrate, is an important feature in dye-sensitized solar cells. Several ruthenium complexes containing substituted groups such as carboxylic acid, dihydroxy, and phosphonic acid on pyridine ligands are described [13,19,70–74]. These functional moieties serve as an anchoring group to immobilize the complex on the nanocrystalline TiO$_2$ films. The immobilized sensitizer absorbs a photon to produce an excited state, which efficiently transfers its electron onto the TiO$_2$ conduction band. To achieve high quantum yields of the excited-state electron-transfer process, the dye ideally needs to be in intimate contact with the semiconductor surface. The ruthenium complexes that have carboxylic acid and phosphonic acid groups show electron-transfer processes in the range of 80–90%. The near-quantitative electron-injection efficiency indicates a close overlap of the ligand π^*-orbitals and the titanium 3d-orbitals [75].

The interaction between the adsorbed sensitizer and the semiconductor surface has been addressed using resonance Raman and Fourier transform infrared (FTIR) spectroscopy. The carboxylic acid functional group could adsorb on the surface in an unidentate, bidentate, or bridging fashion. Yanagida et al. concluded that the sensitizer *cis*-dithiocyanato bis(2,2′-bipyridine- 4,4′-dicarboxylate)ruthenium(II) (**22**) binds to the surface using ester-like chelating linkages [76]. Finnie et al. have reported that the sensitizer (**22**) anchors on the surface of TiO$_2$ as a bidentate or bridging mode using two carboxylate groups per dye [77]. However, Fillinger and Parkinson studied the adsorption behavior of the sensitizer (**22**) and found that the initial binding involves one carboxylate, with subsequent additional binding of two or more carboxylate groups on the surface [78].

Shklover et al. have reported crystal structure and molecular modeling of the sensitizer (**22**) with different anchoring types to the TiO$_2$ anatase surface [79]. In their modeling, the initial attachment of the dye is a single bond A-type (Fig. 10). The main feature of this type of anchoring is the great rotational freedom of the molecule, which leads to immediate capture of another carboxylic group by a neighboring Ti atom resulting in anchoring B- and C-types (Fig. 10). When the sensitizer (**22**) anchors onto the TiO$_2$ surface using both carboxylic groups of the same bipyridyl ligand, this results in a two-bond anchoring of the D-type (Fig. 10). Similar to the transition from A-type to B-type, the B-type anchoring

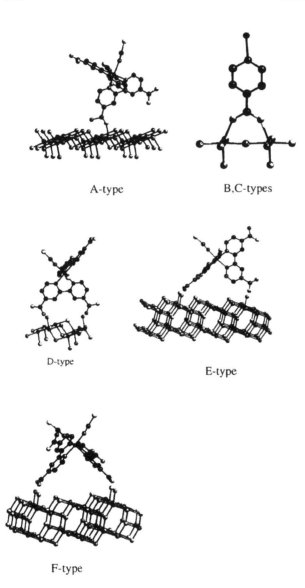

A-type   B,C-types

D-type

E-type

F-type

Figure 10   Possible anchoring modes for a sensitizer (**22**) using the carboxylate functional group.

mode thermal rotation around the axis results in E- and F-types attached via two of its four carboxylic groups (Fig. 10). The carboxylic group either bridges two adjacent rows of titanium ions through bidentate coordination or interact with surface hydroxyl groups through hydrogen bonds. The two remaining carboxylic groups remain in the protonated state. Thermodynamically, the bonding geometry of the E- and -F-types are most favorable. The attenuated total reflection (ATR)–FTIR data discussed in Section III.K using the sensitizer (**22**) and its different protonated forms are consistent with the anchoring of F-type where the sensitizer adsorbs on the surface using two carboxylic groups as a bridging bidentate fashion, which are trans to the NCS ligands.

J. Acid–Base Equilibria of *Cis*-dithiocyanato Bis(2,2-bipyridine-4,4′-dicarboxylate)ruthenium(II) Complex (22)

Understanding of the binding nature of sensitizers onto the TiO$_2$ surface requires detailed knowledge on the pK_a's of the sensitizer. Here we discuss pK_a's of the sensitizer (**22**) and its implications on the adsorption studies. The ground state pK_a values of **22** were determined from the relationship between the change in the optical density or the peak maximum with the pH for a given wavelength [80]. The plot of λ_{max} change versus pH for **22** shows the expected sigmoidal shape, with the pH at the inflection point giving two ground-state pK_a values at 3 and 1.5 $\pm$ 0.1. In complex **22**, there are two 4,4′-dicarboxy-2,2′-bipyridine ligands which could give four separate acid–base equilibria if the dissociation is stepwise. On the other hand, if the dissociation were simultaneous, one would expect one equilibrium constant. The two separate equilibria in complex **22** suggests that the pyridyl subunits are nonequivalent. Scheme 3 presents the simplified two-step equilibrium for complex **22**.

K. Adsorption of Polypyridyl Complexes onto the TiO$_2$ Oxide Surface

The difference in binding properties of the complexes containing carboxyl anchoring groups may stem from the differences in the pK_a values of the complexes. To further understand the pK_a data of complex **22** which contains four protons, two other similar complexes that contains no protons (**8**) and two protons (**56**) were isolated as a tetrabutylammonium salt. The infrared spectra of these three complexes as solid, and after adsorbing onto TiO$_2$ films were recorded in the 4000–400-cm^{-1} region as a solid, using ATR–FTIR. Figures 11 and 12 show a comparison of the solid-state ATR–FTIR spectra of the three free and adsorbed complexes. The infrared spectra of complex **22** as a solid sample shows a broad band at 1709 cm^{-1} due to carboxylic acid groups. However, the complex that contains two protons (**56**) show bands at 1709 and 1610 cm^{-1} due to carboxylic acid and carboxylate groups, respectively. The symmetric stretch of the carboxyl-

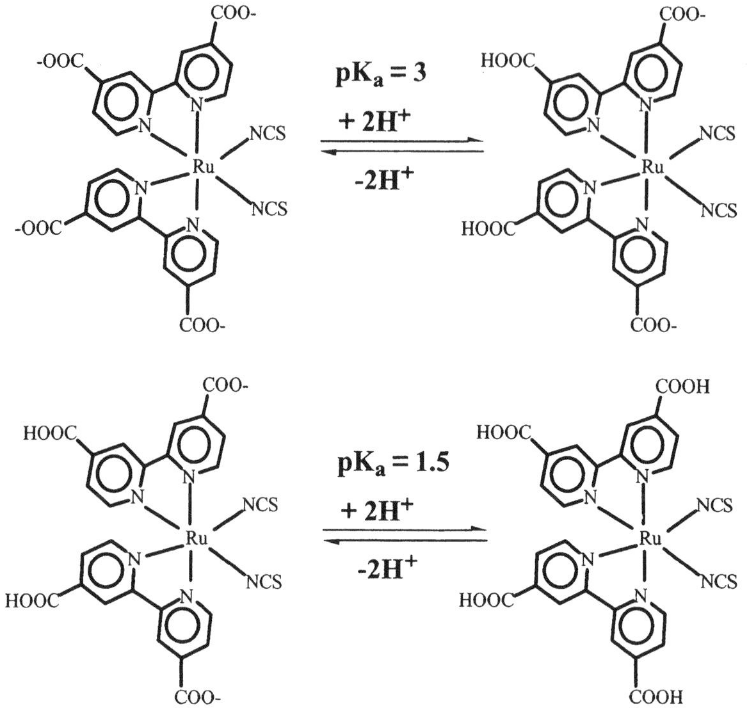

Scheme 3   Simplified two-step acid–base equilibrium of complex **22**.

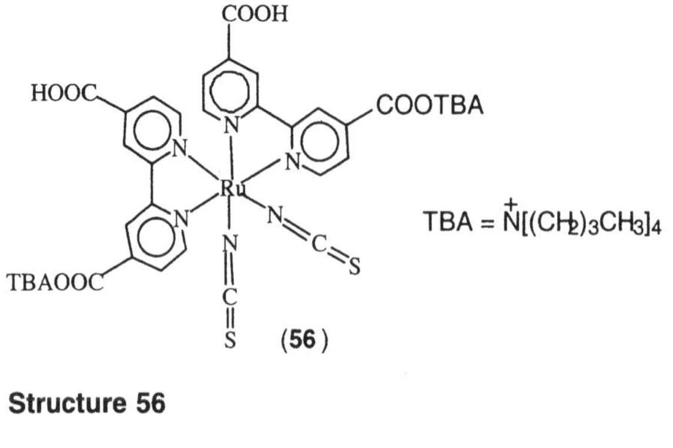

$$TBA = \overset{+}{N}[(CH_2)_3CH_3]_4$$

(**56**)

Structure 56

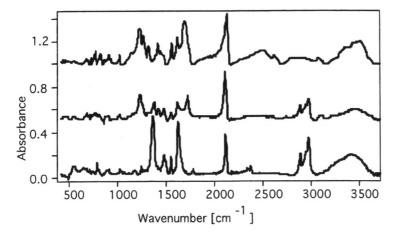

Figure 11 Comparison of the solid-state (dry powder) ATR–FTIR spectra of complexes **8** (bottom), **56** (middle), and **22** (top).

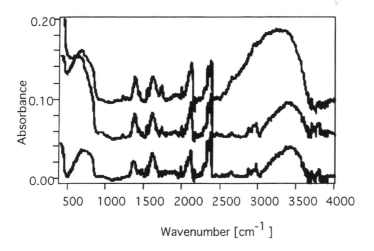

Figure 12 Comparison of the ATR–FTIR spectra of the adsorbed complexes on TiO$_2$: (**8**) (bottom), **56** (middle), and **22** (top).

ate group was observed at 1362 cm^{-1}. The intense peak at 1225 cm^{-1} in both of the complexes is assigned to ν(C—O) stretch. The ν(NC) of the thiocyanate ligand band was observed at 2098 cm^{-1}. The bands at 2873, 2931, and 2960 cm^{-1} are assigned to ν(C—H) of the tetrabutylammonium groups. On the other hand, the FTIR spectra of complex **8** show bands only due to carboxylate groups at 1623 cm^{-1} (—COO$_{as}^{-}$) and at 1345 cm^{-1}(—COO$_{s}^{-}$) and 2105 cm^{-1} (—NCS) [77].

The IR spectrum of complex **22** adsorbed onto the TiO$_2$ surface shows the presence of characteristic bands due to NCS, carbocylic acid, and the carboxylate groups. The presence of carboxylic acid and carboxylate groups indicates that complex **22** is not fully deprotonated on the surface. On the other hand, complex **8** shows the presence of only carboxylate groups, confirming the absence of the ester bond between the complex and the TiO$_2$. Complex **56** contains two protons on the carboxylic groups, which are trans to the NCS ligands, and two carboxylate groups, which are trans to each other. When complex **56** is adsorbed on the surface, it could adsorb using the carboxlic or carboxylate groups. The FTIR spectra show the presence of mainly carboxylate groups, indicating that the complex is being adsorbed on the surface using the two carboxylic groups, which are trans to the NCS ligands (F-type in Fig. 10). The FTIR data are in good agreement with the theoretical modeling of the dye on the TiO$_2$ surface [81].

L.   Effect of Protons Carried by the Sensitizer on the Performance

In order to obtain high conversion efficiencies, optimization of the short-circuit photocurrent (i_{sc}) and open-circuit potential (V_{oc}) of the solar cell is essential. The conduction band of the TiO$_2$ is known to have a Nernstian dependence on pH [13,18]. The fully protonated sensitizer (**22**), upon adsorption, transfers most of its protons to the TiO$_2$ surface, charging it positively. The electric field associated with the surface dipole generated in this fashion enhances the adsorption of the anionic ruthenium complex and assists electron injection from the excited state of the sensitizer in the titania conduction band, favoring high photocurrents (18–19 mA/cm^2). However, the open-circuit potential (0.65 V) is lower due to the positive shift of the conduction-band edge induced by the surface protonation.

On the other hand, the sensitizer (**8**), which carries no protons, shows a high open-circuit potential compared to complex **22**, due to the relative negative shift of the conduction-band edge induced by the adsorption of the anionic complex, although, as a consequence the short-circuit photocurrent is lower. Thus, there should be an optimal degree of protonation of the sensitizer in order to obtain optimum short-circuit photocurrent and open-circuit potential, which determines the power conversion efficiency of the cell.

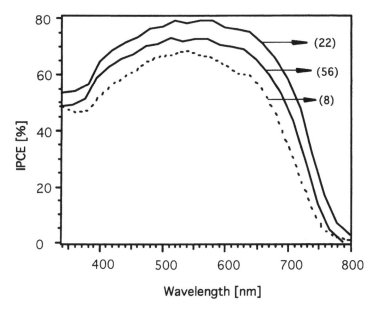

Figure 13 Photocurrent action spectra of nanocrystalline TiO$_2$ films sensitized by complexes **22**, **56**, and **8**. The incident photon-to-current conversion efficiency is plotted as a function of wavelength.

The performance of the three sensitizers **22**, **8** and 56, which contain different degrees of protonation were studied on nanocrystalline TiO$_2$ electrodes [80]. Figure 13 show the photocurrent action spectra obtained with a monolayer of these complexes coated on TiO$_2$ films.

The incident monochromatic photon-to-current conversion efficiency (IPCE) is plotted as a function of excitation wavelength. The IPCE value in the plateau region is 80% for complex **22**, whereas for complex **8**, it is only about 66%. In the red region, the difference is even more pronounced. Thus, at 700 nm, the IPCE value is twice as high for the fully protonated complex (**22**) as compared to the deprotonated complex (**8**). As a consequence, the short-circuit photocurrent is 18–19 mA/cm^2 for complex **22**, whereas it is only about 12–13 mA/cm^2 for complex **8**. However, there is a trade-off inas much as the photovoltage is about 250 mV higher for the latter (0.9 V) as compared to the former sensitizer (0.65 V). Nevertheless, this is insufficient to compensate for the current loss. Hence, the photovoltaic performance of complex **56** carrying two protons is superior to that of compounds **22** and **8**, which contains four and no protons, respectively. The doubly protonated form of the complex is therefore preferred over the other two sensitizers for sensitization of nanocrystalline TiO$_2$ films.

M. Comparison of IPCE Obtained with Various Sensitizers

Figure 14 shows the photocurrent action spectrum of a cell containing various sensitizers. The broad feature covers the entire visible spectrum and extends into the near-IR region up to 920 nm, for complex **24**. The incident photon-to-current conversion efficiency value in the plateau region is about 80% for complex **24**. Taking the light losses in the conducting glass into account, the efficiency of electric current generation is practically 95% over a broad wavelength range extending from 400 to 700 nm. The overlap integral of the absorption spectrum with the standard global AM 1.5 solar emission spectrum for complex **24** yields a photocurrent density of 20.5 mA/cm^2. The open-circuit potential is 720 mV and the fill factor is 0.7. These results were confirmed at the National Renewable Energy Laboratory (NREL), Golden, Colorado (Fig. 15). The complexes [RuL$_3$] (**1**) and [RuL$_2$(NCS)$_2$] (**22** under similar conditions show an IPCE value between 70% and 80% in the plateau region. Although the IPCE values are comparable with that of **24**, the total integrated current decreased significantly due to increasing blue shift of the spectral response of the later complexes.

The cyano-bridged trinuclear complex [RuL$_2$(RuL$'_2$(CN)$_2$)$_2$] (**57**) has found to be an excellent photosensitizer for spectral sensitization of the wide-band-gap

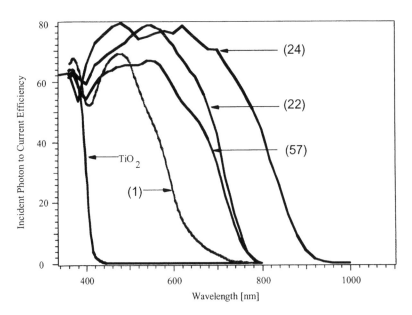

Figure 14   Photocurrent action spectra of bare nanocrystalline TiO$_2$ film, and the sensitizers **1**, **22**, **24**, and **57** adsorbed on TiO$_2$ films. The incident photon-to-current conversion efficiency is plotted as a function of wavelength.

EPFL (Switzerland) nano-crystal dye sensitized cell

Sample: PL0710/2
Oct 30, 1998 10:41 AM
ASTM E 892-87 Global

Temperature = 25.0 $^{\circ}$C
Area = 0.1863 cm^2
Irradiance: 1000.0 Wm^{-2}

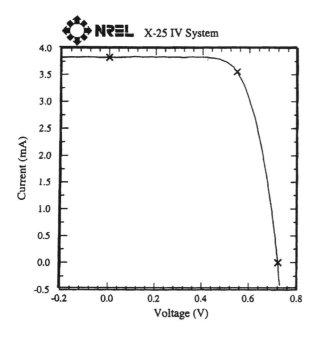

V_{oc} = 0.7210 V

I_{max} = 3.552 mA

J_{sc} = 20.53 mAcm^{-2}

Fill Factor = 70.41%

V_{max} = 0.5465 V

I_{sc} = 3.824 mA

P_{max} = 1.941 mW

Efficiency = 10.4 %

tape aperture, "black" dye

Figure 15   Photocurrent–voltage characteristics of complex **24** measured at NREL.

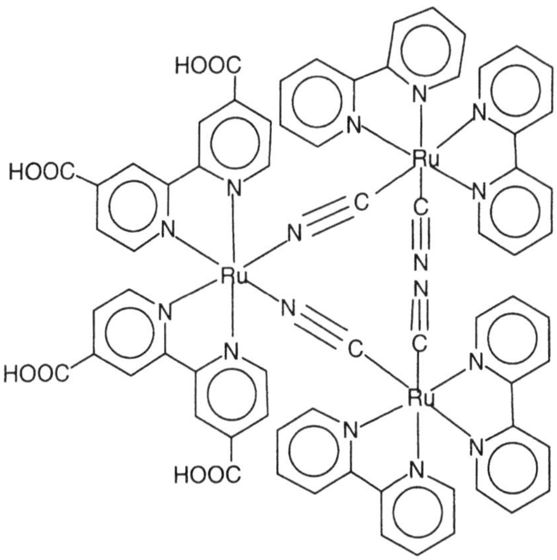

(57)

Structure 57

semiconductor TiO$_2$. Among the possible MLCT excited states associated with the three metal centers, the lowest-energy one is based on the central unit carrying the ligand 4,4'-dicarboxy-2,2'-bipyridine (dcbpy). When coated on a TiO$_2$ electrode, the complex anchors through the carboxyl unit of this central unit. Photons absorbed by the higher-energy peripheral units efficiently transfer excitation energy to this central unit, accounting for its excellent performance. However, the total efficiency of solar cell based on the trinuclear complex (**57**) was lower than the mononuclear complex (**24**)[82,83]. The most likely reason for such behavior is the porosity of the TiO$_2$ semiconductor substrate, where mononuclear complex can access freely into small pores compared to the polynuclear complex. This shows that the ideal sensitizer should be as small as possible and absorb most of the visible light in order to achieve high currents and, thereby, high-energy conversion efficiency.

IV. REDOX COUPLES

The total efficiency of the solar cell depends on the current, voltage, and the fill factor of the cell. Many groups have focused on the development of new sensitizers, which exhibit visible spectral response that matches the solar spectrum [15,20,35,47,50,82,84–86]. The open-circuit photovoltage is one of the key factors that govern the cell efficiency, which is determined by the energy difference between the Fermi level of the semiconductor under illumination and the Nernst potential of the redox couple in the electrolyte (Fig. 3). However, the experimentally observed open-circuit potential using various sensitizers is smaller than the difference between the conduction band and the redox couple due to differences in recombination rates.

Using the triiodide/iodide redox couple and the sensitizers **22** and **56** several groups have reported up to 8–10% solar cell efficiency where the potential mismatch between the sensitizer and the redox couple is around 0.5 V versus SCE. If one develops a suitable redox couple that decreases the potential difference mismatch between the sensitizer and the redox couple, then the cell efficiency could increase by 30%. Toward this goal, Oskam et al. have employed pseudohalogens in place of the triiodide/iodide redox couples, where the equilibrium potential of pseudohalogens is 0.43 V more positive than the iodide/iodide redox couple [87]. Yamada et al. have used cobalt *tris*- phenanthroline complexes as relays in dye-sensitized solar cells [88].

Of course, the design and the development of new redox couples is a challenging task because it has to meet several critical requirements: (1) It should be highly soluble in organic solvents; (2) it should be stable in reduced and oxidized form to undergo millions of turnovers; (3) the redox couple should have a transparent window in the visible region (very low molar extinction coefficient). Toward this goal, we have designed and developed new redox couples (**58–61**), which meet all these stringent requirements and rivals the iodide/iodide redox couple in dye sensitized solar cell. The imidazole and pyridine groups in complex **58** stabilize the oxidized and the reduced state of the metal, respectively. The substituted long aliphatic chains on the imidazole group permits one to dissolve in any polar and nonpolar organic solvents. The molar extinction coefficient of these complexes is typically in the range of 250–350 M^{-1} cm^{-1}, which provides a window for passing most of the light through the electrolyte. For the first time, photovoltaic cells incorporating the relay (**58**) in addition to the iodide/triiodide couple have yielded 65% IPCE and the power conversion efficiency up to 5.5%. On the contrary, the relay that contains **61** yielded 40% IPCE [89].

A. Solid Electrolytes/Hole-Transport Materials

A crucial factor for industrial development of the solar cell is its stability. Several studies have shown good endurance of the various components under accelerated illumination tests. The main problem of the dye-sensitized solar cell is finding a suitable sealing method that can resist organic solvents containing the redox electrolyte for overall long-term stability. A possible solution would be replacement of the liquid electrolyte by a solid electrolyte. Yanagida and others have explored a gel-based electrolyte containing the room-temperature molten salt that

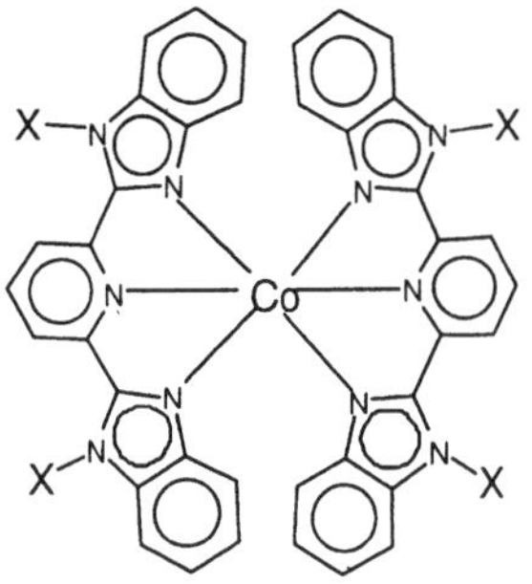

(58)	X	=	—CH$_3$
(59)	X	=	—C$_4$H$_{11}$
(60)	X	=	—C$_{13}$H$_{27}$

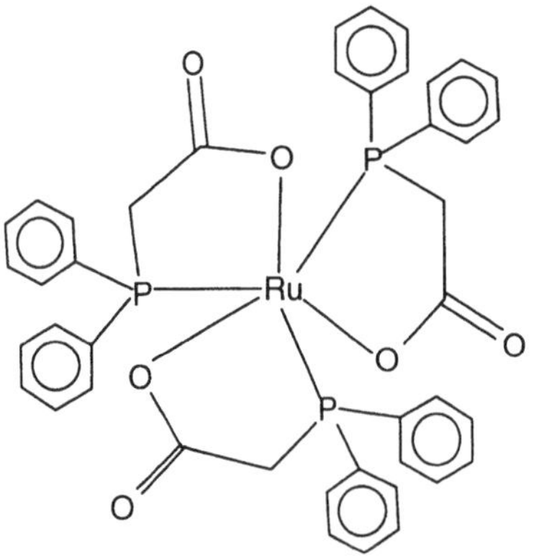

(61)

Structures 58–61

gave light-to-electricity conversion efficiency under AM 1.5 irradiation of 5% [90–94]. Kumara et al have developed solid-state dye-sensitized solar cell based on p-CuI, hole conductors, which gave 3% energy conversion efficiency [95]. However, the short-circuit photocurrent and the open-circuit voltage both decreased due to formation of large crystals that decreases the contact between the dyed TiO$_2$ surface and CuI crystallites. Nevertheless, they have improved stability of the dye-sensitized solar cell dramatically by the addition of small quantity of 1-methyl-3-ethylimidazolium thiocyanate in the hole-conducting CuI solution, which acts as crystal growth inhibitor.

Bach et al. have successfully introduced the concept of a solid p-type semi-conductor (heterojunction), with the amorphous organic hole-transport material 2,2′,7,7′-tetrakis (N,N-di-p-methoxyphenyl-amine)9,9′-spirobifluorene [96]. This hole-conducting material allows the regeneration of the sensitizers after electron injection due to its hole-transport properties. Nevertheless, the incident photon-to-current conversion efficiencies using complex **22** as a charge-transfer sensitizer and spirobifluorene as a hole-conducting material were significantly lower than the redox electrolyte based on iodide/triiodide.

The low efficiencies could be due to lack of intimate contact (interface) between the sensitizer, which is hydrophilic, and spirobifluorene, which is hydrophobic. Moreover, surface charge also plays a significant role in the regeneration of the dye by the electrolyte [97]. In an effort to reduce the charge of the sensitizer and improve the interfacial properties between the sensitizer and spirobifluorene, amphiphilic heteroleptic ruthenium (II) sensitizers (**48–53**) have been used. These sensitizers show excellent stability and interfacial properties with hole-transport materials, which improved the yields of the solid-state solar cell significantly.

V. CONCLUSIONS AND PERSPECTIVES

In this chapter, we have illustrated the working principle of a dye-sensitized solar cell and details of its components. A wide range of complexes has been discussed to emphasize the concept of tunability of the photophysical, redox, and spectral properties. The molecular engineering aspect of the sensitizers for enhanced spectral response across the visible and near-infrared region containing various ligands, which have been judiciously selected to reconcile the tasks of the sensitizer, to afford vectorial electron injection into the semiconductor and efficient solar light harvesting were discussed. Novel amphiphilic ruthenium(II) sensitizers that yield incident photon-to-current conversion efficiency values of about 60% in a heterojunction solar cell due to an enhanced interface between the sensitizer and the hole conductor were illustrated. These sensitizers show excellent stability toward water-induced desorption and opens up a new avenue for the development of low-cost solid-state dye-sensitized solar cells. Also, discussed are the concepts for development of new redox couples, which decreases potential difference mismatch between the sensitizer and the redox couple.

We believe that the development of efficient sensitizers with optimized redox couples could yield cell efficiencies of 13–15%, which would expand our ability to better harvest the sun's energy and decrease our dependency on fossil fuels and nuclear power.

ACKNOWLEDGMENTS

We thank Dr. R. Humphry-Baker for his time and valuable discussions. Financial support from the Swiss Federal Office for Energy (OFEN), the U.S. Air Force Research Office under contract No. F61775-00-C0003, and the Institute for Applied Photovoltaics (INAP, Gelsenkirchen, Germany) is greatly appreciated.

REFERENCES

1. Manabe, S.; Wetherald, T.; Stouffer, R. J. *Clim. Change* **1981**, *3*, 347.
2. Dresselhaus, M. S.; Thomas, I. L. *Nature* **2001**, *414*, 332.

3. Grätzel, M. *Nature* **2001**, *414*, 338.

4. Grätzel, M. *Prog. Photovoltaics* **2000**, *8*, 171.

5. Kalyanasundaram, K.; Grätzel, M. *Coord. Chem. Rev.* **1998**, *77*, 347.

6. Archer, M. D. *J. Appl. Electrochem.* **1975**, *5*, 17.

7. Gleria, R.; Memming, R. *Z. Phys. Chem.* **1976**, *98*, 303.

8. Copeland, A. W.; Black, O. D.; Garett, A. B. *Chem. Rev.* **1942**, *31*, 177.

9. Nazeeruddin, M. K.; Kay, A.; Rodicio, I.; Humphry-Baker, R.; Muller, E., Liska, Vlachopoulos, N.; Grätzel, M. *J. Am. Chem. Soc.* **1993**, *115*, 6382.

10. Hagfeldt, A.; Grätzel, M. *Acc. Chem. Res.* **2000**, *33*, 269.

11. van de Lagemaat, J.; Park, N. -G.; Frank, A. J. *J. Phys. Chem. B* **2000**, *104*, 2044.

12. Kim, K. L.; Benkstein, K. D.; van de Lagemaat, J.; Frank, A. J. *Chem. Mater.* **2002**, *14*, 1042.

13. Yan, S. G.; Hupp, J. T. *J. Phys. Chem.* **1996**, *100*, 6867.

14. Kambe, S.; Nakade, S.; Kitamura, T.; Wada, Y.; Yanagida, S. *J. Phys. Chem. B.* **2002**, *106*, 2967.

15. Hara, K.; Horiuchi, H.; Katoh, R.; Singh, L. P.; Sugihara, H.; Sayama, K.; Murata, S.; Tachiya, S.; Arakawa, H. *J. Phys. Chem. B* **2002**, *106*, 374.

16. Turrion, M.; Macht, B.; Tributsch, H.; Salvador, P. *J. Phys. Chem. B* **2001**, *105*, 9732.

17. Ehret, A.; Stuhl, L.; Spitler, M. T. *J. Phys. Chem. B* **2001**, *105*, 9960.

18. Qu, P.; Meyer, G. J. *Langmuir* **2001**, *17*, 6720.

19. Bignozzi, C. A.; Aragazzi, R.; Kleverlaan, C. J. *Chem. Soc. Rev.* **2000**, *29*, 87.

20. Sauvé, G.; Cass, M. E.; Doig, S. J.; Lauermann, I.; Pomykal, K.; Lewis, N. S. *J. Phys. Chem.* **2000**, *104*, 3488.

21. Chen, C. G.; Chappel, S.; Diamant, Y.; Zaban, A. *Chem. Mater.* **2001**, *13*, 4629.

22. Ferrere, S.; Gregg, B. A. *J. Phys. Chem. B.* **2001**, *105*, 7602.

23. Hagfeldt, A.; Grätzel, M. *Chem. Rev.* **1995**, *95*, 49.

24. Rothenberger, G.; Comte, P.; Grätzel, M. *Solar Energy Mater. Solar Cells* **1999**, *58*, 321.

25. Nazeeruddin, M. K.; Grätzel M. In *Dyes for Semiconductor Sensitization*; Licht, S. Ed. Wiley–VCH: Germany, 2002, p. 407.

26. Barbe, C. J.; Arendse, F.; Comte, P.; Jirousek, M.; Lenzmann, F.; Shklover, V.; Grätzel, *M. J. Am. Ceram. Soc.* **1997**, *80*, 3157.

27. Burnside, S. D.; Shklover, V.; Barbe, C. J.; Comte, P.; Arendse, F.; Brooks, K.; Grätzel, M. *Chem. Mater.* **1998**, *10*, 2419.

28. Cahen, D.; Hodes, G.; Grätzel, M.; Guilemoles, J. F.; Riess, I. *J. Phys. Chem. B.* **2000**, *104*, 2053.

29. Ferber, J.; Luther, J. *J. Phys. Chem. B* **2001**, *105*, 4895.

30. Brooks, K. G.; Burnside, S. D.; Shklover, V.; Comte, P.; Arendse, F.; McEvoy, A. J.; Grätzel M. *Proc. Am. Ceram. Soc.* **2000**, *109*, 115.

31. Nazeeruddin, M. K.; Péchy, P.; Renouard, T.; Zakeeruddin, S. M.; Humphry-Baker, R.; Comte, P.; Liska, P.; Cevey, L.; Costa, E.; Shklover, V.; Spiccia, L.; Decon, G. B.; Bignozzi, C. A.; Grätzel, M. *J. Am. Chem. Soc.* **2001**, *123*, 1613.

32. Haught, A. F. *J. Solar Energy Eng.* **1984**, *106*, 3.

33. De Vos, A. *Endoreversible Thermodynamics of Solar Energy Conversion*, Oxford Science Publishers: Oxford, 1992.

34. Anderson, P. A.; Strouse, G. F.; Treadway, J. A.; Keene, F. R.; Meyer, T. J. *Inorg. Chem.* **1994**, *33*, 3863.

35. Islam, A.; Sugihara, H.; Singh, L. K.; Katoh, H. K. R.; Nagawa, Y.; Yanagida, M.; Takahashi, Y.; Murata, S.; Arakawa, H. *Inorg. Chim. Acta* **2001**, *322*, 7.

36. Juris, A.; Balzani, V.; Barigelletti, F.; Campagna, S.; Belser, P.; von Zelewsky, A. *Coord. Chem. Rev.* **1988**, *84*, 85.

37. Balzani, V.; Juris, A.; Venturi, M.; Campagna, S.; Serroni, S. *Chem. Rev.* **1996**, *96*, 759.

38. Kalyanasundaram, K.; Nazeeruddin, M. K. *Chem. Phys. Lett.* **1992**, *193*, 292.

39. Nazeeruddin, M. K.; Muller, E.; Humphry-Baker, R.; Vlachopoulos, N.; Grätzel, M. *J. Chem. Soc. Dalton Trans.* **1997**, 4571.

40. Ruile, S.; Kohle, O.; Péchy, P.; Grätzel, M. *Inorg. Chimica Acta* **1997**, *261*, 129.

41. Kohle, O.; Ruile, S.; Grätzel, M. *Inorg. Chem.* **1996**, *35*, 4779.

42. Zakeeruddin, S. M.; Nazeeruddin, M. K.; Pechy, P.; Rotzinger, F. P.; Humphry-Baker, R.; Kalyanasundaram, K.; Grätzel, M. *Inorg. Chem.* **1997**, *36*, 5937.

43. Nazeeruddin, M. K.; Pechy, P.; Grätzel, M. *Chem. Commun.* **1997**, 1705.

44. Argazzi, R.; Bignozzi, C. A.; Heimer, T. A.; Castellano, F. N.; Meyer, G. J. *Inorg. Chem.* **1994**, *33*, 5741.

45. Xie, P.-H.; Hou, Y. J.; Zhang, B. W.; Cao, Y.; Wu, F.; Tian, W.-J.; Shen, J. C. *J. Chem. Soc., Dalton Trans.* **1999**, 4217.

46. Nazeeruddin, M. K.; Humphry-Baker, R.; Pechy, P.; Rotzinger, F. P.; Grätzel, M. 10th International Conference on Photochemical Conversion and Storage of Solar Energy, **1997**, p. 201.

47. Hara, S.; Sugihara, H.; Tachibana, Y.; Islam, A.; Yanagida, M.; Sayama, K.; Arakawa, H. *Langmuir* **2001**, *17*, 5992.

48. Yanagida, M.; Sing, L. P.; Sayama, K.; Hara, K.; Katoh, R.; Islam, A.; Sughihara, H.; Arakawa, H.; Nazeeruddin, M. K.; Grätzel, M. *J. Chem. Soc., Dalton Trans.* **2000**, 2817.

49. Kober, E. M.; Meyer, T. J. *Inorg. Chem.* **1983**, *22*, 1614.

50. Heimer, T. A.; Bignozzi, C. A.; Meyer, G. J. *J. Phys. Chem.* **1993**, *197*, 11, 987.

51. Alebbi, M.; Bignozzi, C. A.; Heimer, T. D.; Hasselmann, G. M.; Meyer, G. J. *J. Phys. Chem.* **1998**, *102*, 7577.

52. Nazeeruddin, M. K.; Zakeeruddin, S. M.; Humphry-Baker, R.; Gorelsky, S. I.; Lever, A. B. P.; Grätzel, M. *Coordination Chemistry Reviews* **2000**, *208*, 213.

53. Masood, M. A.; Sullivan, B. P.; Hodges, D. J. *Inorg. Chem.* **1994**, *33*, 5360.

54. Durham, B.; Wilson, S. R.; Hodges, D. J.; Meyer, T. J.; *J. Am. Chem. Soc.* **1980**, *102*, 600.

55. Renouard, T.; Fallahpour, R.-A.; Nazeeruddin, M. K.; Humphry-Baker, R.; Gorelsky, S. I.; Lever, A. B. P.; Grätzel, M. *Inorg. Chem.* **2002**, *41*, 367.

56. Lever, A. B. P. *Inorg. Chem.* **1990**, *29*, 1271.

57. Zakeeruddin, S. M.; Nazeeruddin, M. K.; Péchy, P.; Humphry-Baker, R.; Quagliotto, P.; Viscardi, G.; Grätzel, M. *Langmuir* **2001**, *18*, 952.

58. Tennakone, K.; Perera, V. P. S.; Kottegoda, I. R. M.; Kum, G. R. A. *J. Phys. D: Appl. Phys.* **1999**, *32*, 374.

59. Chandrasekharan, N.; Kamat, P. V. *J. Phys. Chem. B.* **2000**, *104*, 10,851.

60. Gregg, B. A.; Pichot, F.; Ferrere, S.; Fields, C. L. *J. Phys. Chem. B* **2001**, *105*, 1422.

61. Nazeeruddin, M. K.; Moser, J. E.; Grätzel, M. *J. Phys. Chem.* (Communicated).

62. Wöhrle, D.; Meissner, D. *Adv. Mater.* **1991**, *3*, 129.

63. Jaeger, C. D.; Fan, F. F.; Bard, A. J. *J. Am. Chem. Soc.* **1980**, *102*, 2592.

64. Giraudeau, A.; Fan, F. F.; Bard, A. J. *J. Am. Chem. Soc.* **1980**, *102*, 5137.

65. Giraudeau, A.; Ren, F.; Fan, F.; Bard, A. J. *J. Am. Chem. Soc.* **1980**, *102*, 5137.

66. Yanagi, H.; Chen, S.; Lee, P. A.; Nebesny, K. W.; Armstrong, N. R.; Fujishima, A. *J. Phys. Chem.* **1996**, *100*, 5447.

67. Nazeeruddin, M. K.; Humphry-Baker, R.; Murrer, B. A.; Grätzel, M. *J. Chem. Soc., Chem. Commun.* **1998**, 719.

68. He, J.; Benko, G.; Korodi, F.; Polivka, T.; Lomoth, R.; Åkermark, B.; Sun, L.; Hagfeldt, A.; Sundstrom, V. *J. Am. Chem. Soc.* **2002**, *124*, 4922.

69. Nazeeruddin, M. K.; Humphry-Baker, R.; Grätzel, M.; Wöhrle, D.; Schnurpfeil, G.; Schneider, G.; Hirth, A.; Trombach, N. *J. Porphyrins Phthalocyanines* **1999**, *3*, 230.

70. Trammell, S. A.; Moss, J. A.; Yang, J. C.; Nakhle, B. M.; Slate, C. A.; Odobel, F.; Sykora, M.; Erickson, B. W.; Meyer, T. J. *Inorg. Chem.* **1999**, *38*, 3665.

71. Will, G.; Boschloo, G.; Nagaraja Rao, S.; Fitzmaurice, D. *J. Phys. Chem. B.* **1999**, *103*, 8067.

72. Lemon, B.; Hupp, J. T. *J. Phys. Chem. B* **1999**, *103*, 3797.

73. Jing, B.; Zhang, H.; Zhang, M.; Lu, Z.; Shen, T. *J. Mater. Chem.* **1998**, *8*, 2055.

74. Rice, C. R.; Ward, M. D.; Nazeeruddin, M. K.; Grätzel, M. *New J. Chem.* **2000**, *24*, 651.

75. Anderson, S.; Constable, E. C.; Dare-Edwards, M. P.; Goodenough, J. B.; Hamnett, A.; Seddon, K. R.; Wright, R. D. *Nature* **1979**, *280*, 571.

76. Murakoshi, K.; Kano, G.; Wada, Y.; Yanagida, S.; Miyazaki, H.; Matsumoto, M.; Murasawa, S. *J. Electroanal. Chem.* **1995**, *396*, 27.

77. Finnie, K. S.; Bartlett, J. R.; Woolfrey, J. L. *Langmuir* **1998**, *14*, 2744.

78. Fillinger, A.; Parkinson, B. A. *J. Electrochem. Soc.* **1999**, *146*, 4559.

79. Shklover, V.; Ovehinnikov, Y. E.; Braginsky, L. S.; Zakeeruddin, S. M.; Grätzel, M. *Chem. Mater.* **1998**, *10*, 2533.

80. Nazeeruddin, M. K.; Zakeeruddin, S. M.; Humphry-Baker, R.; Jirousek, M.; Liska, P.; Vlachopoulos, N.; Shklover, V.; Fischer, C. H.; Grätzel, M. *Inorg. Chem.* **1999**, *38*, 6298.

81. Nazeeruddin, M. K.; Humphry-Baker, R.; Grätzel, M. *J. Phys. Chem. B* (Communicated).

82. Scandola, F.; Indelli, M. T.; Chiorboli, C.; Bignozzi, C. A. *Topics Curr. Chem.* **1990**, *158*, 73.

83. Nazeeruddin, M. K.; Liska, P.; Moser, J.; Vlachopoulos, N.; Grätzel, M. *Helv. Chim. Acta* **1990**, *73*, 1788.

84. Takahashi, Y.; Arakawa, H.; Sugihara, H.; Hara, K.; Islam, A.; Katoh, R.; Tachibana, Y.; Yanagida, M. *Inorg. Chim. Acta* **2000**, *310*, 169.

85. Hou, Y.-J.; Xie, P.-H.; ZJing, B.-W.; Cao, Y.; Xiao, X.-R.; Wang, W.-B. *Inorg. Chem.* **1999**, *38*, 6320.

86. Lees, A. C.; Kleverlaan, C. J.; Bignozzi, C. A.; Vos, J. G. *Inorg. Chem.* **2001**, *40*, 5343.

87. Oskam, G.; Bergeron, B. V.; Meyer, G. J.; Searson, P. C. *J. Phys. Chem. B* **2001**, *105*, 6867.
88. Wen, C.; Ishikawa, K.; Kishima, M.; Yamada, K. *Solar Energy Mater. Solar Cells* **2000**, *61*, 339.
89. Nusbaumer, H.; Moser, J. E.; Zakeeruddin, S. M.; Nazeeruddin, M. K.; Grätzel, M. *J. Phys. Chem. B* **2002**, *105*, 10,461.
90. Kubo, W.; Kitamura, T.; Hanabusa, K.; Wada, Y.; Yanagida, S. *Chem. Commun.* **2002**, 374.
91. Kubo, W.; Murakoshi, K.; Kitamura, T.; Wada, Y.; Hanabusa, K.; Shirai, H.; Yanagida, S. *Chem. Lett.* **1998**, 1241.
92. Nogueira, A. F.; De Paoli, M. A.; Montanari, I.; Monkhouse, R.; Nelson, J.; Durrant, J. R. *J. Phys. Chem. B* **2001**, *105*, 7517.
93. De Paoli, M.-A.; Machado, D. A.; Nogueira, A. F.; Longo, C. *Electrochim. Acta.* **2001**, *46*, 4243.
94. Matsumoto, M.; Miyazaki, H.; Matsuhiro, K.; Kumashiro, Y.; Takaoka, Y. *Solid State Ionics* **1996**, *89*, 263.
95. Kumara, G. R. A.; Konno, A.; Shiratsuchi, K.; Tsukahara, J.; Tennakone, K. *Chem. Mater.* **2002**, *14*, 954.
96. Bach, U.; Lupo, D.; Comte, P.; Moser, J.-E.; Weissörtel, F.; Salbeck, J.; Spreitzer, H.; Grätzel, M. *Nature* **1998**, *395*, 583.
97. Pelet, S.; Moser, J.-E.; Grätzel, M. *J. Phys. Chem. B* **2000**, *104*, 1791.

11

Applications of Semiconductor Electro-Optical Properties to Chemical Sensing

Anne-Marie L. Nickel

Milwaukee School of Engineering, Milwaukee, Wisconsin, U.S.A.

Jeng-Ya Yeh, Gordon A. Shaw,
Luke J. Mawst, Thomas F. Kuech, and
Arthur B. Ellis

University of Wisconsin–Madison, Madison, Wisconsin, U.S.A.

I. INTRODUCTION

There is a growing need to sense molecules for applications as diverse as waste management, explosives detection, and disease prevention. Goals of chemical sensor development are to create devices that require little power and that are robust, sensitive, selective, fast, compact, and inexpensive. In this chapter, we describe optical techniques based on semiconductor luminescence that are promising methods for chemical sensing applications.

The application of semiconductor luminescence to chemical sensing can rely on the chemical, electrical, and optical properties of II–VI and III–V semiconductors [1]. These properties provide the binding capability, transducing mechanism, and signal required for chemical sensing. The diverse chemical compositions of semiconductor materials provide a range of surfaces for molecular binding.

345

Customarily, semiconductor surfaces are chemically or physically prepared to optimize their chemical and/or electro-optical properties. For chemical sensing applications, a freshly etched surface often provides greater chemical sensitivity. A Br_2/MeOH etch of n-CdSe, for example, has typically yielded larger luminescence responses to analytes than have polished samples. Additionally, transducing films have been used to modify semiconductor surfaces to enhance the selectivity of CdSe for particular analytes [2].

The transducing mechanism of semiconductor luminescence involves the modification of the semiconductors' surface electrical properties through molecular adsorption. Changes in solid-state electro-optical properties result from adsorption of the molecule of interest onto the semiconductor surface.

Semiconductor band-gap luminescence results from excited electrons recombining with electron vacancies, holes, across the band gap of the semiconductor material. Electrons can be excited across the band gap of a semiconductor by absorption of light, as in photoluminescence (PL), or injected by electrical bias, as in electroluminescence (EL). Both types of luminescence have been used in chemical sensing applications [1,3].

II. SEMICONDUCTOR LUMINESCENCE

Changes in intensity of semiconductor PL or EL can be used to detect molecular adsorption onto semiconductor surfaces [1,3]. PL occurs most efficiently when ultra-band-gap radiation excites electrons from the valence band to the conduction band of a direct-band-gap semiconductor and the electrons recombine radiatively with the holes left behind in the valence band.

Electroluminescence in semiconductor devices typically uses the interface of a p-type and an n-type semiconductor, forming a so-called p–n junction. Establishment of thermal equilibrium between the two materials results in a built-in voltage across the unilluminated junction. The equilibrium is disturbed by the application of an external voltage or bias. If electrons and holes are provided, sufficient energy to cross the potential energy barrier separating the n-side and p-side, then the charge carriers can recombine in large quantities to produce easily measured luminescence.

III. PHOTOLUMINESCENCE FOR CHEMICAL SENSING

The semiconductor surface and bulk can have quite different electronic properties. Because the regularity of the bulk crystal lattice is lost at the surface, there is often a large number of dangling or unsaturated bonds. This can lead to a large density of surface-localized energy levels within the band gap, called surface states. Electrons occupy surface states at energies up to the Fermi level, creating an electric field in the near-surface region of the solid. The region supporting

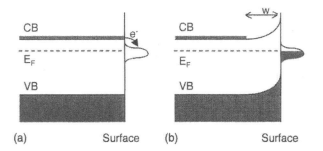

Figure 1 Idealized band diagrams of *n*-type semiconductors showing the conduction-band (CB) edge, valence-band (VB) edge, and Fermi level (E_F). A hypothetical distribution of surface states is represented by the "hump" at the surface. Shading represents filled electronic states. (a) Band structure before thermal equilibrium in the dark is reached; (b) band structure after thermal equilibrium in the dark is established; surface states are filled up to E_F with electrons. The electric field present in (b) is represented by the bending of the band edges near the surface.

the electric field is called the depletion region and has a thickness W that is typically on the order of hundreds to thousands of angstoms, depending on carrier concentration, as shown in Fig. 1 [4]. The electric field causes the conduction and valence bands to bend up at the surface of an *n*-type semiconductor and to bend down for a *p*-type semiconductor. Because the near-surface electric field promotes separation rather than recombination of electron–hole pairs, this region is considered nonemissive and is called a "dead layer." The dead-layer thickness D is on the order of the depletion width.

The depletion width can play a role in analyte-induced modulation of the semiconductor PL [4]. As molecules adsorb onto the surface of the semiconductor, the dead-layer thickness can change, resulting in what can be described as a "luminescent litmus test": When Lewis bases adsorb onto the semiconductor surface, they donate electron density to the solid, which decreases the electric field and thus decreases the dead-layer thickness. The reduction in D causes an enhancement in the PL intensity from the semiconductor. Figures 2a and 2b present typical PL enhancements observed from an etched *n*-CdSe substrate: Relative to a nitrogen reference ambient, adsorption of the Lewis bases ammonia and trimethylamine cause a reversible increase in PL intensity. In contrast, when Lewis acids adsorb onto the surface, they can withdraw additional electron density, causing the electric field to increase and the PL intensity to decrease. Such effects have been observed with gases like sulfur dioxide [5].

The quantitative form of the dead-layer model relates PL intensity to dead-layer thickness, which is assumed to approximate the depletion width W [6,7]:

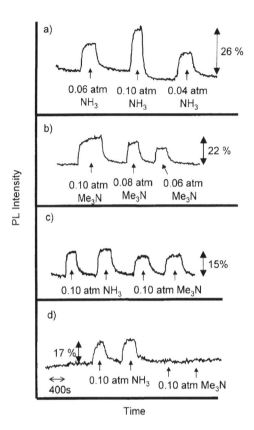

Figure 2 Typical PL enhancements due to exposure to amines: (a) etched n-CdSe surface exposed to ammonia; (b) etched n-CdSe surface exposed to trimethylamine; (c) n-CdSe coated with a trimethylamine-imprinted PAA film exposed sequentially to ammonia and trimethylamine; and (d) n-CdSe coated with an ammonia-imprinted poly(acrylic acid) (PAA) film exposed sequentially to ammonia and trimethylamine. Nitrogen gas was used as the reference ambient between exposures to the amine-containing ambients with the indicated partial pressures. Samples were excited at 633 nm and the PL was monitored at 720 nm.

$$\frac{PL_{ref}}{PL_x} = \exp(-\alpha'\Delta D) \tag{1}$$

In this equation, PL_{ref} represents the PL intensity in a reference ambient (typically, vacuum or nitrogen for gas phase analytes and pure solvent for solution analytes); PL_x represents the PL intensity in the presence of the analyte; $\alpha' = (\alpha + \beta)$ is the sum of the absorptivities of the semiconductor for the exciting and emitted light, and $\Delta D = (D_{ref} - D_x)$ is the change in the dead-layer thickness caused by adsorption. This form of the dead-layer model assumes that the surface recombination velocity S is unaffected by adsorption or that it remains very large in the absence and presence of the analyte ($S \gg L/\tau$ and $\alpha L^2/\tau$, where L and τ are the minority carrier diffusion length and lifetime, respectively). The model can be tested by exciting at several different wavelengths to determine whether a constant value of ΔD is observed. For the data in Fig. 2a, a ΔD value of ~ 100 Å was calculated from the PL changes [8].

In general, the magnitude of the PL response to the adsorption of analytes increases with concentration between threshold and saturation concentrations. At saturation, all of the available surface sites are filled and, therefore, no change in PL intensity occurs beyond that point. This concentration dependence allows for an estimation of the analyte-binding constant using the Langmuir adsorption isotherm model. The quantitative form of the model is shown in Eq (2), where the fractional surface coverage, θ, is estimated from the fractional changes in the dead-layer thickness [9]:

$$\frac{1}{\theta} = 1 + \frac{1}{KP} \tag{2}$$

In this equation, K represents the equilibrium constant for binding of the analyte to the surface and P is the analyte partial pressure or concentration. When PL changes are at their saturated intensity, PL_{sat}, then $\theta = 1$; the PL intensity of the reference ambient, PL_{ref}, corresponds to $\theta = 0$, and the PL intensity at intermediate surface coverages, PL_x, allows θ to be defined as follows:

$$\theta = \frac{\ln(PL_{ref}/PL_x)}{\ln(PL_{ref}/PL_{sat})} \tag{3}$$

The Langmuir model has been used to estimate binding constants for a variety of molecules in solution and in the gas phase, which cause adsorbate-induced PL responses with CdSe [1] Figure 3a shows typical data for the PL response of an etched n-CdSe crystal to ammonia at various concentrations. The inset of the figure displays the same data as a double reciprocal plot, corresponding to Eq. (2) and yielding a binding constant K of $750 \pm 250 \; M^{-1}$.

IV. CHEMICAL SELECTIVITY THROUGH MOLECULARLY IMPRINTED POLYMER FILMS

Molecularly imprinted polymer (MIP) films can be utilized in conjunction with CdSe PL changes to enhance the selectivity of the sensor. MIPs have been developed to mimic the selective binding of molecules to biological substrates, as in

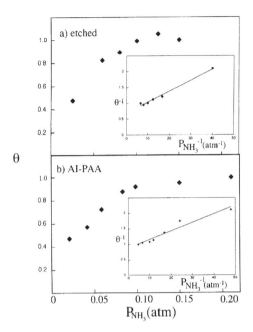

Figure 3 Fractional surface coverage θ, determined using Eq. (3), as a function of ammonia pressure for (a) an etched n-CdSe crystal and (b) a n-CdSe crystal coated with an ammonia-imprinted PAA film. The insets show the same data as double reciprocal plots, yielding equilibrium constants for (a) and (b) of $750 \pm 250\ M^{-1}$ and 1000 ± 500 M^{-1}, respectively. In both sets of experiments, the PL was excited at 633 nm and monitored at 720 nm. (Reprinted with permission from *Chem. Mater.* **2001**, *13*, 1391. Copyright 2001, American Chemical Society.)

the specific binding that occurs between antigens and antibodies. Polymerization of a monomer with a linked analyte molecule creates a polymer film with an imbedded analyte. Removal of the analyte yields a polymer film with specific binding sites for the target molecule, as shown schematically in Fig. 4. In addition to chemical sensing, typical applications of molecular imprinting include catalysis and separations [10–12].

The use of MIP films with emissive semiconductor substrates can potentially impart selectivity to PL and EL responses. To illustrate, in addition to the PL enhancements caused by ammonia and trimethylamine (Figs. 2a and 2b), other amines like methylamine and dimethylamine elicit a similar PL response when adsorbed onto the bare surface of etched n-CdSe. These amines cannot generally be reliably distinguished with this technique because of the similarity of their PL

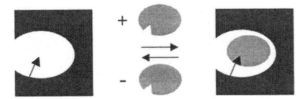

Figure 4   Schematic representation of a MIP binding and releasing a target molecule. (Reprinted with permission from *Chem. Mater.* **2001**, *13*, 1391. Copyright 2001, American Chemical Society.)

signatures. However, as described below, we have shown that MIP films on CdSe can improve the selectivity of this PL-based detection method [8].

Poly(acrylic acid) (PAA), a MIP film candidate, has been shown to bind to the bare CdSe surface from methanol solution with considerable affinity [13]. Placement of drops of a PAA–methanol solution on the surface of CdSe and evaporation of the solvent leaves a PAA film on the semiconductor surface. Once coated with this PAA film, CdSe shows no change in PL intensity in the presence of amines. Despite the lack of a PL change, the deprotonation of the carboxylic acid could be observed by the shifting of the infrared (IR) carboxylic acid peak to lower frequencies characteristic of the carboxylate anion upon amine binding, as shown in Fig. 5. The reaction chemistry is:

$$[—CH_2CH(COOH)—]_n(s)$$
$$+ \ xNR_3(g) \leftrightarrow \{[—CH_2CH(COO)—]_n^{x-} \cdot xHNR_3^+\} \ (s) \quad (4)$$

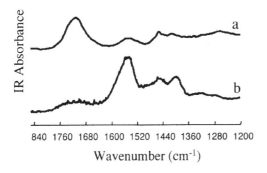

840 1760 1680 1600 1520 1440 1360 1280 1200

Wavenumber (cm⁻¹)

Figure 5   Infrared absorption spectra for a PAA film deposited onto *n*-CdSe in the presence of (a) nitrogen and (b) ammonia, both at 1 atm. (Reprinted with permission from *Chem. Mater* **2001**, *13*, 1391. Copyright 2001, American Chemical Society.)

Imprinting of the PAA film can be achieved by bubbling amines into a solution of PAA in methanol, depositing the solution onto the semiconductor surface, evaporating the methanol, and then removing the amine gas *in vacuo.* Coating n-CdSe with imprinted PAA films allows molecules used for imprinting to penetrate and bind to the surface, thus altering the PL intensity. For example, trimethylamine-imprinted PAA films deposited onto n-CdSe allowed trimethylamine and ammonia to bind to the semiconductor surface, as evidenced by PL responses to both analytes (Fig. 2c). In contrast, an ammonia-imprinted PAA film on CdSe yields a PL enhancement in response to ammonia, but not to the larger trimethylamine molecules (Fig. 2d); for ammonia, a value of ΔD comparable to that found with the etched semiconductor was observed.

Binding constants for the ammonia-imprinted, PAA-coated sensors were estimated with the Langmuir adsorption isotherm model (Fig. 3b). These binding constants for ammonia adsorption onto ammonia-imprinted, PAA-coated n-CdSe were comparable to those determined by PL for ammonia binding to the bare, etched n-CdSe surface, suggesting that the imprinted polymer film acts as a size-selective sieve. These results identify molecular imprinting as a potential strategy for distinguishing among analytes with semiconductor luminescence-based chemical sensors.

Photoluminescence experiments with III–V wafers of $In_{0.50}(Ga_{0.90}Al_{0.10})_{0.50}P$ were conducted [14]. The Lewis basic gaseous analytes ammonia, methylamine, dimethylamine, and trimethylamine all yielded reversible PL enhancements. The Lewis acid sulfur dioxide, in contrast, caused reversible quenching of the semiconductor's PL intensity. These PL intensity changes were consistent with analyte-induced modifications of the dead-layer thickness.

V. ELECTROLUMINESCENCE

Another type of semiconductor luminescent signal, EL, can be obtained from III–V semiconductor p–n junctions, as in light-emitting diodes (LEDs). LEDs are revolutionizing the lighting and display industries by providing more efficient, durable, and robust methods for producing light [15–18]. In addition, they have chemical sensing applications. The EL intensity from LEDs formed from III–V semiconducting materials has been shown to be modified by the adsorption of analytes onto the device's surface [3,14], LEDs provide a promising means of chemical sensing, not only due to their small size, low cost, energy efficiency, and longevity but also because the techniques used to grow the devices, such as metal organic chemical vapor deposition (MOCVD), provide the wherewithal to customize the material by preparing it virtually one atomic layer at a time [15,19]. This control over the LED composition may provide beneficial surface chemistries for chemical sensing. The comparative stability of III–V semiconductors in air and water relative to n-CdSe is also of value for chemical sensing applications in common ambients.

VI. CHEMICAL SENSING WITH LIGHT EMITTING DIODES (LEDs)

A series of LEDs with different active-layer thicknesses was grown by low-pressure MOCVD, with the structure given in the schematic in Fig. 6. The LEDs used were double-heterostructure, edge-emitting devices wherein p-type and n-type semiconductors sandwich an undoped, low-band-gap energy semiconductor (active layer). The p–n junction double-heterostructure is more efficient at trapping electrons and holes within the active layer for recombination, enhancing EL efficiency.

These devices showed EL enhancements to ammonia, methylamine, dimethylamine, trimethylamine, and sulfur dioxide that increased in magnitude with concentration until saturation was reached [14]. The LEDs with larger active layers produced the greatest change in EL intensity with exposure to sulfur dioxide and the amines. Intensity changes were attributed principally to surface recombination velocity effects, as the significant forward biases employed should eliminate the depletion width.

VII. INTEGRATED DEVICES

As described earlier, when a p–n junction is forward biased, it can emit light. When it is reverse biased, it can serve as a photodiode to measure light intensity. It occurred to us that by carving a trench in a diode, one portion could be forward biased and the other portion reverse biased so that the light emitted from one region of the structure could be detected by the other region, as shown in the

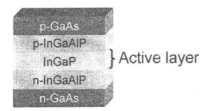

Figure 6   Schematic diagram of the double-heterostructure p–n junction diode used for chemical sensing experiments. Electrical contact is made to the top and bottom surfaces with metal films. (Adapted from Ref. 3.)

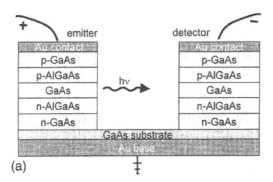

(a)

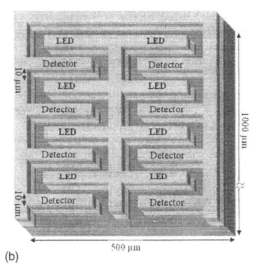

(b)

Figure 7   Schematic diagrams of a trench structure carved in a *p–n* diode: (a) side view of the trench structure that illustrates how light is produced by the forward-biased diode (LED) and detected by the reverse-biased diode (photodiode); (b) top view of these LED/photodiode pairs on a wafer. (Adapted from Ref. 14.)

schematic in Fig. 7. Moreover, arrays of such self-aligned diode structures could be easily fabricated using standard microelectronics lithographic techniques. Such LED/photodiode trench structures provide a means to increase sensor surface-to-volume ratio and thus enhance surface adsorption effects on semiconductor electro-optical properties. They also provide a route to completely integrated sensor structures, as the light emitter, light detector, and sensor element are all combined on a single chip.

To implement this strategy, multilayered semiconductor structures were grown by MOCVD and then processed using lithographic techniques to create trenches of 10–20 μm. Trenches of 10 μm were used to create arrays of 34 interdigitated LED/photodiode pairs, as shown in Fig. 8. As molecules adsorb onto the surfaces of these semiconducting materials, the electronic properties of the surfaces can be altered and thus changes in current can be observed when molecules such as ammonia and sulfur dioxide adsorb onto the surfaces of the diodes.

Both $Al_{0.4}Ga_{0.6}As/GaAs/Al_{0.4}Ga_{0.6}As$ and $In_{0.49}(Ga_{0.50}Al_{0.50})_{0.51}P/In_{0.49}Ga_{0.51}P/In_{0.49}(Ga_{0.50}Al_{0.50})_{0.51}P$ devices produce changes in photodiode current that signal the adsorption of gases such as ammonia and sulfur dioxide onto the surfaces of the integrated devices. A reproducible and reversible current enhancement of ~30% with respect to N_2 was found for the devices in response to SO_2, as shown in Fig. 9. Identical flow rates of SO_2 and of N_2, the reference gas, were used to minimize thermal variations (*vide infra*). The responses were measured with reference to the photocurrent of the devices. These photocurrent changes could be due to some combination of changes in the depletion width and surface recombination velocity [14]. In evaluating these responses, we wanted to estimate the relative contributions of adsorption onto the emitter and detector diode surfaces. As shown in Fig. 10, we found small current changes in response to SO_2 with an external excitation source (HeNe laser, $\lambda = 633$ nm), which suggests that adsorption onto the surfaces of both components of the integrated device— the LED and photodiode—contribute to the observed current changes. The current enhancements are reversible and reproducible; thus, these integrated devices possess promising chemical sensing characteristics.

Figure 8 Optical image of an integrated LED/photodiode wafer like that shown schematically in Fig. 7b.

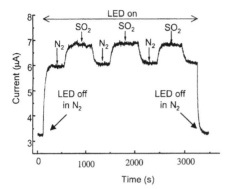

Figure 9 Photocurrent from LED–photodiode integrated device (active-layer thickness is ~500 nm, trench width is ~20 μm) for three cycles of $N_2 \rightarrow SO_2 \rightarrow N_2$ at 1 atm each. A reproducible and stable enhancement of ~30% was observed with respect to the N_2 ambient. (From Ref. 14.)

VIII. TEMPERATURE EFFECTS

Temperature plays an important role in semiconductor luminescence. In general, as temperatures decrease, band-gap luminescence increases in intensity and decreases in wavelength by ~0.2–0.3 nm/°C [20]. Joule heating in the LEDs can result in significant thermal effects that will limit their usefulness as chemical

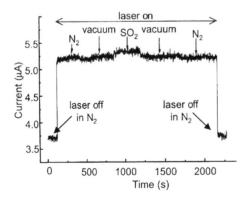

Figure 10 Photocurrent of the integrated device's photodiode (active layer thickness is ~250 nm, trench width is ~15 μm) in N_2 (1 atm), vacuum (< 0.001 atm), and SO_2 (1 atm) ambients upon excitation with an external 633-nm HeNe laser. (From Ref. 14.)

sensors. Using the dimensions of our experimental setup, finite-element thermal modeling was employed to study the steady-state temperature distribution within the LED under various heat-sinking configurations. Under these conditions, the surface of an LED operating in a flow of nitrogen will increase in temperature more than 10°C above room temperature, whereas the surface of an LED affixed to a thermoelectric cooler (TEC), set at 15°C for our experiments, only increases by about 2°C. These data indicate that temperature control from a TEC permits adsorption effects to be decoupled from thermal effects in sensor operation.

IX. CONCLUSION

Variations of semiconductor PL and EL intensities resulting from analyte adsorption are promising techniques for chemical sensing. When coupled with films such as MIPs, the selectivity of such structures may be improved. Integrated devices in which forward- and reverse-biased diodes are juxtaposed using microelectronics fabrication methods provide an opportunity to create completely integrated sensor structures on a single chip and to prepare arrays of such structures.

ACKNOWLEDGMENTS

We thank Kimberly Rickert for assistance in obtaining the optical microscopy image. We are grateful to the National Science Foundation for generous support of this work (CTS-9810176 and BES-9980758).

REFERENCES

1. Seker, F.; Meeker, K.; Kuech, T. F.; Ellis, A. B. *Chem Rev.* **2000**, *100*, 2505.
2. Ivanisevic, A.; Reynolds, M. F.; Burstyn, J. N.; Ellis, A. B. *J. Am. Chem. Soc.* **2000**, *122*, 3731.
3. Ivanisevic, A.; Yeh, J.-Y.; Mawst, L.; Kuech, T. F.; Ellis, A. B. *Nature* **2001**, *409*, 476.
4. Ellis, A. B.; Brainard, R. B.; Kepler, K. D.; Moore, D. E.; Winder, E. J.; Kuech, T. F.; Lisensky, G. C. *J. Chem. Ed.* **1997**, *74*, 680.
5. Meyer, G. J.; Luebker, E. R. M.; Lisensky, G. C.; Ellis, A. B. In *Studies in Surface Science and Catalysis*; Anpo, M.; Matsuura, T., Eds. Elsevier Science: Amsterdam, 1989.
6. Burk, A. A.; Johnson, P. B.; Hobson, W. S.; Ellis, A. B. *J. Appl. Phys.* **1986**, *59*, 1621.
7. Mettler, K. *Appl. Phys.* **1977**, *12*, 75.
8. Nickel, A.-M. L.; Seker, F.; Ziemer, B. P.; Ellis, A. B. *Chem. Mater.* **2001**, *13*, 1391.
9. Atkins, P. *Physical Chemistry*; W. H. Freeman: New York, 1998.
10. Dickert, F. L.; Hayden, O. *Adv. Mater.* **2000**, *12*, 311.
11. Haupt, K.; Mosbach, K. *Chem. Rev.* **2000**, *100*, 2495.

12. Wulff, G. *Angew. Chem. Int. Ed. Engl.* **1995**, *34*, 1812.
13. Seker, F.; Ellis, A. B. *Macromolecules* **1999**, *33*, 582.
14. Yeh, J.-Y.; Rusli, S.; Pornsuwan, S.; Ivanisevic, A.; Nickel, A. M.; Ellis, A. B.; Kuech, T. F.; Mawst, L. J. *Proc. SPIE* **2001**, *4285*, 69.
15. Condren, S. M.; Lisensky, G. C.; Ellis, A. B.; Nordell, K. J.; Kuech, T. F.; Stockman, S. A. *J. Chem. Ed.* **2001**, *78*, 1033.
16. Haitz, R.; Kish, F.; Tsao, J. Y.; Nelson, J. S. *The Case for a National Research Program on Semiconductor Lighting*; Washington DC, 1999.
17. Savage, N. *Technol. Rev.* **1999**, *1103*, 38.
18. Ponce, F. A.; Bour, D. P. *Nature* **1997**, *386*, 351.
19. Steranka, F. In *High Brightness Light Emitting Diodes*; Stringfellow, G. B.; Craford, M. G., Eds. Academic Press: San Diego, CA, 1997.
20. Varshni, Y. P. *Physica* **1967**, *34*, 149.

Index

Printed and bound by CPI Group (UK) Ltd, Croydon, CR0 4YY

25/10/2024

01779227-0001